Environment at Crossroads

Challenges and Green Solutions

Prof. Arun Arya
Head (Retd.)
Department of Environmental Studies
The Maharaja Sayajirao University of Baroda
Vadodara

SCIENTIFIC
PUBLISHERS

Published by

SCIENTIFIC PUBLISHERS (INDIA)
5 A, New Pali Road, P.O. Box 91
Jodhpur 342 001 (INDIA)
E-mail: info@scientificpub.com
Website : www.scientificpub.com

ISBN: 978-93-89184-93-8
eISBN:

Printed in India

In the loving memory of Er. Shri O.P. Arya (1-12-1935 – 15-3-2005)

हस्तियाँ मिट जाती हैं
आशियाँ बनाने में,
बहुत मुश्किल होती है
अपनों को समझाने में,
एक पल में किसी को भुला मत देना
जिन्दगी लगा दी जिन्होंने
समाज को बनाने में,
हम सब को प्रगति के पथ पर
बढ़ाने में।
नगरों को बनाने
और बागों को सजाने में।

Shubhanginiraje Gaekwad
CHANCELLOR

सत्यं शिवं सुन्दरम्

The Maharaja Sayajirao University of Baroda

FOREWORD

Earth's climate has changed during its evolutionary history, resulting into five mass extinctions. Scientists have observed loss of more than 75% species, in these extinctions. Now the environmental pollution has threatened the survival of plants, animals and human lives. Climate directly affects the survival of biodiversity. The warmer world may induce geographic expansion of certain group of organisms like insects, fungi etc. potentially changing the landscape of infectious diseases. Cases of vector-borne diseases the dengue, malaria and swine flu have increased in recent years and are responsible for large number of human lives.

Air pollution is a global problem that causes premature mortality, damage to crops and changes in ecosystem. It is estimated that in developing world, air pollution alone contributes to the death of more than 800 thousand people every year. Municipal solid waste, used as land fill, can be converted into useful compost or may be converted into source of energy. Disseminating recent scientific knowledge is important in order to sensitive scientists, students and public. In this regard an edited volume entitled, "Environment at Crossroads: Challenges and Green Solutions: by Prof. Arun Arya, Ex-Head Department of Environmental Studies. The Maharaja Sayajirao University of Baroda is a welcome academic effort.

The present volume encompasses 25 well-written chapters on various aspects of pollution management and bio-diversity conversation. Chapters include pollution caused by vehicles, dust and cement dust, food allergy as well as diseases caused by fungi. An article on pollution of the river Vishwamitri, in Vadodara, is included. The volume includes chapters dealing with molecular aspect of stress related genes in plants.

I am sure that edited volume will attract the attention of environmental scientists, biologists, biotechnologists and city planners. It will be useful to educate students, protect organic objects from deterioration, reduce pollution and make life better, I compliment Prof. Arya on this book and I am hopefully it will solicit the response it deserves.

Shubhanginiraje Gaekwad
CHANCELLOR

Contents

B: Climate Change and Sustainable Development

C: Environmental Problems and Solutions

Preface

The increase of global human population and the enhanced world wide environment pollution has been a grave concern in each corner of the planet. It seems that we have been challenging the testing limits of the planet with ever increasing threshold of environmental toxicants, pollutants, mutagens, carcinogens, teratogens and clastogens. In pursuit of greater profit margins we have been cutting corners without realizing that our activities will come back and haunt us for the repeated mistakes we have been making in mishandling environmental issues. In other words, we have been consciously sweeping the burning problems under the carpet and just been looking towards God. We were flooded by calamity of natural and man made disaster-floods throughout the country. This shows our poor planning after 70 years of independence. We have perfectly acted as ignorant individual for a long time and were not ready to come face to face with the crude and ugly truth hovering behind our back. However, everything has a breaking point and finally the house of cards starts falling apart in front of those same eyes which were so professionally been trained to ignore their warning symptoms across the past decades.

This is one of the realities that we are currently facing world-wide, because of the short term gains, we have ignored the long term sustainable aspect of our fragile ecosystems and the environment. Be it loss of precious natural resources or biodiversity or extensive pollution halting an industrial operation or destruction of a virgin forest due to slash and burn method agriculture; the final impact is always the same. The impact being loss of livelihood, employment opportunities, social and economic instabilities, poor health and vengeance full local environment and a hopeless future. We choose to be ignorant when we should have responded and the nature takes its toll in due time. Hence it is important to understand and appreciate the challenges of extensive environment pollution and ecological degradation and to take responsibility and leadership to cope with such challenges rather than avoiding them.

Students and researchers in areas of Environmental Sciences and Applied Environmental Studies and will cater to the needs of academics and researchers involved in finding sustainable solutions to global environmental issues including loss of biodiversity, intelligent traffic and health foods. The author gratefully acknowledges the kind support, help, cooperation of all the authors and peer reviewers involved in the project for their contributions and constructive criticisms,

The author(s) are solely responsible for the thoughts, opinions, views and information provided in different chapters of this volume. Any copyright infringement or plagiarism act is the sole responsibility of the authors. Hope the readers will enjoy the volume.

Arun Arya
Ex- Head, Env. Studies
The Maharaja Sayajirao University of Baroda,
Vadodara (Gujarat), India

A: Environmental Pollution and Microbes

स्वच्छ पर्यावरण

कुछ ऐसा हम करें
कि चारों ओर हो हरियाली,
वनों में सिंह गरजें
बागों में नाचें मोर
जैव विविधता न बने कोरा नारा,
हम भावी पीढ़ी के लिये छोड़ें
वातावरण न्यारा।
रोकें उपयोग रसायनिक खादों का
जैव उर्वरकों का हो बोलबाला
कमी हो रसायनिक कीट नाशकों की,
हम बनायें वातावरण सुन्दर
जैव उर्वरकों का प्रयोग
न केवल धरती की उर्वरता बढ़ायेगा
एक नया स्वर्णिम अध्याय
गांव–गांव में लिखा जायेगा,
अपना भारत फिर स्वच्छ
एवं स्वस्थ बन जायेगा।

Chapter - 1

Use of fast growing white wood rotting fungus *Ganoderma lucidum* (L.) Karst. for Cultural and Bioremediation Studies

Arun Arya, Chitra Arya and Pradyut Dhar[1]

Environmental Science, Faculty of Science,
The Maharaja Sayajirao University of Baroda, Vadodara-390002, India.
[1]Department of Botany, Faculty of Science,
The Maharaja Sayajirao University of Baroda, Vadodara-390002, India.
E mail: sarojarun10arya@rediffmail.com,

ABSTRACT

A new development in the field of environmental biotechnology is the microbe based sorbents for the removal and recovery of heavy metals from the industrial waste water. Heavy metals as widespread pollutants is of great environmental concern as they are non-degradable and persistent. Many of these metals are toxic and hazardous. The term biosorption is defined as a process in which solids of natural origin are used.

Pollution from metals is very common, as they are used in many industrial processes such as electroplating, textiles, paint and leather. The wastewater from these industries is often used for agricultural purposes, so besides the immediate damage to the ecosystem it is spilled into, the metals can enter far away creatures and humans through the food chain. Mycoremediation is one of the cheapest, most effective and environmental-friendly solutions to this problem. Many fungi are hyperaccumulators, that means they are able to concentrate toxins in their fruiting bodies for later removal. This is usually true for population that have been exposed to contaminants for long time, and have developed a high tolerance, and happens via biosorption on the cellular surface, which means that the metals enter the mycelium in a passive way with very little intracellular uptake. A variety of fungi, such as *Pleurotus*, *Aspergillus*, *Trichoderma* have proven to be effective, in removal of toxic metals from wastewater and on land. Not all the individuals of a species are effective in the same way in the accumulation of toxins. The single individuals are usually selected from an old-time polluted environment, such as sludge or wastewater, where they had time to adapt to the circumstances, and the selection is carried on in the laboratory. A dilution of the water can drastically improve the ability of biosorption of the fungi. The capacity of certain fungi to extract metals from the ground also can be useful for bioindicator purposes, and can be a problem when the mushroom is an edible one. For example, the shaggy ink cap (*Coprinus comatus*), a common edible north-hemisphere mushroom, can be a very good bioindicator of mercury, and accumulate it in its body, which can also be toxic to the consumer. Chapter presents the account of biosorption potential of *Ganoderma* mushroom

Keywords: White wood rotting fungus, *Ganoderma lucidum*, Cultural and Bioremediation Studies

1.1 INTRODUCTION

Analysis of heavy metals in wastewater is important because of their significant role in various complex processes, such as surface soil and water loading, bioaccumulation in living organisms, it acts as atmospheric catalyst and results in increase in the frequency of air and water-borne diseases. There is continuous increase in concentration of trace metals in the atmosphere. The metals in the hydrosphere are of environmental importance because of their interactions with solid geological materials, their influence on biological processes and their interactions with the atmosphere by evaporation processes. Biosorption is a technology that represents an alternative to conventional water treatments for heavy metal recovery. This technology allows the reuse of agricultural and industrial residue. Basically the term biosorption describes the removal of polluting agents from aqueous solutions by using biomass.

The commonly used procedure for metal removal are

Reverse Osmosis: It is a process in which heavy metals are separated by semi permeable membrane at a pressure greater than osmotic pressure caused by dissolved solids in wastewater. The disadvantage is that the method is expensive.

Electrodialysis: In this process, the heavy metals are separated through the use of semipermeable ion selective membranes. Application of an electrical potential is done between two electrodes. The main disadvantage of this process is formation of metal hydroxides, which clog the membrane.

Ultrafiltration: filtration is facilitated by membrane operations that us porous membranes for the removal of heavy metals. The main disadvantage of this process is generation of sludge.

Ion-exchange: In this process, metals ions from dilute solution are exchanged with ions held by electrostatic forces on the exchange resin. The disadvantages include, high cost and partial removal of certain ions.

Chemical Precipitation: Precipitation of metals is achieved by the addition of coagulants as alum, lime, iron salts etc. The large amount of sludge containing toxic compounds produced during the process is main disadvantage.

Phytoremediation; use of certain plants to clean up toxic metal pollutants.

Analysis of heavy metals in wastewater is important because of their significant role in various complex processes, such as surface soil and water loading, bioaccumulation in living organisms, it acts as atmospheric catalyst and results in increase in the frequency of air and water-borne diseases. There is continuous increase in concentration of trace metals in the atmosphere. The metals in the hydrosphere are of environmental importance because of their interactions with solid geological materials, their influence on biological processes and their interactions with the atmosphere by evaporation processes. Biosorption is a technology that represents an alternative to conventional water treatments for heavy metal recovery. This technology allows the reuse of agricultural and industrial residue. Basically the term biosorption describes the removal of polluting agents from aqueous solutions by using biomass.

New technologies have evolved for the removal of toxic metals from waste waters. Biosorption is based on metal binding capacities of various biological materials. Biosorption can be defined as ability of biological materials to accumulate heavy metals from waste water through metabolically mediated or physic-chemical pathways of uptake (Fourest and Roux, 1992). Algae, bacteria and fungi have proved to be potential metal biosorbents (Volesky, 1986). The major advantages of biosorption over conventional treatment methods include (Kratochvil and Volesky, 1998), low cost, high efficiency and no additional nutrient requirements.

1.2 ROLE OF REISHI MUSHROOM IN HUMAN HEALTH

A considerable number of studies in countries like Japan, China, USA, and UK in the past 30 years have shown that the consumption of red Reishi has been linked to the treatment of a vast range of diseases, common ailments, and conditions. From asthma to zoster, the applications of red Reishi seem to be related to a multitude of body organs and systems.

However, most of the scientific research that has been conducted appears to strongly support red Reishi's role as a nutraceutical - a nutritional supplement that can yield medical benefits through its normalization and regulation of the body's organs and functions. The role of red Reishi in maintaining a healthy lifestyle can best be explained through the Traditional Chinese Medicine (TCM) point of view because none of the known active components taken alone is as more effective than the consumption of Reishi itself. Whereas, Western medicine focuses on the “cure” after the disease has already occurred, TCM, established through over 2,000 years of human observation, focuses on disease prevention by sustaining the right balance within the body through proper nutrition, exercise, and meditation. Reishi is an important adaptogenic herb in TCM in helping the body to maintain this balance and also restore the balance, when one is sick (Gopalkumar *et al.* 2010).

Another major active ingredient found in red Reishi are triterpenes , called ganoderic acids. Preliminary studies indicated that ganoderic acids help alleviate common allergies by inhibiting histamine release, improve oxygen utilization and improve liver functions. Triterpenes are bitter in taste and the level of the triterpene content contained in a product can be determined by the bitterness (Tim *et al.* 2004).

Regular consumption of red Reishi can enhance our body's immune system and improve blood circulation, thus improving better health conditions. Commonly known as Reishi mushroom is highly ranked in Oriental folklore. *Cordyceps, Lentinus edodes* and *Ganoderma* are three important medicinal mushrooms. In China, medicine reishi has been considered as a panacea for all types of diseases. The fruiting bodies of this mushroom contain a variety of chemical substances (Tim). The fungus has been used to treat a variety of diseases including chronic hepatopathy, hypertension, neurasthenia, insomnia, bronchitis, gastric ulcer, diabetes, hyperglycemia and cancer (Zhang, *et al.* 2002).

1.3 MOLLICULAR ACTION OF HEAVY METAL COPPER ON CELLS

A heavy metal like copper shows antimicrobial mechanisms on microbes. The antimicrobial mechanisms are very complex and take place in many ways, both inside cells and in the interstitial spaces between cells. Examples of some of the molecular mechanisms noted by various researchers include the following:

The 3-dimensional structure of proteins can be altered by copper, so that the proteins can no longer perform their normal functions. The result is inactivation of bacteria or viruses. Copper may disrupt enzyme structures, and functions by binding to sulfur- or carboxylate-containing groups and amino groups of proteins.

Copper may interfere with other essential elements, such as zinc and iron. Copper facilitates deleterious activity in superoxide radicals. Repeated redox reactions on site-specific macromolecules generate OH- radicals, thereby causing "multiple hit damage" at target sites. Copper can interact with lipids, causing their peroxidation and opening holes in the cell membranes, thereby compromising the integrity of cells. This can cause leakage of essential solutes, which in turn, can have a desiccating effect. Copper damages the respiratory chain in *Escherichia coli* cells and is associated with impaired cellular metabolism.

Faster corrosion correlates with faster inactivation of microorganisms. This may be due to increased availability of cupric ion, Cu2+, which is believed to be responsible for the antimicrobial action. Microbes require copper-containing enzymes to drive certain vital chemical reactions. Excess copper, however, can affect proteins and enzymes in microbes, thereby inhibiting their activities. Researchers believe that excess copper has the potential to disrupt cell function both inside cells and in the interstitial spaces between cells, probably acting on the cells' outer envelope.

1.4 REMEDIATION OF POLLUTANTS BY FUNGI

Fungi are amongst the primary saprotrophic organisms in an ecosystem, as they are efficient in the decomposition of matter. Wood-decay fungi, especially white rot, secretes extracellular enzymes and acids that break down ligninan, cellulose, the two main building blocks of plant fiber. These are long-chain organic (carbon-based) compounds, structurally similar to many organic pollutants. They do so using a wide array of enzymes. In the case of polycyclic aromatic hydrocarbons (PAHs), complex organic compounds with fused, highly stable, polycyclic aromatic rings, fungi are very effective also in marine environments. The enzymes involved in this degradation are ligninolytic and include lignin peroxidase, versatile peroxidase, manganese peroxidase, general lipase, laccase and sometimes intracellular enzymes, especially the cytochrome P450.

There are several factors that affect the rate and ability of fungi to break down the pollutants. The first of which is the physical nature of the pollutants. Generally, contaminants that are molecularly simpler are easier to break down than more complicated ones. Temperature also controls the rate at which contaminants are broken down. Low temperatures slow the process and warm temperatures speed it up. Fungi prefer a pH of 4 to 5, that is they prefer a pH which is slightly more acid than a neutral pH. Oxygen is also essential to fungal metabolism. The initial step of hydrocarbon degrading involves adding oxygen to the hydrocarbon, and so a lack of oxygen in the environment slows the process.

In an experiment conducted in conjunction with Dr. S.A. Thomas, a major contributor in the bioremediation industry, a plot of soil contaminated with diesel was inoculated with mycelia of oyster mushrooms; traditional bioremediation techniques (bacteria) were used on control plots. After four weeks, more than 95% of many of the PAH (polycyclic aromatic hydrocarbons) had been reduced to non-toxic components in the mycelial-inoculated plots. It appears that the natural microbial community participates with the fungi to break down contaminants, eventually into carbon dioxide and water. Wood-decay fungi are particularly effective in breaking down aromatic pollutants (toxic components of petroleum), as well as chlorinated compounds (certain persistent pesticides) (Thomas *et al.,* 1999). In 2007 a similar method was practiced in San Francisco. Oil had contaminated the shoreline after a cargo ship spilled 58,000 gallons of heavy fuel oil. Hair mats, the size of a doormat, acted as sponges to soak up spilled oil. An experiment was designed that collected and layered with oyster mushrooms and straw: the mushrooms broke down the oil and after several weeks the resulting soil was clean enough to be used for roadside landscaping. Mycofiltration is a similar process, using fungal mycelia to filter toxic waste and microorganisms from water in soil.

1.5 EFFECT OF CULTURE MEDIA ON GROWTH

Most mycologists develop preferences for certain types of media based on experience and peculiarities of the type of fungi that are routinely grown. Media will affect colony morphology and color, whether particular structures are formed or not, and may affect whether the fungus will even grow in culture. For example, some fungi lack the necessary enzymes to utilize different carbon sources. All fungi require several specific elements for growth and reproduction.

Following broth media were tried

GMY	Glucose 20 g. Malt Extract 20 g. yeast Extract 1g
AMY	Antibiotic (chloramphenicol) Malt Extract 20 g. yeast Extract 1g
STMY	Starch, 20g Malt Extract 20g Yeast Extract 1g
SMY	Sucrose 20 g. Malt Extract 20 g. Yeast Extract 1 g
MMY	Mannose 20 g. Malt Extract 20 g. Yeast Extract 1 g
MY	Malt Extract 20 g. yeast Extract 1g

Different carbohydrates are essential to provide energy and are used by fungi with different rates. Basidiomycetous fungi prefer maltose than glucose. Detailed results are presented in Table 1 and Fig 1

Table 1. Increase in fungal biomass of two strains of *Ganoderma* in six different media

Media	Strain	Mycelial Dry Weight		
		5 Day	10 Day	15 Day
GMY	GN	155	223	230
	G29	150	200	210
AMY	GN	222	222	315
	G29	170	230	280
STMY	GN	168	285	415
	G29	170	280	350
SMY	GN	247	227	186
	G29	180	240	250
MMY	GN	210	220	395
	G29	200	220	380
MY	GN	160	150	120
	G29	180	200	200

It is evident from the above table that strain GN grew better in STMY medium followed by MMY and AMY. While strain G29 grew better in medium MMY followed by STMY and AMY.

Effect of two culture media PDA and PMA was observed on the growth of two strains GN and G29 of *Ganoderma lucidum*. The growth was more in G29 on PDA and of GN strain on PMA detailed results are shown in Fig. 1.

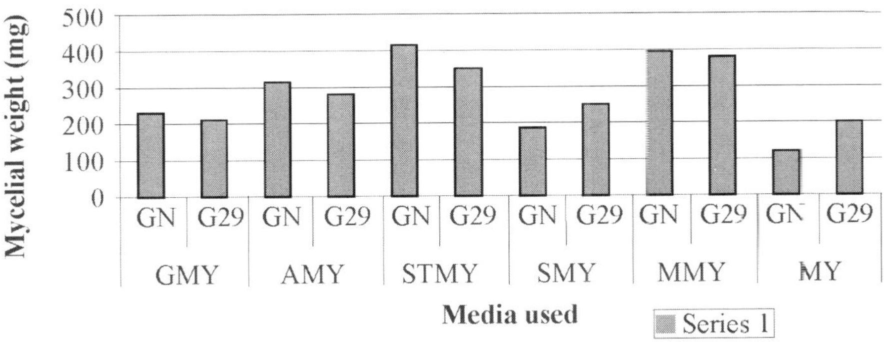

Fig. 1. Histogram showing growth (mg) of *G. lucidum* after 15 days on six different media

1.6 REMEDIATION STUDIES

Dry fruiting bodies of *Ganoderma lucidum* are used as medicine. A part of the fruiting body was crushed in a grinder. 2g of the powdered mass of *Ganoderma lucidum* was weighed and mixed with the 100ml of 100,500 and 1000 ppm of Cd salt. The cadmium salt solution was prepared from the Cadmium Sulphate (extrapure) salt i.e. $3CdSO_4.8H_2O$ in 100ml of distilled water in three different concentrations.

The mixture of *Ganoderma lucidum* and $CdSO_4$ solution was mixed and placed in Rotary shaker for 20, 40, 60 and 120 min. and after that the solution was filtered using Whattman filter paper no. 1. After filtration the biomass was kept in the Owen for drying at 72°C for 24 hr.

The dried mass was analyzed by the instrument RDX-500 XRF X-ray fluorescence spectrometer for reduction in con. of $CdSO_4$

The experimental results revealed that reduction was fast in 100 ppm conc at 40 sec. although further reduction in conc. of heavymetal salt was observed at 50 and 100 ppm conc. Similar results have been observed by Suseem and Saral (2014) studied the efficacy of *Pleurotus eous* mushroom, which has analgesic, anti-inflammatory and antipyretic properties, in biosorption of heavy metals like Lead, Chromium and Nickel. The experiment conducted on mushroom bodies obtained from Kerala Agriculture University, Trivandrum resulted in pH 5, 3, and 7 respectively for removal of Pb, Cr, and Ni respectively. The percentage removal of all the heavy metals was found to be biosorbent dependent and found to increase with increase in biosorbant dosage. Biosorption varies with metals. This mushroom is a potential source of raw material for the development of biosorbent. Terragon oyster mushroom *P. eous* exhibited maximum uptake of Lead (93.2%) among the other metals Chromium (27.6%) and Nickel (38.3%).

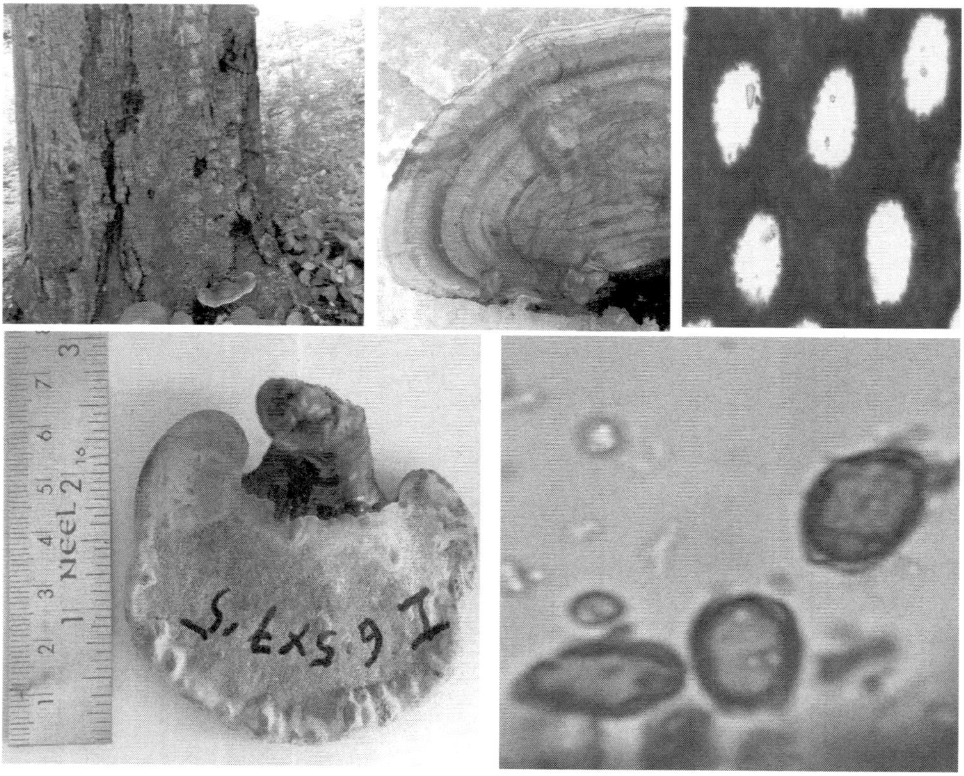

Plate 1.

Immobilized *A. bitorquis* had higher heavy metal ions remediation potential as compared to other two strains. The maximum potential of live immobilized *A. bitorquis* for remediation of Cr(III), Pb (II), Cr (VI), and Cu (II) was 226.6, 208.5, 207.3, and 205.1 mg/g, respectively (Hanif and Bhatti, 2013).

Acknowledgements

We thank UGC for funding the program of UGC- DRS to Botany Dept. and providing financial assistance in one of the major research projects to Prof. A. Arya for funding the project on Endophytes.

REFERENCES

Fourest, E. and Roux, J.C. (1992) Heavy metal biosorptton by fungal mycehal byproduct mechanism and influence of pH. Appl. Mlcrobiol Blotechnol, **37**, 399–403

Ghosh, S. (2009). Phytoremediation and bioremediation: Energy technologies for pollution control. In: Ecodegradation to air pollution, eds. A. Arya, S.J. Bedi and V.S Patel, *Sci. Pub. India*, Jodhpur. pp. 293-309.

Gopalkumar G., Martin, F., Antony, S.K., Pillai, T.G. and Nair, CKK. (2010). Preclinical studies on the use of medicinal mushroom *Ganoderma lucidum* as an adjuvant in radiotherapy of cancer. *Curr. Sci.* 99: 1084-1090.

Hanif M.A. and Bhatti N. (2013) Remediation of heavy metals using easily cultivable, fast growing, and highly accumulating white rot fungi from hazardous aqueous streams. Pages 238-248 | https://doi.org/10.1080/19443994.2013.848413

Hikino, H. and Mizuno, T. (1989). Hypoglycemic actions of some heteroglycans of *Ganoderma lucidum* fruit bodies. *Plants Med*, 55: 385.

Kimura, Y., Taniguchi, M. and Baba, K. (2002). Antitumor and antimetastatic effects on liver of triterpenoids fractions of *Ganoderma lucidum*: mechanism of action and isolation of an active substance. *Anticancer Res*, 22: 3309-3318.

Kratochvil, D. and Volesky B. (1998) Biosorption of Cu from ferruginous wastewater by algal biomass Water Research. 32(9): 2760-2768.

Kumar, V.V. and Kaladharan, P. (2006). Biosorption of metals from contaminated water using sea weed, *Curr. Sci.* 90: 1263-1267.

Maruyama, H.K. and Murofushi, A. (1989). Antitumour activity of Sarcodon aspratus, *J. Pharmacobiodyn*, 12: 118-123

Singh, R.P., Verma, R.C., Arora, R.K., Mishra, K.K., Bhanu, C. and Singh, M. (2007). Medicinal mushrooms of Uttaranchal with reference to *Ganoderma, Auricularia* and *Cordiceps sinensis*. In: *Mushroom Biology and Biotechnology* (R.D. Rai, S.K. Singh, M.C. Yadav and R.P. Tewari, eds.). Mushroom Society of India, Solan. pp. 321-324

Suseem S.R. and Saral M.A.(2014) Biosorption of heavy metals using mushroom *Pleurotus eous*. J of Chemical and Pharmaceutical Research 6(7): 2163-2168.

Tim L., Yihuai, G. and Shufeng, Z. (2004). Global marketing of medicinal ling 2hi mushroom *G. lucidum* products and safety concerns, Int. J. Med. Mushroom, 6: 189-194

Thomas SA, P Becker, MR Pinza, and JQ Word. (1999). Mycoremediation: A Method for Test- to Pilot-Scale Application. PNL-SA-30881, in Proceedings, Battelle In Situ and On-Site Bioremediation, Fifth International Symposium, April 19-22, 1999, San Diego, California.

Volesky, B. (1986) Biosorbent Materials, Biotechnol. Bioeng Symp., 16: 121-126 (1) 4. C. Ercole, F. Veglio, L. Toro, G. Ficara and A. Lepidi

Zhang G.L., Wang, Y.H., Ni, W., Teng, H.L. and Lin, Z.B. (2002). Hepatoprotective role of *Ganoderma lucidum* polysaccharide against BCG-induced immune liver injury in mice. *World J Gastro*, 8: 728-733.

Chapter - 2

A Look into Dark Side of Fungal agents: Researches on Human Diseases

Varsha Raimalani and Brinda Panchamia

Department of Biochemistry, Faculty of Science
The Maharaja Sayajirao University of Baroda, Vadodara, 390002 India
E mail; varsharai1512@gmail.com, brindapanchamia@yahoo.in

ABSTRACT

From so many years fungi have been a gift to mankind because of their resourceful nature. They perform various essential roles in maintaining balance of nature whether it is decomposition of organic matter or in nutrient cycle. The first word everyone remembers is "Penicillin" when they hear of fungi. Since, the discovery of penicillin they are useful for production of many antibiotics and enzymes used in industries. Also, they are being used tremendously in food industry for production of wine, beer, bread, etc. Moreover, they act as a direct source of food such as mushrooms and truffles. Besides providing all these advantages they still can be a threat to living organisms. Many species of fungi produce bioactive compounds that are toxic to animals including humans. This paper deals with negative side of fungi which can be lethal to humans. Diseases like Rhinitis, fungal allergy, dermatophytosis, mycosis, Aspergillosis, and Candidiasis are described.

Keywords: Fungi, Rhinitis, Allergy, Dermatophytosis, Mycosis, Aspergillosis, Candidiasis

2.1 INTRODUCTION

"Fungi" is a trivial term nowadays. These are eukaryotic microorganisms, such as unicellular yeasts and filamentous moulds, as well as fascinating mushrooms. Speaking of which these are beautiful, colourful and eye captivating organisms. They exist in two forms either as parasites or as symbionts. Those which act as parasites, prey on insects, plants and animals which help in breakdown of dead organic matter. Lichens are the best example for those which dwell as symbionts. Both ecologically and economically fungi are important organisms. For example, they help in maintain the continuity of nutrition cycle by breaking down dead organic matter in the ecosystem. Fungi are everywhere. There are approximately

1.5 million different species of fungi on Earth, but only about 300 of those are known to make people sick. Fungal diseases are often caused by fungi that are common in the environment. Fungi live outdoors in soil and on plants and trees as well as on many indoor surfaces and on human skin. Most fungi are not dangerous, but some types can be harmful to health. They act as a source for a number of supplements like,

a. Drugs such as penicillin, and lentinan from *Lentinus* mushroom (Arya and Arya, 2003)

b. Endophytes *Pestalotiopsis* is known to enhance Taxol production

c. Food such as single cell proteins and edible mushrooms

d. Certain fungi and mushrooms are used as Nutraceuticals

e. Bubbles in bread, champagne and beer are produced via the process of fermentation.

But as it's rightly said, "Every coin has dual sides- a positive and a negative." So is the case with fungi. Yes, they do have negative side i. e., in spite of having such exquisite morphology and resourceful nature they are responsible for causing a number of plant and animal diseases (including human). This chapter deals with fungus which causes various skin diseases (termed mycosis) and neurodegenerative diseases in humans which can prove to be fatal if untreated. Some of them are Aspergilloses, candidiasis, cryptococcosis and many more. Moreover, persons having immuno-deficiencies are particularly susceptible to diseases caused by genera such as *Aspergillus* and *Candida*, etc. Since fungi can reproduce by asexual method by conidia formation, their growth is more pronounced in presence of humidity and suitable temperature. Same is true for vectors like insect and mosquitoes and diseases are more severe during rainy season, while the allergy cases are more during rainy and spring season, when fungal spores and pollen grains are also present in large numbers. Most of the health concerns regarding global warming are focused on vector-borne and parasitic diseases, as they have been long considered tropical-region-associated diseases that might extend to previously more temperate regions (Paaijmans *et al.* 2009). However, the potential effects of global warming on the fungi have not been considered.

Global warming could have a significant effect on fungal populations. First, a warmer climate could change the distribution of heat-tolerant and susceptible species by favoring those that are more thermotolerant, and by creating conditions for more environmental fungi to spread and enter into closer contact with human populations, as postulated for *P. brasiliensis* and *C. gattii* (Casadevall and Perfect 1998,Greer and Fisman 2008). Second, under strong selective pressure, the prevalence of species adapted to heat tolerance may increase. There is some evidence for this in the finding that some fungi in urban areas grow faster at warmer temperatures than their rural counterparts (McLean *et al.*, 2005).

For every 1°C gained in body temperature in the range of 30 to 42°C, approximately 6% of the fungal species are excluded as potential pathogens (Robert and Casadevall, 2009). Global warming means narrowing of the thermal gradient between ambient and mammalian temperatures. The current gradient is approximately 22°C, and consequently, every degree increase in the global average temperature reduces the gradient by about 5%. We hypothesize that with current global warming, the prevalence of fungal diseases will increase by the mechanisms previously discussed. As thermotolerance is more commonly found within the basidiomycetes (Robert and Casadevall, 2009), this group may be the major contributor of new fungal pathogens.

2.2 DIFFERENT TYPES OF MYCOSIS

Mycoses (plural) are classified according to the tissue levels initially colonized.

a. **Superficial mycosis:** These are usually confined to the outer layers of skin, hair, and nails.

b. **Cutaneous mycosis:** Cutaneous mycoses extend deeper into the epidermis, and also include invasive hair and nail diseases. These diseases are restricted to the keratinized layers of the skin, hair, and nails. Unlike the superficial mycoses, host immune responses may be evoked resulting in pathologic changes expressed in the deeper layers of the skin. The organisms that cause these diseases are called dermatophytes. The resulting diseases are often called ringworm

c. **Subcutaneous mycosis:** Subcutaneous mycoses involve the dermis, subcutaneous tissues, muscle and fascia. These infections are chronic and can be initiated by piercing trauma to the skin which allows the fungi to enter. These infections are difficult to treat and may require surgical interventions such as debridement.

d. **Systemic mycosis:** Systemic mycosis are subdivided into :

   (i) Systemic mycosis due to primary pathogens originate primarily in the lungs and may spread to many organ systems. Organisms that cause systemic mycoses are inherently virulent.

   (ii) Systemic mycosis due to opportunistic pathogens are infections of patients with immune deficiencies who would otherwise not be infected. Examples of opportunistic mycoses include Candidiasis, Aspergillosis.

2.3 CANDIDIASIS

Candida albicans is a versatile pathogen of humans. The reason for this is its ability to survive as a commensal at distinct places in the body. *Candida* spp. can cause more number of diseases to that of other commensal organisms (example *E. coli*). An overgrowth of yeast leads to yeast infection. There are more than 20 species of *Candida*, the most common being *Candida albicans*. Under certain conditions, they can become so numerous and cause infections, particularly in warm and moist areas. Examples of such infections are vaginal yeast infections, thrush (infection of tissues of the oral cavity), skin and diaper rash, and nailbed infections and in warm moist body areas, such as underarms.

2.3.1 Causes of Candidiasis

A healthy immune system and a normal microbial flora are sufficient to protect our body from *Candida* infections. However, some events such as extensive antibacterial treatment or dysfunction of the immune system may enable this fungus to outnumber the microbial flora on mucosal surfaces. Oral and vaginal infections with *C. albicans* are extremely common in even mildly immunocompromised individuals (Rechard *et al.,* 2001).

As mentioned above, the yeast is already present. You don't catch it. A number of factors can increase the chance of the yeast growing out of control. The leading cause is overuse of antibiotics. Yeast must compete for the right to live on us with various other organisms, many of them bacteria. These bacteria, which live on the skin and in the intestine and vagina, among other places, are harmless but good at fighting off yeast. When we take antibiotics to deal with less friendly bacteria, we kill off these harmless ones as well. Yeast, which is unaffected by antibiotics, moves into the vacated spots once occupied by bacteria, and starts to grow and multiply (Berhard, 2004).

Therefore, *C. albicans* is able to survive and proliferate in radically changing environments with drastic changes in oxygen and carbon dioxide, pH, osmolarity, availability of nutrients, and temperature. In addition, it has to resist the constant surveillance of the immune system. Only few pathogenic microorganisms have the potential to cause infections at such a broad range of body sites (Richard *et al.* 2001; Odds, 1988). Steroids and some cancer medications weaken the immune system and can allow yeast to flourish. *C. albicans* infections of the mouth (known as *oral thrush*) most often develop in people with diseases such as cancer and AIDS. Hot weather and tight clothing are also risk factors, as they create the ideal environment for *Candida*. Other conditions that tend to encourage yeast include obesity and pregnancy. Yeast generally infects intertriginous areas, that is, areas where skin contacts skin. Overweight people have more folds in their skin. They also sweat more and *C. albicans* is fond of moist skin. Pregnancy causes temporary obesity and may weaken the immune system, increasing the risk of yeast infections.

2.3.2 Severity of Candidiasis

Treating and caring for patients with fungal diseases is becoming an increasing economic burden. Estimated numbers of patients who will need expensive antifungal drugs for some of the life-threatening fungal illnesses are summarized here.

Candidaemia infections occur can be predicted at around 300,000 worldwide per year - with a mortality of 30-55%. Oral, oesophageal and vulvovaginal candidiasis (thrush) cases are more. Oral thrush occurs in ~13.3 million people worldwide based on ~90% of HIV/AIDS patients1 not taking but needing anti-retroviral therapy, estimated by the WHO in 20112. Oral thrush also occurs in normal babies, people taking inhaled steroids for asthma, following radiotherapy to the head and neck for cancer, in denture wearers and in some leukaemia and transplant patients. Candida infection of the oesophagus (gullet) affects an estimated ~3 million people as ~20% of HIV/AIDS patient's not on anti-retroviral therapy, and ~0.5% if on antiretroviral therapy develop it. Repeated attacks of vulvovaginal candidiasis affect at least 80 million women annually as 5–9% have at least 4 attacks annually. The impact of on quality of life is substantial. About 70% of all premenopausal women develop thrush at some point in their lives. Invasive and life-threatening fungal infection causes Candida infection. *Candida peritonitis* occur both those undergoing long term peritoneal dialysis for renal failure (CAPD) and post-surgical patients, usually in intensive care. In a large multicentre study in 101 French intensive care units (ICU), hospital-acquired *Candida peritonitis* was documented in 73 patients over 8 months, compared with 123 patients with candidaemia without *Candida peritonitis*; 26 patients had both15. Assuming this is generalizable to other populations, this suggests a ratio of 1 patient with hospital-acquired (almost all post-operative) *Candida peritonitis* for every 2 patients with candidaemia, in ICU. As between 30 and 50% of candidaemia cases occur in ICU, and there are about 400,000 episodes of candidaemia globally, this suggests about 60,000 - 100,000 cases of *Candida peritonitis* each year. The mortality of *Candida peritonitis* was 38%. In those with end stage renal disease worldwide (~1.7M) CAPD is used in about 50%. Patients get 1 infection per 18 months on average and ~0.05 episodes per patient year are attributable to *Candida* spp., equivalent to ~42,500 cases annually. Around 150,000 of the 7.5 million patients admitted to intensive care (ICU) in Europe, USA and Japan each year grow *Candida* in their urinary system.

2.4 ASPERGILLOSIS

Apergillosis is an opportunistic infection caused by inhaling spores of the mould *Aspergillus*, commonly present in the environment; the spores invade blood vessels, causing haemorrhagic necrosis and infarction. Symptoms may be those of asthma, pneumonia, sinusitis, or rapidly progressing systemic illness. Diagnosis is primarily clinical but may be aided by imaging, histopathology, and specimen staining and culture. Treatment is with voriconazole, amphotericin B (or its lipid formulations), caspofungin, or itraconazole. Invasive infections are usually acquired by inhalation of spores or, occasionally, by direct invasion through damaged skin.

Major risk factors include

- Neutropenia when prolonged (typically > 7 days)
- Long-term high-dose corticosteroid therapy
- Organ transplantation (especially bone marrow transplantation with graft-vs-host disease [GVHD])
- Hereditary disorders of neutrophil function (*e.g.*, chronic granulomatous disease)

Aspergillus spp. tends to infect open spaces, such as pulmonary cavities caused by previous lung disorders (*e. g.*, bronchiectasis, tumor, TB), the sinuses, or ear canals (otomycosis). Such infections tend to be locally invasive and destructive, although systemic spread sometimes occurs, particularly in immunocompromised patients. However, aspergillosis is unusual in those with HIV infection. *A. fumigatus* is the most common cause of invasive pulmonary disease; *A. flavus* most often causes invasive extra pulmonary disease, probably because these patients are more severely immunosuppressed than patients infected with *A. fumigatus*.

Over 10 million patients in Europe, USA and Japan are at risk of invasive aspergillosis (IA) each year because of corticosteroid or other therapies. Over 50% of patients with IA die, even with treatment. Over 200,000 patients develop IA annually. Key groups include ~10% of acute leukaemia (300,000 new cases annually) (30,000 IA cases) and stem cell and other transplants. 7% of the global number of moderate and severe COPD = 65M (WHO)) (60,000 confirmed and IA cases). There are 11.9M COPD admissions in China each year and a mean rate in OECD countries of 198/100,000 (range 364 (Ireland) to 71 (Portugal). A recent large survey of IA in liver failure in China documented a 5% rate, with a 95% mortality. These patients probably account for 50-65% of patients, with those admitted to intense care (ICU), with lymphoma or chronic leukaemia and various immunological disorders and treatments accounting for the remainder. Under diagnosis is a major problem in this disease.

Aspergilloma

The infection sometimes forms fungal ball (aspergilloma), a characteristic growth of tangled masses of hyphae, with fibrin exudate and few inflammatory cells, typically encapsulated by fibrous tissue. Occasionally, there is some local invasion of tissue at the periphery of the cavity, but usually the fungus just resides within the cavity with no appreciable local invasion. A chronic form of invasive aspergillosis occasionally occurs, particularly in patients taking corticosteroids long-term and those with chronic granulomatous disease, which is characterized by a hereditary phagocytic cell defect. Primary superficial aspergillosis is uncommon but may occur in burns; beneath occlusive dressings; after corneal trauma (keratitis); or in the sinuses, mouth, nose, or ear canal. Allergic bronchopulmonary aspergillosis is a hypersensitivity reaction to *A.*

fumigatus that results in lung inflammation unrelated to fungal invasion of tissues.

In severely immunocompromised patients, extrapulmonary invasive aspergillosis begins with skin lesions, sinusitis, or pneumonia and may involve the liver, kidneys, brain, and other tissues; it is often rapidly fatal. Aspergillosis in the sinuses can form an aspergilloma or cause allergic fungal sinusitis or a chronic, slowly invasive granulomatous inflammation with fever, rhinitis, and headache. Patients may have necrosis cutaneous lesions overlying the nose or sinuses, palatal or gingival ulcerations, signs of cavernous sinus thrombosis, or pulmonary or disseminated lesions.

- Usually fungal culture and histopathology of tissue samples
- Galactomannan antigen test on serum and bronchoalveolar lavage fluid

Because *Aspergillus* spp. are common in the environment, positive sputum cultures may be due to environmental contamination or noninvasive colonization in patients with chronic lung disease; positive cultures are significant mainly when sputum is obtained from patients with increased susceptibility due to immunosuppression or when there is high suspicion due to typical imaging findings. Conversely, sputum cultures from patients with aspergillomas or invasive pulmonary Aspergillosis are often negative because cavities are often walled off from airways and because invasive disease progresses mainly by vascular invasion and tissue infarction.

Chest CT is far more sensitive and should be done if patients are at high risk (i.e., neutropenic). CT of sinuses is done if sinus infection is suspected. A movable fungus ball within a cavity lesion is characteristic on both, although most lesions are focal and solid. Sometimes imaging detects a halo sign (a hazy shadow surrounding a nodule) or cavitation within a necrotic lesion. Diffuse, generalized pulmonary infiltrates are seen in some patients. Suggested treatment includes, use of Voriconazole and Amphotericin B (including lipid formulations) Echinocandins as salvage therapy, Sometimes surgery for Aspergilloma.

2.5 OTHER COMMON FUNGAL DISEASES

Following is the brief account of certain common fungal diseases:

i. Cryptococcal Meningitis: Over 1 million people develop cryptococcal meningitis in AIDS, each year, resulting in ~600,000 deaths, of which ~70% are in sub-Saharan Africa. Meningitis is an infection and inflammation of the meninges. Your meninges are the membranes that cover your spinal cord and your brain. Meningitis can be caused by many different germs, including bacteria, fungi, and viruses. Cryptococcal meningitis (CM) is one of the exceptions. It is caused by two types of fungus: *Cryptococcus neoformans* and *Cryptococcus gattii*. This disease is rare in normally healthy people. It is more common in people who have compromised immune systems, such as AIDS patients. CM is caused by a fungus called *C. neoformans*. This fungus is found in soil all over the world. It is usually found in soil that contains bird droppings. Symptoms of CM usually come on slowly. Within a few days to a few weeks of contact, an infected person may develop the following symptoms:

- headache
- nausea and vomiting
- mental changes such as: confusion, hallucinations, and/or personality changes
- lethargy
- sensitivity to light

If left untreated, CM may lead to more serious symptoms such as: brain damage, hearing loss, coma, and hydrocephalus ("water on the brain") (Bicanic and Harrison, 2005). Prior to about 1979, *Cryptococcal meningitis* was very rare worldwide (Berger, 2011).

If you have CM, you will be given antifungal drugs. The most common choice is amphotericin B. You will need to take the drug daily. Your doctor will monitor you closely while you are on this drug to watch for nephrotoxicity (meaning the drug is toxic to your kidneys). Amphotericin B is typically administered intravenously. This means that it is administered directly into your veins. One can probably also take flucytosine, another antifungal medication, while you are taking the amphotericin B. The combination helps treat the condition more quickly than amphotericin B alone. You will need to have your spinal fluid tested repeatedly during treatment. After your tests have come back negative for CM for two weeks, you will generally stop taking amphotericin B and flucytosine. You will typically switch to taking only fluconazole, which you will generally keep taking for about eight weeks.

ii. Pneumocystis Pneumonia : Peumocystis Pneumonia (PCP) is an illness caused by the fungus *Pneumocystis jirovecii*. PCP is one of the most frequent and severe opportunistic infections in people with weakened immune systems, particularly people with HIV/AIDS. All of the ~14.8 million HIV/AIDS infected patients2 who should be receiving anti-retroviral therapy are at risk of Pneumocystis pneumonia (PCP), as well as many other immunocompromised patients, unless taking oral antifungal prophylaxis with cotrimoxazole. The rate of PCP as an AIDS indicator disease is very variable. In Nigeria 40% of new AIDS cases in children develop PCP but many rates are lower than this. Pneumocystis pneumonia has a 10-30% mortality in the USA and UK24,25. Ø Precise estimates of annual incidence are difficult because of diagnostic deficiencies but case numbers certainly exceed 400,000 globally per year.

The symptoms of PCP are fever, dry cough, shortness of breath, and fatigue. In people with weakened immune systems, PCP can be very serious, so it is important to see a doctor if you have these symptoms. In HIV-infected patients, PCP usually presents sub-acutely, and symptoms include a low-grade fever. In

HIV-uninfected patients, symptoms of PCP tend to develop more quickly and patients typically experience a high fever.

PCP is extremely rare in healthy people. Most people who get PCP have weakened immune systems due to HIV/AIDS, cancer treatments, or organ transplants. Other groups of people who are at risk for PCP include:

- HIV-exposed but uninfected children
- People who are receiving immunosuppressive therapies, such as organ transplant patients
- People with connective tissue diseases or chronic lung diseases

There is no vaccine to prevent PCP. Some groups of people who are at high risk of developing PCP may need to take a medication called TMP-SMX to prevent the illness from occurring. If your doctor thinks you are at risk for developing PCP, he or she might prescribe this medicine for you. TMP-SMX prophylaxis is currently recommended for:

- All HIV-infected patients with CD4 < 350 cells / μL
- Infants born to HIV-infected mothers
- Children with a history of PCP
- Stem cell transplant patients

iii. Histoplasmosis : About 25,000 cases of symptomatic histoplasmosis are estimated in the USA annually. Histoplasmosis may be a life-threatening infection in new borne and AIDS patients and is common in Central and South America. Most people who are exposed to the fungus Histoplasma never have symptoms. Other people may have flu-like symptoms that usually go away on their own. Fever is the common symptoms of histoplasmosis include:

- Fever
- Cough
- Fatigue (extreme tiredness)
- Chills
- Headache
- Chest pain
- Body aches

Symptoms of histoplasmosis may appear between 3 and 17 days after a person breathes in the fungal spores. For most people, the symptoms of histoplasmosis will go away within a few weeks to a month. However, some people have symptoms that last longer than this, especially if the infection becomes severe. In some people, usually those who have weakened immune systems, histoplasmosis can develop into a long-term lung infection, or it can spread from the lungs to other parts of the body, such as the central nervous system (the brain and spinal cord).

iv. Allergic Fungal Sinusitis : Allergic fungal sinusitis and rhinitis affects ~12 million people at any time. Allergic sinusitis and rhinitis affects ~15% of the world's population or around 900 million people and ~1.3% is predominantly fungal allergy (assuming a 50% endoscopic examination rate).

Fungi act as causative agents in allergic rhinitis and rhinosinusitis, Ambient mold spores are widely distributed in nature, and an estimated 3 to 10% of the world's population is allergic to molds. There are compelling epidemiologic links between mold (fungal) allergy and illnesses such as asthma and "asthma with allergic rhinitis." Fungal allergy is more prevalent in areas of high ambient mold spore concentrations. However, epidemiologic studies have failed to demonstrate a direct relationship between fungal allergy and allergic rhinitis either via outdoor or indoor exposure. Fungal allergy is clearly linked to a subset of chronic rhinosinusitis (CRS) known as allergic fungal rhinosinusitis (AFRS). This condition represents an intense allergic response against colonizing fungi giving rise to formation of allergic (eosinophilic) mucin, mucostasis, and sinus opacification. A broader role for colonizing fungi has been postulated in CRS owing to the demonstration of fungi in mucus in the vast majority of cases of CRS, and *in vitro* studies demonstrating that certain fungi, particularly *Alternaria*, elicit a "modified" allergic response in patients with CRS that is independent of Ig.

Allergic fungal disease include Allergic bronchopulmonary aspergillosis (ABPA). Approximately 4.8 million people develop ABPA among the 193 million adults with active asthma worldwide. Individual country estimates of asthma for 70 countries are now available, superseding older estimates. ~ 15% of people with cystic fibrosis develop ABPA, ~9,000 affected. Severe Asthma with Fungal Sensitisation (SAFS) is predicted to affect 3.25-13 million adults worldwide depending on the frequency of severe asthma (5-20% of all asthmatics).

v. Fungal Kiratitis: Eye infection: Central microbial keratitis is one of the major causes of ophthalmic morbidity and visual loss. Worldwide, the reported incidence of fungal keratitis is 17% to 36%, caused by fungi, most commonly Fusarium and *Aspergillus* and other less common species like *Candida, Curvularia,* and *Monosporidium*, among others (Mravicic, 2010). Organisms belonging to *Drechslera* species are rare causes of human infection and are considered opportunistic pathogens; however, it has been seen in healthy hosts (Sheridan, 1976). Only 10 cases of human infection with these organisms have been reported in the literature (Rolston, 1985). A new fungal keratitis was reported due to *Drechslera* spp. treated with topical voriconazole.

vi. Athlete's Foot : *Tinea pedis* : It is also known as "ringworm of the foot", *tinea pedum*, and moccasin foot is a common and contagious disease that causes itching, scaling, flaking, and sometimes blistering of the affected areas. Its medical name is *Tinea pedis*, a member of the group of diseases or conditions known as tinea, most of which are dermatophytoses (fungal infections of the skin), which in turn are mycoses (broad category of fungal infections). Globally, athlete's foot affects about 15% of the population. Tinea pedis is caused by fungi such as *Epidermophyton floccosus* or fungi of the *Trichophyton* genus including *T. rubrum* and *T. mentagrophytes*. These fungi are typically transmitted in moist communal areas where people go barefoot, such as around swimming pools or in showers, and require a warm moist environment like the inside of a shoe to incubate. Fungal infection of the foot may be acquired (or reacquired) in many ways, such as by walking in an infected locker room, by using an infested bathtub, by sharing a towel used by someone with the disease, by touching the feet with infected fingers (such as after scratching another infected area of the body), or by wearing fungi-contaminated socks or shoes. Infection can often be prevented by keeping the feet dry by limiting the use of footwear that enclose the feet, or by remaining barefoot.

However, the spread of the infection is not limited to skin. Toe nails become infected with fungi in the same way as the rest of the foot, typically by being trapped with fungi in the warm, dark, moist inside of a shoe. Fungal infection of the nails is called tinea unguium, and is not included in the medical definition of "athlete's foot", even though toe nails are part of the foot. Fungi are more difficult to kill inside and underneath a nail than on and in the skin. But if the nail infection is not cured, then the fungi can readily spread back to the rest of the foot. The fungi can also spread to hair, grow inside hair strands, and feed on the keratin within hair, including the hair on the feet, the hair of one's beard, and the hair on one's head. From hair, the fungi can spread back to skin. To effectively treat athlete's foot, it is necessary to treat the entire infection, wherever it is on the body, until the fungi are dead and the skin has fully healed.

vii. *Tinea unguium* (nails): Dermatophytic onychomycosis is a fungal infection of the nail. It is the most common disease of the nails and constitutes about half of all nail abnormalities. This condition may affect toenails or fingernails, but toenail infections are particularly common. It occurs in about 10% of the adult population.

viii. *Tinea manuum* : *Tinea manuum* (or *tinea manus*) is a fungal infection of the hand. It is typically more aggressive than tinea pedis but similar in look. Itching, burning, cracking, and scaling are observable and may be transmitted sexually or otherwise, whether or not symptoms are present.

ix. *Tinea cruris* (groin) : *Tinea cruris*, also known as "crotch itch", "crotch rot", "Dhobie itch", "*eczema marginatum*", "gym itch", "jock itch", "jock rot", "scrot rot" and "ringworm of the groin" is a dermatophyte fungal infection of the groin region in any sex, though more often seen in males. In the German sprachraum this condition is called *tinea inguinalis* (from Latin *inguen* = groin) whereas *tinea cruris* is used for a dermatophytosis of the lower leg (Latin *crus*). Tinea cruris is similar to, but different from *C. intertrigo,* which is an infection of the skin by *Candida albicans*. It is more specifically located between intertriginous folds of adjacent skin, which can be present in the groin or scrotum, and be indistinguishable from fungal infections caused by *tinia*. However, candidal infections tend to both appear and disappear with treatment more quickly. It may also affect the scrotum.

x. *Tinea corporis* *Tinea corporis* (also known as "ringworm", *tinea circinata*, and *tinea glabrosa*) is a superficial fungal infection (dermatophytosis) of the arms and legs, especially onglabrous skin; however, it may occur on any part of the body.

xi. *Tinea capitis* (scalp) *Tinea capitis* (also known as "Herpes tonsurans", "Ringworm of the hair,"Ringworm of the scalp, "Scalp ringworm", and "Tinea tonsurans") is a superficial fungal infection (dermatophytosis) of the scalp. The disease is primarily caused by dermatophytes in the *Trichophyton* and *Microsporum* genera that invade the hair shaft. The clinical presentation is typically single or multiple patches of hair loss, sometimes with a 'black dot' pattern (often with broken-off hairs), that may be accompanied by inflammation, scaling, pustules, and itching. Uncommon in adults, tinea capitis is predominantly seen in pre-pubertal children, more often boys than girls. At least eight species of dermatophytes are associated with tinea capitis. Infections from *Microsporum* species are mainly in South America, Southern and Central Europe, Africa and the Middle East. The disease is infectious and can be transmitted by humans, animals, or objects that harbor the fungus. Griseofulvin is the most commonly used drug, but other newer antimycotic drugs, such as terbinafine, itraconazole, and fluconazole have started to gain acceptance.

xii. *Tinea faciei* (face) It generally appears as a red rash on the face, followed by patches of small, raised bumps. The skin may peel while it is being treated. Tinea faciei is contagious just by touch and can spread easily to all regions of skin.

xiii. *Tinea barbae* (beard) *Tinea barbæ* (also known as Barber's itch, "Ringworm of the beard," and "Tinea sycosis") is a fungal infection of the hair. Generally, the infection occurs as a follicular inflammation, or as a cutaneous granulomatous lesion, i.e. a chronic inflammatory reaction. It is one of the causes of folliculitis. It is most common among agricultural workers, as the transmission is more common from animal-to-human than human-to-human. The most common causes are *Trichophyton mentagrophyts* and *T. verrucosum*.

2.6 ROLE OF FUNGI IN NEURODEGENERATIVE DISEASE

Has anyone ever given a thought to “fungi” as one of the causal reason for Parkinson’s?

Parkinson’s disease is a condition in which part of the brain becomes progressively damaged over many years. A person suffering from Parkinson’s goes through various physical and psychological conditions. The three main symptoms of Parkinson’s disease are:

- tremor (involuntary shaking of particular parts of the body)
- slow movement
- stiff and inflexible muscles

Parkinson's disease is caused by a loss of nerve cells in part of the brain called the substantia nigra. This leads to a reduction in the amount of a chemical called **dopamine** in the brain.

Dopamine plays a vital role in regulating the movement of the body by acting as a messenger between the parts of the brain and nervous system that help control and co-ordinate body movements. Thus, loss of nerve cells leads to reduction in dopamine levels and thereby is responsible for many of the symptoms of Parkinson's disease. This means that the part of the brain controlling movement cannot work as well as normal, causing movements to become slow and abnormal.

A recent study shows that a compound named, 1-octen-3-ol produced by fungi can be one of the factors for Parkinson's. Researchers from The State University of New Jersey, New Brunswick and Emory University, Atlanta carried out a study on this compound and tested it on flies. The study revealed that Drosophila flies when exposed to the molecules had difficulty with movement; loss of dopamine neurons, reduced levels of dopamine and died earlier than flies which were not exposed. As the flies were affected they further tested it with human cell lines and found that exposure to the molecules also caused difficulty in the dopamine system on human embryo kidney cells (Inamdar *et al.* 2013). Although this study provides insights to understanding the neurological system a more in depth but this doesn't proves that living in mouldy environments can cause one a Parskinson's. This still requires more studies correlating the exposure of the compound and risk of development towards Parkinson's

2.7 CONCLUSIONS

Fungi play a pivotal role in maintaining the existence of life on earth. But their existence is beneficial if they are present to a certain extent within living organisms. Their over limitation is a check point as once they penetrate inside human's body, their existence proves to be fatal to life as they can cause numerous diseases. These can be identified by Serologic tests (mainly for *Aspergillus*, *Blastomyces*, *Candida*, *Coccidioides*, *Cryptococcus,* and *Histoplasma*).

The human pathogen *Candida albicans* is successful both as a commensal and as a pathogen. It is clear that *C. albicans* utilizes several genes whose expression is required for virulence.

Recent findings suggest that the fungus senses local environmental changes and rapidly responds to these changes by modifying its transcriptional profile. Transcriptional regulation may not only help the fungus to adapt to the host environment but also to actively change the local microenvironment. In vivo transcriptional profiling will help further to light on fungal strategies and adaptations. Still, there is a lot of scope to improve the strategies for identifying and improving the impact of these diseases in order to fight back better.

REFERENCES

Arya, A. and Arya, C. (2003) Production of Shiitake mushroom on tree waste Presented in 12th World forestry Congress Quebec City Canada 2003

Berger, J.R. (2011) Cryptococcal meningitis in Medmerits http://www.medmerits.com/ index. php/article/cryptococcal_meningitis/P0

Bernhard, H. (2004). From commensal to pathogen: stage- and tissue-specific gene expression of *Candida albicans*. *Current Opinion in Microbiology* 7:336–341

Bicanic, T. and Harrison, T.S. (2005) *Cryptococcal meningitis* British Medical bulletin http://bmb.oxfordjournals.org/content/72/1/99.full

Casadevall, A. and Perfect, J. R. (1998). *Cryptococcus neoformans*. ASM Press, Washington, DC.

Greer, A.N.V. and Fisman, D. (2008). Climate change and infectious diseases in North America: the road ahead. CMAJ 178:715–722.

Inamdar, A.A., Hossain, M.M., Bernstein, A.I. Miller, G.W. Jason, R. Richardson, J.R. and Bennett, J.W. (2013). Fungal-derived semiochemical 1-octen-3-ol disrupts dopamine packaging and causes neurodegeneration. PNAS 110 (48) 19561–19566

Odds, F.C. (1988) : Candida and Candidiasis edn 2. London: Bailliere Tindal.

Paaijmans, K.P., Read, A.F. and Thomas, M.B. (2009). Understanding the link between malaria risk and climate. Proc. Natl. Acad. Sci. U. S. A. 106:13844–13849.

McLean, M.A. Angilletta, J.M.J. and Williams, K.S. (2005) If you can't stand the heat, stay out of the city: thermal reactions norms of chitinolytic fungi in an urban heat island. *J. Thermal. Biol.* 30: 384-391

Mravicic, I. Dekaris, I. and N. Gabric, (2010) "An overview of fungal keratitis and case report on trichophyton keratitis", Collegium *Antropologicum*, 34, (2): 271–274.

Richard A., Calderone C. and William A. F. (20010.Virulence factors of *Candida albicans. TRENDS in Microbiology*. 9 (7) July 2001

Robert, V.A. and Casadevall, A. (2009). Vertebrate endothermy restricts most fungi as potential pathogens. J. Infect. Dis. 200:1623–1626.

Rolston, K.V. Hopfer, R.L. and Larson, D.L. (1985). "Infections caused by *Drechslera* species: case report and review of the literature," Reviews of Infectious Diseases, 7, (4): 525–529,

Sheridan, J.E. (1976) "*Drechslera* spp. and other seed-borne pathogenic fungi in New Zealand cereals," *New Zealand Journal of Agricultural Research*, 20: 91–93.

■■■■

Chapter - 3

Occurrence of Fungal Endophytes inhabiting the leaves of three *Terminalia* spp.

Pradyut Dhar and Arun Arya

Department of Botany, Faculty of Science,
The Maharaja Sayajirao University of Baroda, Vadodara- 390002
Email: pradyutdhar@gmail.com, sarojarun10arya@rediffmail.com

ABSTRACT

It has been found that fungal endophytes both bacteria and fungi reside in internal tissues of the living plants without causing any immediate damage to the plant. Grass endophytes are known to provide protection against herbivory and pathogens that increase their fitness. Endophyte members belonging to family Xylariaceae produce compounds of high biological activity. Such increase in secondary metabolites has led to screen a large number of plants for the occurrence of fungal endophytes. These organisms form a potential source to produce pharmaceutically active compounds and discovery of novel drug compounds. Though host specificity among endophytes is expressed at the family level, yet some endophytic genera such as *Phomopsis* and *Phyllosticta* occur on a wide variety of host range.

A study was conducted to isolate various endophytes from the leaves of five different *Terminalia* spp. and their extracellular enzyme production *in vitro*. The observations revealed the presence of maximum 15 endophytes from leaves. The isolation from petiole and leaf tip also resulted in presence of several fungal endophytes. Five of these fungi were further studied to ascertain their potential to produce extracellular enzymes.

Key-words: Fungal endophytes, Enzymes, *Terminalia*, *Phomopsis*, Colonization frequency.

3.1 INTRODUCTION

Fungal organisms settle on foliar surface and easily colonize leaves and twigs, first as epiphytes and later lead the life as biotrophs and necrotrophs. The term endophyte has been used to denote a particular type of systemic, nonpathogenic symbiosis. The grass endophytes provide their hosts with a number of benefits, such as protection against herbivory and pathogens, that increase their fitness (Clay, 1990, Saikkonen *et al.,* 1998). They can live there for all part of their life cycle (Petrini *et al.*, 1992). Fungal endophytes reside in healthy tissues and grow asymptomatically within aerial plant tissues such as leaves, stems (Bills and Polishook, 1992) and inner bark (Tejesvi, 2005). of all terrestrial plant taxa studied to date and are diverse and abundant in tropical woody angiosperms. Endophytic fungi were studied from *T. arjuna*, an important ethnopharmacological plant extensively used for heart diseases. Tejasvi *et al.* (2005) reported 22 fungal genera from 5 plants growing in 3 different locations. *Pestalotiopsis, Chaetomium* and *Myrothecium were* most common fungi.

Fungal endophytes have been found in healthy plant tissues like stem and leaves and energe to sporulate at plant or host tissue senescence. Scientist recognized Clavicipitaecous endophytes and 3 groups of Non clavicipitaecone (Nc) endophytes. NC endophytes can be further differentiated in one groups which may colonize below and above ground part and 2 groups present only above ground parts. Endophytes may be diverse at an exceedingly small scale; a sing conifer needle may harbor several dozen species. Many endophytes are highly host – or tissue – specific. Endophytes may represent a substantial number of undiscovered fungi (Arnold *et al.,* 2003). Fungal pathogens, parasites, and symbionts may exploit the host cell wall metabolism to support the colonization of their hosts. The cumulative evidence suggests that the plant cell wall Degrading enzymes play a control role in the softening and remodeling of the cell. wall during symbiosis.

Endophytes may enhance the ecophysiology of host plants and enable plants to counter abiotic stresses such as drought (Arechavaleta *et al.,* 1989) and metal contamination (Malinowski and Belesky, 2000). Endophytes increase plant biomass and act as biocontrol agent (Danielsen and Jensen, 1999; Campanile *et al.*, 2007). Symbiosis between *Curvularia protubereta* and *Dichanthium lanuginosum* allowed both the partners to tolerate temperatures up to 65°C (Marquez *et al.*, 2007). Endophytic *Colletotrichum musae* and *Fusarium moniliforme* impair the photosynthetic efficiency in banana and maize.

Presence of foliar endophytes have resulted into a score of novel and / or active compounds (Schulz *et al.*, 2002), they have attracted increasing interest in recent years. Isolation of cytotoxic podophyllotoxins (Cragg and Newman, 2004), taxol and paclitaxel from yew tree bark has opened new potential of plant based metabolites. Total 313 bioactive compounds isolated from endophytes. *Terminalia* an important Genera of family Combretaccae consists of – species. Three species *T. arjuna, T. bellerica* and *T. chebula* are extensively used in Ayurvedic medicine Tejesvi *et al.*(2005) reported endophytic fungal organisms from twig and inner bark of *T. arjuna*.

Pathogenic fungi capable of symptomless occupation of their hosts during a portion of the infection cycle “quiescent infections” (Williamson, 1994), and strains with impaired virulence can be considered endophytes (Schardle *et al.,*

1994). Most biologists agree that species composition of the internal mycobiota is distinct for various hosts, organs and tissues.

3.2 MATERIALS AND METHODS

Study Site

Study site selected was at Gora Forest Range, Narmada District of Gujarat (21°49'40.85"N, 73°44'18.87"E, Elev. 779 ft), Sagai Forest Range (21°43' 41.46"N, 73°38'57.89"E, Elev. 1297 ft) and Arboretum, The M.S. University of Baroda, Vadodara (22°19'15.93"N, 73°10'45.96"E, Elev. 129 ft).

1. Different culture media listed below were used for isolation of Endophytes from *Terminalia* plants

1. Modified Asthana and Hawker's Medium 'A'		2. Potato Dextrose Agar Medium		3. Czapek's Medium	
Glucose	10.0 g	Peeled Potato	200.0 g	Sucrose	30.0g
KNO_3	3.5g	D-Glucose	20.0 g	$NaNO_3$	3.0g
KH_2PO_4	1.75g	Agar	20.0 g	K_2HPO_4	1.0g
$MgSO_4.7H_20$	0.75g	Distilled water	1000 ml	$MgSO_4.7H_2O$	0.50g
Agar	20.0 g	pH	6.5	KCl	0.50g
Distilled Water	1000 ml			$FeSO_4.7H_20$	0.01 g
pH	6.0			Agar	20.0 g
				Distilled Water	1000 ml
				pH	6.2

2. Isolation of Endophytic fungi

Leaf samples were collected from survey area. First the leaves were washed in running tap water for 10 minutes to drain off the dust particles present on leaf surface. The leaves were cut into 0.5×0.5 cm^2 and surface sterilized in NaOCl solution (1% available Cl_2) for 2 min. The leaves were washed in sterile distilled water for 2 minutes. Then the leaves were inoculated in suitable culture media. The growth of the organisms was observed after 7 and 21 days of inoculation.

Data Analysis

The percent frequency of colonization (Khan *et al.,* 2007 and Khan *et al.,* 2010) of 5 selected fungal endophytes in leaf isolates was calculated to find out the diversity of fungal endophytes in the 3 different species of *Terminalia*.

$$\% \text{ CF} = \frac{\text{No. of tissue segments colonized by the fungus}}{\text{Total number of tissue segments plated}} \times 100$$

3.3 RESULTS AND DISCUSSION

A total of 197 isolates belonging to 24 taxa was obtained from 2400 leaf pieces of six plants of the three species of *Terminalia*. Maximum 37/197 × 100 = isolates were from plants of *T. bellerica* growing in arboratums. Percentage of *T. crenulata* was more in arboratum as compared to *T. chebula* (Harde) and *T. catappa* in Gora forest range of Narmada district. In three plants the number of endophytes was less in trees growing in Sagai forest range no plant of *T. catappa* was found in this range.

Fungal endophytes like *Rhizopus stolonifer, Aspergillus niger, Chaetomium globosum Pestalotiopsis* sp., *Alternaria alternata, Cladopsporium* sp. *Fusarium oxysporum* and *Nigrospora sphaerica* were present in different *Terminalia* leaves. Species of *Pestalotiopsis Myrothecium* and *Trichoderma* were recorded from the twigs of *T. arjuna* (Tejesvi *et al.,* 2005). *Pestalotiopsis microspora* was obtained as an endophyte from *I. morobensis* from Papua New Guinea.

Two different species of *Fusarium* were isolated from five different locations in three spp. It was not recorded from the leaves of *T. cattapa*. Two spp. of *Chaetomium* were present in the leaves similar to twigs of *T. arjuna* (Tejesvi *et al.,* 2005).

Mycologists found that in panama where 100% of mature leaves of diverse trees and shrubs typically contain endophytes, individual leaves may harbor up to one isolate per 2 mm of leaf tissue and frequently contain dozens of species. Different leaves on the same tree may have quite distinctive assemblages of endophytic fungi (Arnold *et al.,* 2003). Individual plants may harbor hundreds of species.

Table 1. Showing fungal endophytes isolated from Leaf lamina of three *Terminalia* spp.

Sl. No.	Fungal organisms	Presence of Endophytes		
		T. bellerica	***T. chebula***	***T. crenulata***
1.	*Alternaria alternata* Fr. Keiss.	+	+	+
2.	*Aspergillus niger* van Tiegh	+	+	+
3.	*Aspergillus flavus* Link.	+	+	+
4.	*Chaetomium* sp.	-	-	+
5.	*Curvularia prasadii* R.L. & B.L. Mathur	_	_	+
6.	*Curvularia pallescence* Boedijn	-	-	+
7.	*Emericella* sp.	_	_	+
8.	*Fusarium oxysporum* Schltdl.	+	-	+

Sl. No.	Fungal organisms	Presence of Endophytes		
		T. bellerica	*T. chebula*	*T. crenulata*
9.	*Fusarium roseum* Link.	-	+	-
10.	*Gloeosporium* sp.	-	+	-
11.	*Lasiodiploidia theobromae* (Pat.) Griffon& Maubl.	+	-	-
12.	*Nigrospora sphaerica* (Sacc.) E.W. Mason	+	-	-
13.	*Pestalotiopsis* sp.	+	-	-
14.	*Phomopsis* sp.	+	+	+
15.	*Rhizopus stolonifer* (Ehrenb.) Vuill.	+	-	-
16.	*Trichoderma viride* Pers.	-	-	+

Table 2. Showing fungal endophytes isolated from petiole of three *Terminalia* spp.

S. No.	Fungal Organisms	Presence of Endophytes		
		T. bellerica	*T. chebula*	*T. crenulata*
1.	*Alternaria alternata* Fr. Keiss.	-	+	+
2.	*Aspergillus flavus* Link	+	-	-
3.	*Aspergillus niger* van Tiegh	+	-	-
4.	*Chaetomium* fusisporale J.N. Rai & Mukerji	-	-	+
5.	*Curvularia* sp.	-	+	-
6.	*Emericella* sp.	-	-	+
7.	*Fusarium roseum* Link.	-	+	-
8.	*Fusarium* sp.	-	-	+
9.	*Gloeosporium* sp.	-	+	-
10.	*Nigrospora sphaerica* (Sacc.) E.W. Mason	+	-	-
11.	*Pestalotiopsis* sp.	+	-	-
12.	*Phomopsis* sp.	+	+	+
13.	*Trichoderma viride* Pers.	-	-	+

Table 3. Showing fungal endophytes isolated from Leaf tip of three *Terminalia* spp.

Sl. No.	Name of Fungal organism	Presence of Endophytes		
		T. bellerica	***T. chebula***	***T. crenulata***
1.	*Alternaria alternata* Fr. Keiss	-	+	+
2.	*Aspergillus flavus* Link	+	+	+
3.	*Aspergillus niger* van Tiegh	-	+	+
4.	*Chaetomium* sp.	+	-	-
5.	*Emericella* sp.	-	-	+
6.	*Fusarium roseum* Link	-	+	-
7.	*Phomopsis* sp.	+	+	+
8.	*Rhizopus stolonifer* (Ehrenb.) Vuill.	+	-	-
9.	*Trichoderma viride* Pers.	-	-	+

The three locations differed in species richness, location 1, arboratum resulted into more isolates from *T. bellerica* and *T. crenulata* whole location 2 Narmada was more suitable for isolates in *T. chebula* and *T. catappa*. The assemblage was location and host specific *T. bellerica* harbored 20 fungi while *T. chebula* only 7 fungi. Two spp. of *Phomopsis* were recorded from the leaves of *T. bellerica*, while *Cladosporium herbarium* was recorded from *T. bellerica* and *T. crenulata*. Similar to occur of taxa in *T. arjuna* (Tejesvi, 2005) *Chaetomium globosum* and *Pestalotiopsis* of was recorded from different location in three *Terminalia* while *Myrothecium verrucaria* was recovered from leaf segments of *T. bellerica* and *T. crenulata.*

The broad diversity of these fungi as well as their easy access makes them interesting for natural product screening programme. How these fungi enter in the plant and produce cellulose degrading enzymes is our next study target.

Acknowledgements

We are grateful to University Grants Commission (UGC) for providing financial assistance in the form of Major Research Project 41-458/2012 (SR) Dated 16/07/2012.

REFERENCES

Arechavaleta, M., C.W. Bacon, C.S. Hoveland and D.E. Radcliffe. (1989). Effect of the tall fescue endophyte on plant response to environmental stress. *Agron. J.* 81233.

Arnold A.E., Mejia L.C., Kyllo D., Rojas E.I., Maynard Z., Robinns N. and Herre E.A., (2003). Fungal endophytes limit pathogen damage in a tropical tree. PNAS USA 100, 15649-15654.

Bills G.F. and Polisshook J.D. (1994). Abundance and diversity of microfungi in leaf litter of a lowland rain forest in Costa Rica. Mycologia 86, 187-198.

Campanile G, Ruscelli A and Luisi N. (2007). Antagonistic activity of endophytic fungi towards Diplodia corticola assessed by in vitro and in planta tests. European Journal of Plant Pathology 117: 237–246.

Carroll, G.C. and Carroll, F.E. (1978). Studies on the incidence of coniferous needle

Clay, K. (1990). Fungal Endophytes of Grasses. *Ann. Rev. of Ecology and Systematics,* 21: 275-297.

Cragg, G.M. and D. J. Newman (2004/Rev.2006), Plants as a source of anti-cancer agents in Ethnopharmacology, [Eds. Elaine Elisabetsky, Nina L. Etkin], in Encyclopedia of Life Support Systems (EOLSS), Developed under the Auspices of the UNESCO, Eolss Publishers, Oxford.

Danielsen S and Jensen DF. (1999). Fungal endophytes from stalks of tropical maize and grasses: isolation, identification, and screening for antagonism against *Fusarium verticillioides* in maize stalks. *Biocontrol Science and Technology* 9: 545–553.

Khan, R., Shahzad, S., Choudhary, M.I., Khan, S.A. and Ahmad, A. (2007). Biodiversity of the endophytic fungi isolated from *Calotropis procera* (AIT.) R. BR., *Pak. J. Bot.,* 39(6): 2233-2239

Khan, R., Shahzad, S., Choudhary, M.I., Khan, S.A. and Ahmad, A. (2010). Communities of Endophytic fungi in medicinal plant *Withania somnifera*, *Pak. J. Bot.,* 42(2): 1281-1287.

Malinowski DP and Belesky DP. (2000). Adaptations of endophtye-infected cool-season grasses to environmental stresses: mechanisms of drought and mineral stress tolerance. *Crop Science* 40: 923–940.

Márquez, L.M., Redman, R.S., Rodriguez, R.J. and Roossinck, M.J. (2007). A virus in a fungus in a plant – three way symbiosis required for thermal tolerance. Science 315: 513–515.

Petrini, L. (1987). Endophytic fungi of alpine Ericaceae. The endophytes of *Loiseleuria procumbens*. In: G.A. laursen, J.F. ammirati and S.A. Redhead (ed.), *Arctic and alpine mycology of leaves*. Springer-verlag, New York. pp-71-77.

Redman, R.S., Dunigan, D.D. and Rodriguez, R.J. (2001). Fungal symbiosis: from mutualism to parasitism, who controls the outcome, host or invader? New Phytologist 151: 705–716

Saikkonen, K., Faeth, S.H., Helander, M. and Sullivan, T.J. (1998). Fungal endophytes: a continuum of interactions with host plants. *Annual Review of Ecology and Systematics,* 29: 319-343.

Saikkonen, K., Lehtonen, P., Helander, M., Koricheva, J. and Faeth, S.H. (2006). Model systems in ecology: dissecting the endophyte-grass literature. *Trends in Plant Sci.* 11: 428-433.

Schardl, C.L., Leuchtmann, A.H., Tsai, F., Collett, M.A., Watt, D.M. and Scott, D.B. (1994). Origin of a fungal symbiont of perennial rye-grass by interspecific hybridization of a mutualist with the ryegrass choke pathogen, *Epichloe typhina. Genetics,* 136: 1307-1317.

Tejasvi, V., Basavanna, M., Monnanda, S.N., Harishchandra, S.P., Kukkundoor, R.K., Subbiah, V. and Hunthrike, S.S. (2005). Endophytic fungal assemblages from inner bark and twig of *Terminalia arjuna* W. & A. (Combretaceae). *World Journal of Microbiology & Biotechnology*. 21: 1535-1540.

Schulzt B. (2002). Endophytic fungi: a sources of novel metabolites. In British Mycological Society, International Symposium Proceedings on Bioactive fungal metabolites – Impact and Exploitation at University of Wales.

Stone J.K., Polishook J.D. and White J.F. (2004). Endophytic fungi. In: Biodiversity of fungi. Inventory and monitoring methods (Mueller G.M., Bills G.F., Foster M.S., eds). Elsevier Academic Press, San Diego, USA. pp. 241-270.

Williamson, P.R. (1994). Biochemical and molecular characterization of the diphenol oxidase of *Cryptococcus neoformans*: identification as a laccase. *J Bacteriol,* 176: 656–664.

■■■■

Chapter - 4

Role of microbes in Biodeterioration of Museum objects in Gujarat

Arun Arya

Department of Environmental Studies, Faculty of Science,
The Maharaja Sayajirao University of Baroda, Vadodara, Gujarat-390002, India
Email: sarojarun10arya@rediffmail.com

ABSTRACT

Baroda museum and picture gallery is known for its valuable cultural heritage including Egyptian mummy, skeleton of blue whale and world famous European oil paintings. Microorganisms particularly fungi produce different affects and discolouration of museum objects. For preservation of valuable artefacts it is necessary to study the indoor aeromycoflora and enzyme producing capabilities of these biodeteriogens. Fungus like *Emericella nivea* was associated with fingers of mummy causing white spots. A detailed account of these fungi in damage of miniature paintings, mummy, manuscript and herbarium sheets is presented.

Keywords: Fungi, Biodeterioration, Museum objects, paper, mummy, coins

4.1 INTRODUCTION

Museums are not merely storehouses but they are centers of education in the service of Society. Museums have collections of cultural, artistic, ethnological archaeological, historical and scientific significance. Today museums have also become centers of non-formal education. According to Foster (1986),"A good museum attracts, entertains, arouses curiosity, leads to questioning and thus promotes learning. It is an educational institution that is set up and kept in motion, that it may help the members of the community to become happier, wiser and more effective human beings" (Dana, 1909).

The word MUSEUM is derived from the Greek word, Mouseion, which was a place of contemplation, a philosophical institution or a temple of the Muses. ICOM (1990) defines a museum as a non-profit making, permanent institution in the service of society and its development, which communicates and exhibits the objects for purpose of study, education and enjoyment, material evidence of people and their environment.

- Aerobiology is a branch of biology that studies organic particles, such as bacteria, fungal spores, very small insects, pollen grains and viruses, which are passively transported by the air (Spieksma, 1991).
- Aerobiologists have traditionally been involved in the measurement and reporting of airborne pollen and fungal spores as a service to allergy sufferers (Larsson, 1993). P.H. Gregory (1973) has written a book on aerobiology.

The Baroda museum and picture gallery is famous for its collection of European oil paintings all over Asia. The buildings of Baroda museum are situated in pastoral environment of Sayaji Park (Plate I). This Museum was built in the Indo-Saracenic style of architecture in 1984. The department of Museology also had a good museum. This is a pioneer institution in the field of Museum training at the postgraduate level, associated with the M. S University. The department was established fifty years back in the Baroda Museum and picture gallery. The department is presently housed in a new three storeyed building in the campus of the Faculty of Fine Arts, Opposite Sayajibaug, near the river Vishwamitiri. Although the department started its new building in 1976, the museum was thrown open to the public in the year 1989 only. Museum of the Department possesses different types of objects, such as tribal musical instruments, woods carving, objects made of paper mache, paintings and paper manuscripts, etc.

The Museology and Baroda Museum are situated within a radius of 1 km in Sayajigunj area of Baroda City. Two other spots, Oriental Institute and Fatehsingh Museum are situated 7 km apart. The Oriental Institution is a part of the M. S. University of Baroda. It has a collection of a large number of books and old Journals, paper and palm leaf manuscripts. Now it is situated on Palace road. Earlier it was housed in Smt. Hansa Mehta Library (Plate I) near Faculty of Science.

Biodeterioration can be defined as any undesirable change in materials of economic importance, brought about by the activities of living organisms. Biodeteriogens may be plants, microbes and animals including human beings. Biodeteriogens like fungi, bacteria and actinomycetes pose a severe threat to museum objects. Microbes gain entry into the indoor atmosphere through wind currents and settle on various objects by impaction (Gregory, 1973). Sudden Jumps-introductory phases required Aerospora form a normal component of the earth's atmosphere. Aerobiologists have correlated such data with clinical allergy. These studies are helpful in assessing the deterioration and management of the environment. The aerobiological data may be used for the formation of disease forecasting system of plants (Mishra, 1992) and human beings (Davis and Smith, 1973; Spieksma, 1980). Role of certain fungi present in indoor air and Biodeterioration caused by them is discussed in the resent paper.

4.2 STUDY OF INDOOR AEROMYCOFLORA

It is noteworthy that twenty-four spp. of fungi belonging to fifteen genera were isolated from Baroda Museum in 1999 and twenty-two spp, belonging to seventeen genera were isolated in 2000 (Table 1). The presence of viable spores of *Aspergillus japanicus*, A. *wentii, Chaetomium atrobrunneum, C, gangleg-arum, Lewia infectoria, Monascus rubur* and *Phoma nebulosa* is noteworthy in indoor air. These biodeteriogens have the potential causing plant as well a animal disease, and biodeterioring cultural property at the same time. According to Nair (1972), fungi that can grow on any suitable substrata and play a prominent role in the deterioration of museum materials. Tilak and Kulkarni (1972) and Tilak *et al* (1970-1972) failed to detect such fungi during their surveys in Ajanta caves. Conidia of *Aspergilli* and Fungi imperfecti dominated in their slides. They concluded that the excreta of bats and birds residing in these caves serve as feeding source for the microorganisms and this acts as a secondary source of aerospora. Air in the Mummy chamber harboured different mycoflora (Table 1), out of which *Aspergillus* species dominated; in all twenty-eight colonies were produced in 5 petriplates

Table 1. Presence of indoor aeromycoflora in two different galleries of Baroda Museum and Picture gallery, Baroda in 2005

Sr. No.	Fungi isolated	No. of Fungal colonies			
		PG*	EG*	Total	%
1.	*Alternaria alternata*	2	2	4	1.22
2.	*Aspergillus flavus*	5	10	15	4.42
3	*A. niger*	20	10	30	9.42
4	*A. sydowii*	8	7	15	4.91
5	*A versicolor*	20	10	30	9.42
6	*Cladosporium herbarum*	-	5	5	1.10
7	*Chaetomium gangligerum*	10	-	10	2.45
8	*Curvularia lunata*	5	10	15	5.15
9	*Emericella nidulans*	10	-	10	3.68
10	*Fusarium moniliforme*	-	10	10	3.68
11	*F. oxysporum*	5	10	15	5.52
12	*Monilia sitophila*	5	-	05	3.07
13	*Penicillum citrinum*	20	-	20	7.98
14	*Rhizopus stolonifer*	20	10	30	9.42
15	*Trichoderma lignorum*	5	5	10	5.52

Occurrence of fungal flora in 5 Petriplates.
Note: PG= Picture gallery, EG= Egyptian Gallery

Plate 1. Baroda museum and Picture gallery and Smt. Hansa Mehta library. Fungal biodeteriogens, *Aspergillus, Penicillium* and *Curvularia* Miniature Painting and Egyptian mummy

Plate 2. Biodeteriogens *Chaetomium, Rhizopus and Nigrospora* causing damage to miniature painting and herbarium sheet of Lygodium

4.3 DETERIORATION OF MUSEUM OBJECTS

Museum objects included miniature paintings, paper objects, wooden objects, herbarium sheets, cloths with embroidery designs. A survey of different miniature paintings resulted into isolation of *Aspergillus niger*, *A. flavus* and *Chaetomium gangligerum* and *C. globosum*. Objects like herbarium specimens revealed the presence of *Curvularia prasadi* and *Penicillium citrinum*

Table 2. Symptoms of biodeterioration on different Kinds of museum materials

Sl No	Symptoms	Example
1.	Different Stains and Discoloration	Caused by surface growth of Fungi on paper etc. and production of colouring compounds
2.	Disfigurement	Positive changes in the surface of a painting due to fungi
3.	Fibrilization	Loss of coherence in fibrous material as in wool when attacked by Microorganisms.
4.	Penetrations	Invasion of the lumen of certain natural fibers by fungi.
5.	Changes in mechanical properties	Loss of weight and strength of wood on reduction or breaking of Lignin.

4.4 DETERIORATION OF MUMMY IN BARODA AND KOLKATA MUSEUM

Pyramids of Giza are made up of 2 million blocks of stone. Weight of each stone is 2 tons. These were cut and then brought here. Preservation of body from decomposition was practiced by Egyptian society. As they believed in existence of God, Living being (Soul - **Ba**) and dead body. The human being has existence even after death hence his/ her body should not deteriorate. Mummy is derived from Greek word Mumiyah which means Bitumen. It helped in maintenance of personal identity and Creation of a form resembling to God after his/ her embalmment. The entire requirement had to be satisfied in order to ensure the survival of dead in the next world. (David, 1978). Lopez- Martinez *et al.* (2007) isolated a total of 469 fungi from 12 mummies. Among the isolated fungi were *Penicillium, Cladosporium, and Aspergillus* spp. Fungi were present in keratinized tissue like nails, and hairs etc. (Howard, 1983).

In Baroda Museum and Picture gallery there is Egyptian gallery, which houses, the 154 cm long mummy is kept in wooden coffin box made up of *Sekamore* tree. This mummy is 230 BC old. The internal body parts are preserved separately. The body was washed in palm wine and then kept in Natrum for 70 days. In Baroda Museum the deterioration of Egyptian Mummy was recorded during year 2000. The two fingers of right foot showed the presence of white coloured growth. On isolation from the spot using Potato dextrose agar medium the presence of fungus *Emericella nivea* was recorded. This is perfect state of *Aspergillus*. The presence of white coloured conidia caused white coloured patches in fingers, which spread to other fingers as well.

4.5 CONCLUSION

Baroda museum is a gift to citizens by its visionary ruler maharaja Sayajirao Gaekwad III. It was built in 1894 and picture gallery was inaugurated in 1921. The museum houses a large number of objects, these are exposed to environment where different fungi can grow on organic materials like paper, cloth, wood and dried and preserved animals and Egyptian mummy. The study showed a number of fungi like species of *Aspergillus, Penicillium, Rhizopus, Cladosporium, Alternaria* etc. *Emericella nivea* is reported from fingures of Egyptian mummy which was brought by Maharaja from New York and *Cheatomium gangligerum* and *C. globosum* from miniature paintings. *Chaetomium* is a cellulose degrading fungus. There is need for periodic inspection of organic objects and maintain the humidity and temperature in the galleries to reduce the fungal proliferation and conserve the valuable cultural heritage for the younger generations.

REFERENCES

Davis R.R. and Smith L.P. (1973). Aspects of human allergy and fungal spores. Clinical Allergy, 3: 23-267

David R. (1978). Mystries of the mummies Macmillan Pub. Co. Inc. New York pp58-81

Fatmi A.S. and Zarrin A. (2014) Egyptian mummies in Indian Museums. In : biodeterioration of Cultural Property- 7 95-104

Foster R.A. (1986). The larger provincial museums and art galleries. In : Manual of curatorship. Butterworths, London, 66-75.

Gregory P.H. (1973) The microbiology of the atmosphere. Leonord hill books, Aylesbury, England 2nd ed. Pp 377.

Howard D.H. (1983) Fungi pathogenic for human and animals. Pathogenicity and detection, Marcel and Decker New York 267-271

http://www.ehow.com/way_5149770_listerine-cure-toenail fungus. html#ixzz2qr5p1FkY

ICOM 1990 ICOM Statutes code of professional ethics, Paris pp. 3

James TGH (1979) An Introduction to Ancient Egypt British Museum Pub. Ltd. London 286

Lopez- Martinez R. Henandez- Harnandez F. Millan Chu B.E. and Manzano-Gyayosen L.J. (2007) Effectiveness of Imazalil to control the effect of fungal deterioration on mummies at mexico city museum. Revista Ibe de Micologia pp 283-288.

Maspero G. (1908) New Light on ancient Egypt T. Fisher Unwin, London, pp 315

Mishra G. and Mishra K.B. (1992) Aeromycoflora of a mica mine. In: National symposium on Management of microbes in service of mankind. Allahabad Univ. 19-21.

Nair S.M. (1972) In: Conservation in tropics. Pros. of AST- Pacific seminars on Con. of Cul. Properties. (ed. Agrawal O.P.) Rome 150-158.

Spieksma, F.T.M. (1980) Importance of aerobiological studies for the prediction of human diseases. In: Aerobiological studies for the prediction of human diseases. Aerobiol. Symp. Munich 307-315.

Tilak S.T., Sharma B.R. N., Sengupta S.R. and Kulkarni R.L. (1970) The deterioration of Ajanta and Ellora paintings. In Con. of Cul. Prop. In India Proceedings of V seminar, New Delhi 77-82.

■■■■

Chapter - 5

Effect of cement dust pollution on the colonization of leaf surface microorganisms

Arun Arya and Ankita Bhatt

Department of Environmental Studies, Faculty of Science, The Maharaja Sayajirao University of Baroda, Vadodara- 390002.

Email : sarojarun10arya@rediffmail.com, bhattankita04@gmail

ABSTRACT

Many physical, chemical and biological factors bring about causative changes in composition of aeromycoflora of an area and different fungal species are restricted to that particular areas with specific environmental conditions. As aerobiology deals in large parts with bio-particles present in air, it contributes a lot in enumeration of types of bio-particles present. Among all the air borne bio-particles, fungal spores constitute the greatest and most important portion in air (Salvaggio and Lars, 1981). On this basis, of the recent aerobiological investigations, it can be broadly classified into two categories as Indoor or Intramural aerobiology and Outdoor or Extramural aerobiology.

Cement industry may cause increase in SPM along the distance from the point source with an annual average of about 175.64 $\mu g/m^3$. SPM conc. was more in winter and less in rainy season. This increased pollution load and presence of pollens may increase the concentration of fungal pollutants. The pollutant dust may be responsible for reduced photosynthesis and less protein content.

Keywords: Cement dust, Leaf surface, Microorganisms

5.1 INTRODUCTION

Atmospheric pollution causes serious health hazards. Volcanic eruption housing construction activities and moving traffic causes dust pollution. Pollution of smoke and indoor dust can increase the probability of Asthma and respiratory problems. Chemical industry produce volatile chemical pollutants and cause health problems. In 2018 in New Delhi during Deepawali air was polluted and due to SMOG visibility was poor. Children and elderly people were not advised to go for walk during morning and evening hours.

Indian Cement industry has been the major force behind the nation's construction industry for nearly ten decades. From small beginning in the year 1914, this has grown into a major industry in India. By 1999 the total production is likely to exceed 90 million tonnes (Mehta and Dubey, 2003). Lot of dust is released by cement industry. The dust is responsible for both visible and latent consequences on ecosystem. The pollutant dust may reduce the photosynthetic area and can reduce the yield of plants. The openings or pores may be blocked reducing the gaseous exchange and rate of transpiration.

Studies on the phylloplane micro flora of safflower revealed that the bacterial population was more than the fungal population in both the varieties at various growth stages of crop. The mean fungal population was highest at flowering stage ($5.32 \times 102/cm^2$ leaf area) whereas the bacterial population was more at the stage of physiological maturity ($32.15 \times 102/cm^2$ leaf area). Fungal and bacterial populations were highest on phylloplane of variety 'Bhima' (Manndhar and Suryawanshi, 2009).

5.2 OCCURRENCE OF PHYLLOSPHERE MICROORGANISMS

Phyllosphere, being a rich and complex habitat for microorganism, is a larger space for both agricultural and environmental important microbes. According to a group of researchers a large array of culture dependent and independent bacteria, fungi, yeast and algae are found on the phylloplane, known as epiphytic microbes. Different studies have been revealed that bacteria are the most dominant microbes on phyllosphere, average population is 106-107 CFU cm^{-2} (up to 108 CFU per gram) of leaf (Hirano and Upper, 2000; Andrews, and Harris 2000). The growth of phylloplane microbes generally depends on various biotic and abiotic factors.

Many physical, chemical and biological factors brings out causative changes in composition of aeromycoflora of an area and different fungal species are restricted to that of particular area with specific environmental conditions (Bajwa *et al.*, 1997).

5.2.1 Morphological Characteristics

The phylloplane inhabitants possess morphological characteristics for their survival. These include the development of pigment in their mycelia, spores, pycnidia, apothecia, cleistothecia for protection against strong light and desiccation. The dark pigments act as light screen. These pigments are often referred to as melanin.

5.2.2 Physiological Characteristics

Phylloplane fungi possess a number of Physical and Physiological characteristics, which are discussed as below:

(A) Nutrition: Phylloplane microfungi have ability to decompose cellulose by producing cellulases. In addition, pectinases, cutinases, proteases have also been estimated in many fungi e.g. *Alternaria alternata, Aureobasidium pullulans, Botrytis cineria, Cladosporium herbarum.*

(B) Radiations: Besides certain exceptions, high intensity light inhibits mycelial growth, while light (of normal intensity) is not harmful to phylloplane fungi. The UV portion of spectrum plays a significant role. Fungi containing melanin pigment are resistant to UV exposure as compared to those containing hyaline mycelia.

Hyaline spores of Aureobasidium and Sporobolomyces are killed by exposure of UV for five minutes, whereas dark spores of *Alternaria* and *Epicoccum* survive even after exposure for 35 minutes. High intensity of UV radiation becomes lethal to microorganisms.

(C) Temperature: Generally, the phylloplane fungi are mesophilic growing at temperature 20-25°C. Even some fungi (*e.g. Alternaria alternata, Aureobasidium pullulans, Cladosporium herbarum*, and *Botrytis cinerea*) can grow below 0°C.

(D) Relative humidity: During rain and dew formation, high relative humidity on leaf surface is found. But when there is wind the microenvironment is reduced. Therefore, phylloplane fungi get advantages at low levels of relative humidity. Generally, a faster growth rate of the germ tube and mycelium of some fungi occurs at 100% relative humidity. Even some could not grow below 93% of relative humidity.

5.3. PHYLLOPLANE AND PHYLLOSPHERE MICROFLORA

The term phyllosphere was first coined by Last, (1955) to denote the leaf surface environment. The phylloplane, the surface of plant leaves is a complex terrestrial habitat that is characterized by a variety of microorganisms including bacteria, filamentous fungi and yeast. Phylloplane fungi are the mycota growing or the surface of leaves. There are two groups of phylloplane fungi: residents and casuals. Residents can multiply on the surface of healthy leaves without noticeably affecting the host. Whereas, casuals land on the leaf surface but cannot grow. Phylloplane fungi have been poorly studied as compared to endophytes, saprobes and pathogenic fungi.

Lot of investigations have been carried out on the phylloplane flora of leaf surfaces of several plants growing in garden or cultivated in many parts of the world by several researchers (Abdel-Fattah *et al.*, 1977; Eicker, 1976; Nagaraja, 1991; Sharma, 1974; Dalal 2014). El-Said A.H.M. (2001) reported the fungi from leaf surfaces (phyllosphere and phylloplane) and observed were basically similar on the two types of media and the most common fungi were *Alternaria, Aspergillus, Chaetomium, Cladosporium, Cochliobolus, Curvularia, Gibberella, Memnoniella, Mycosphaerella, Setosphaeria* and *Stachybotrys*. The indoor aeromycology includes the study of indoor aeromycoflora of libraries (Nitrini *et al.*, 1992),

5.4 HEALTH EFFECTS OF PHYLLOPLANE MYCOFLORA

The researches done on the study of airborne fungal spores and pollen grains and its impact on human health, animal beings and plants and other flora and fauna reveal that the airborne fungal spores cause the respiratory disorders like asthma, allergic rhinitis, skin diseases, ear diseases etc. The airborne fungal spores and their concentration vary from place to place. The aeromycoflora of an area and different fungal species are restricted to that particular areas with specific environmental conditions. The phylloplane mycoflora may change according to host plants and pollutant like SPM or cement dust. These spores present in air and inhaled by various living beings like human beings, animals, birds. and also they settled on the plant surfaces, where they germinate and produce mycelia or they may produce toxins.

On getting the suitable substratum, they germinate, forms mycelia and complete their life cycle or multiply themselves and causes the diseases to the objects like human skins, eyes, heads, throats, lungs etc. and causes the infections to the plants and living organisms too. Such fungi which deposited on leaf surfaces and the study of such leaf surface environment is called the phyllosphere.

5.5 POLLUTION DUE TO CEMENT INDUSTRY

Air pollutants of major concern are particulate matter and gaseous pollutants. Particulate matter may be cement dust, stone crusher fugitive dust, coal dust, fly ash, lead particles, oxides of magnesium and iron etc. Sources of particulate pollutants are a variety of natural and anthropogenic activities. Air pollution problems in India show a highly irregular pattern due to a nonuniform pattern of industrialization. The gaseous pollutants as well as fly ash particulates cause various injury symptoms on the leaves. These injuries adversely affect the physiological functions especially the photosynthetic activity of the plants (Trivedy and Singh, 1995). Pandit and Prasannakumar (2009) reported that dust holding capacity of leaves varies according leaf structure, its surface geometry, epidermal and cuticular features. The Protein and chlorophyll content decreased from 500 m to 50m.

Scientists reported that the pollutants from stone crushers may produce several diseases. Silicosis is an important disease caused by crystalline silica. A variety of dust particles can help to increase the mycoflora on leaves.

5.6 ISOLATION OF PHYLLOSPHERE MICROFLORA

Isolation and Identification of phyllosphere mycoflora A modified leaf washing technique was adopted to estimate phylloplane mycoflora of healthy and virus infected tomato plants. Disc of 3 mm diameter were cut randomly from leaves of each variety with sterile cork borer. Fifty disc were placed in 250 ml conical flask containing 100 ml sterile distilled water and shake for 20 minutes to get a homogenous suspension. One ml suspension was pipette out in to sterilized petriplates. The plates were poured with peptone dextrose agar medium containing rosebengal and streptomycin medium (For Fungi) and mixed thoroughly and kept undisturbed in dust free chamber at room temperature for 7 days. On incubation, the fungal colonies were observed and identified on the basis of morphological and reproductive characters.

$$\text{Population per cm}^2 = \frac{\text{No. of colonies / ml washing}}{\text{Area of one disc} = \pi r2} \times 100$$

where 'r' is the radius of disc in cm

$$\text{Relative abundance} = \frac{\text{No. of colonies of the fungal form}}{\text{Total number of fungal colonies}} \times 100$$

Table 1. Occurrence of different Fungal species on *Calotropis* and *Croton* leaves

Sr. No.	Name of Fungi isolated	*Calotropis*	*Croton*
1	*Mucor hemalis*	0.50	1.0
2	*Rhizopus stolonifer* (Ehrenb ex Fr.) Linder.	0.77	2.0
3	*Aspergillus niger* Van Tiegh	4.66	6.80
4	*Aspergillus flavus*	2.33	3.50
5	*Aspergillus terreus* Thom.	7.00	4.77
6	*Alternaria alternata* (Fr.) Keissler	2.00	5.00
7.	*Curvularia clavata* Walker	2.00	2.00
8.	*Fusarium oxysporum*	1.40	1.77
9.	*Trichoderma viride* Pes.	0.55	0.33
10.	*Penicillium citrinum*	1.33	0.33
11.	*Cladosporium herbarum*	1.0	0.55
	Total		

Data based on three replicates of leaf per cm^2 of leaf

The number of fungal colonies of *Aspergillus terreus* . *Aspergillus niger, Alternaria alternata Curvularia clavata* were more in *Calotropis procera*. The colonies of *Aspergillus terreus*. *Aspergillus niger* and *Alternaria alternata* were more in *Croton bonpadelium.*

Shinde *et al.* (2004) studied the grain mold fungi in relation to physical and nutritional parameters of Sorghum grains. Garud (1992) reported resistance sources, mechanisms and resistance screening techniques for grain molds. Magar (2003) reported the occurrence of mold flora at different grain development stages in Sorghum. These spores present in air and inhaled by various living beings like human beings, animals, birds, and also they settled on the plant surfaces, where they germinate and produce mycelia or they may produce toxins which are allergic in nature, sometimes they cause diseases of incurable forms to living beings. Are allergic in nature, sometimes they cause diseases of incurable forms to living beings.

Plants with high APTI were recommended for use in domestic, industrial and urban land scapping both for their aesthetic values and phytoremediation potential of air polluted environment, while the plants with low APTI (Sensitive) were to be used as bio-indicators of environmental air quality. Air pollution Tolerance Index (APTI) is used to determine the degree of sensitivity of plants to environmental air pollution. Singh (1991) suggested a method of determining Air Pollution Tolerance Index (APTI) by synthesizing the values of four different biochemical parameters-leaf extract pH, ascorbic acid, total chlorophyll and relative water content. If APTI value for a plant is high it will be more resistant to diseases.

The result revealed that different plants respond differently to air pollution. This variation of the APTI can be attributed to the variation in any of the four biochemical parameters which govern the computation of the index. The relative water content (RWC) ranged between 70% to 99.29% in *Ficus*, while ascorbic acid ranged from 3.5mg/l in *Ixora.* In the computation of APTI using the formula, A (T+P) + R,/10, A, which is the ascorbic acid is the multiplication factor and therefore determines to a great extent what the value of APTI would be. The same thing applies to T (total chlorophyll).

REFERENCES

Abdel-Fattah, H.M., Moubasher, A.H. and Abdel-Hafez, S.I.I. (1977). Fungus flora of root and leaf surface of broad bean cultivated in Oases, Egypt. Naturalia Monspeliensia Ser. Bot. 27: 167-177.

Andrews, J.H. and Harris, R.F. (2000) The ecology and biogeography of microorganisms of plant surfaces, Annu Rev Phytopathol, 38, 145–180.

Bajwa, R., M.H. Shah, A. Javaid and Z. Tasneem.(1997). Aeromycoflora of Lahore. I. Seasonal variation in air mycoflora of highly commercialized, thickly populated areas. Pak. J. Pl. Sci., 3(1): 17-24.

Dalal. L.P. (2014) Study of Phylloplane Flora of College Premises of Wardha. International Journal of Scientific & Engineering Research, Volume 5, Issue 2, February-2014 672 ISSN 2229-5518 IJSER © 2014 http://www.ijser.org

Eicker, A. (1976). Non-parasitic mycoflora of the phylloplane and litter of Panicum coloatum. Trans. Brit. Mycol. Soc. 67: 275- 281.

El-Said A.H.M. (2001). Phyllosphere and Phylloplane Fungi of Banana Cultivated in Upper Egypt and their Cellulolytic Ability. Mycology. 29(4):210-217

Garud, T.B. 1992. Resistance sources, mechanisms and resistance screening techniques for grain molds. Proceedings of XXII Annual Sorghum Workshop, Surat, Gujarat, India, 2-4 Apr. 1992:26.

Hirano, S.S. and Upper, C.D. (2000) Bacteria in the leaf ecosystem with emphasis on Pseudomonas syringae — a pathogen, ice nucleus, and epiphyte, Microbiol Mol Biol Rev, 64, 624–653.

Magar, S. J. 2003 Occurrence of mold flora at different grain development stages in sorghum. M.Sc. (Agri.) dissertation submitted to Marathwada Agriculture University, Parbhani (M.S.) PP: 85

Mehta N. and Dubey P.S. (2003) Air Quality in three different cement producing areas. In Air Pollution Development at what cost? (eds.) Jasrai Y.T. and Arya A. Daya Publishing House, Delhi 231 pp.

Nagaraja, T. G. (1991). Rhizosphere and phyllosphere studies in *Strychnos nuxvomica* Linn. Ad. Plant Sci. 4: 171-173.

Nitrini, S.M. de O., Matle, G.R. and Matle, M.H. (1992). Monitoring the number of moulds in Library submitted to an air sterilizer apparatus. In: Preprints of 2nd Int. Conf. on Biodeterioration of Cultural Property -2. Yokohama (Japan). pp.221-227.

Pandit B.R. and Prasannakumar P.G. (2009) Reduction in protein and chlorophyll content in *Calotropis gigantean* (L.) R.Br. growing near stone crusher. In: Eco-degradation Due to air pollution (eds.) Arya A. Bedi S.J. and Patel V.S. Scientific Pub. Jodhpur India 151-158.

Salvaggio, J. and Lars, A. (1981) Mold induced asthma. J. of Aller. and Clin. Immunol. 68: P 327-364. Sharma, P.D. 1974. Experimental studies of some micro-fungi from decayi

Singh S.K. and Rao D.N. (1983) Evaluation of plants for their tolerance to air pollution. In Proc. Symp. On Air Pollution Control, New Delhi, India ; Indian assoc. for air Pollution Control ; 218-241.

Shinde, P.V., T.B. Garud, S.D. Somwanshi, P.H. Ghante and B.M. Shinde, 2004. Effect of two mold causing fungi on physical and nutritional properties of kharif sorghum grains. Int. Sorghum Millets Newslett., 45: 33-34.

Chapter - 6

Studies on Seed Borne Diseases of *Cicer arietinum* (Chick pea) and Trials with Fungus *Trichoderma* as a Bio-control Agent against *Fusarium* spp.

Niharika Nema[1], and Arun Arya[2]

[1]Cellular and Molecular Biology Programme,
The Maharaja Sayajirao University of Baroda,Vadodara, Gujarat-390002, India
[2] Department of Environmental Studies,
The Maharaja Sayajirao University of Baroda,Vadodara, Gujarat-390002, India
Email: niharikanema@gmail.com; sarojarun10arya@rediffmail.com

ABSTRACT

Our balanced diet consists of two major components, cereals and pulses besides vegetable and fruits. Seed health is being recognized as one of the important criteria in evaluating seed quality. According to current estimates, 10% of food grain is lost in storage due to microbial spoilage and insect attack. Storage fungi grow faster in 70-90% humidity. *Cicer arietinum* L. (Chickpea) commonly called as gram (Chana) is most popular pulse and is important for making gram flour (besan). The reduction in yield and poor germination of gram seeds may be due to deterioration caused by various fungal organisms and insect pests. After the harvest many fungal problems result into loss of about 20% of produce. This results in to loss of seed viability, production of toxins and loss of food value.

To find out the occurrence and solution of seed borne infection on gram an effort was made to analyse fungal flora on gram (local variety) in control sets using three wet blotter papers, and with treatment of 1% NaOCl in PDA medium. The results showed occurrence of seven different fungi. Two species of *Aspergillus* were present. The paper deals with control of *A. flavus* and *Fusarium* sp. with *Trichoderma viride* a member belonging to Mitosporic fungi.

Keyword: Seed borne diseases, *Cicer arietinum*, Chick pea, Fungi, *Trichoderma, Fusarium,* Bio-control

6.1 INTRODUCTION

Plant pathology is the science dealing with the diseases of plants. These diseases cause significant losses in the crop yields. Plant suffers from three enemies: the pathogenic microorganisms causing diseases; insects and animals. When the ability of plant to carry out its function properly is interfered with by either a pathogenic microbes or an adverse environmental factor, the cells malfunction or die, and the plant is said to become diseased.

Some pathogens survive as dormant mycelium in the seeds or other propagative structures of host plants. The dormant mycelium becomes active and grows as germination takes place, affecting the plants as the seeds develop. The real story starts when seed formation takes place. The seeds of infected plants may never be formed or if they develop they carry the disease which is shown in its morphology later when they are sown for germination. Such types of diseases can be called the seed borne diseases. Among the pathogens associated to the seeds are viruses, bacteria, nematodes and fungi (Richardson, 1979)

The genus *Cicer* L., formerly regarded as a member of tribe Vicieae Alef, nine species including the cultivated chickpea, *Cicer arietinum*, are annuals and the rest are perennial shrubby species. The earliest remains of chickpea seeds are present in archaeological excavations from Cayonu, Turkey, dates back to 7500-6800 BC. Chickpea was introduced to India as early as 2000 BC (Vishnu-Mittre, 1974) after which it has become an important integrated part of the Indian diet as pulse and for making gram flour (besan). Neergaard (1979) a famous mycologist had described many seed borne pathogens from pulses. Fungal colonization in seed tissues reduces the germination, viability, longevity and vigour of seeds. Zaidi *et al.* (1991) and Arya and Chauhan (1995) had described seed mycoflora associated with chickpea.

Microbes are also used as bio-control agents i.e. the microorganisms to fight as pesticides, insecticides, etc. Microorganisms have indirect positive effects on plants affecting adversely for population density, dynamics, and metabolic activities of soil borne pathogens mainly through competition, antibiosis, lysis and hyperparasitism. Antibiosis is a biological interaction between the 2 or more organisms that is detrimental to at least one of them; it can also be an antagonistic association between an organism and the metabolic substances produced by another.

Hyperparasitism is documented for *Trichoderma*. While using *Trichoderma* as the bio-control agent, it produces compounds which are inhibitory to storage mold growth. The hyphae of this organism coils around the hyphae of the pathogen. The cellulases digest the pathogen's cell wall. This way it is an important weapon in integrated pest management. As an interesting fact *Trichoderma* is also found in the seed mycoflora. In the present investigation blotter method was employed and efforts were made to control *Fusarium* wilt *in vivo* using *Trichoderma viride.* Seed borne diseases of chickpea were studied by Haware *et al.* (1986)

6.2 MATERIALS AND METHODS

1. Chickpea seeds were treated with NaOCl (1% available chlorine) for 5 min and kept on 3 wet blotter papers (treated).
2. Chickpea seeds were kept on 3 wet blotter papers (control).
3. The mycoflora associated with germinating seeds was isolated and identified.
4. In order to observe the biocontrol activities of *Trichoderma viride,* maize-meal medium was used. The isolated *Trichoderma viride* and *Fusarium oxysporum* species was separately grown.
5. In several pots the chickpea seeds were grown with *T. viride*, and *Fusarium.*
6. Control sets were also maintained.

Media used

A. Modified Asthana and Hawker's medium 'A'

3.5g	KNO_3
1.75g	KH_2PO_4
0.75g	$MgSO_4.7H_2O$
10.0g	Dextrose
20.0g	Agar
1000ml	distilled water

B. Maize meal medium

100g sand (washed many times with tap water and dried)

20g maize grains (broken)

Water for moisture

Add all the contents in a 150ml flask and autoclave for 45 min at 15lbs psi.

C. Potato Dextrose Agar medium

200g potato (washed, peeled, diced and boiled)

20g agar

20g dextrose

1000ml distilled water

Boil the potato in 500 ml of distilled water. Filter the contents. Take the filtrate and add agar. Heat while dissolving agar and continuously stirring. Add dextrose and make the final volume to 1000 ml by adding distilled water.

Transfer the medium in 150 ml conical flasks and culture tubes (as per the requirement). Medium was then autoclaved.

6.3 RESULTS AND DISCUSSION

Seed mycoflora of chickpea showed the presence of *Fusarium, Trichoderma, Phoma, Aspergillus, Rhizopus,* and *Cladosporium.* Detailed results are shown in Table 1 and Fig. 3. Untreated seeds showed more number and types of fungi. Seed borne infection of fungal pathogens is important not only for its association with seeds but also contamination of soil by permanently establishing its inocula (Bello and Sisterna, 2010).

Table 1. Results of blotter paper test showing % seed infection, % seed germination and radicle length.

S.No.	% seed infection	% Seed germination	Radicle length after 5 days (cm)
1.	100	25	4.5
2.	90	25	5.0
3.	20	100	4.3
4.	100	10	3.7
5.	80	40	2.0
6.	10	100	5.0
7.	90	10	1.5
8.	10	50	2.0
9.	80	30	1.6

Species of *Aspergillus, Penicillium,* and *Rhizopus* may produce aflatoxins and other mycotoxins which affect the seed viability, germination and vigour. These are called the storage fungi, which mainly consists of several species of *Aspergillus.* Species of *Penicillium* are encountered at times. Usually in lots of grain stored at low temperatures and with moisture contents above 16%. The storage fungi do not invade grains to any appreciable degree or extent before harvest. Field fungi require humidity more than 90% (Arya and Arya, 2006, Christensen and Kaufmann, 1965). *Cladosporium, Fusarium,* and *Phoma,* are a few examples which grow with the germinating seed and hinder the outcome of the crop and are called the field fungi.

Scientists have found various kinds of fungi growing on chickpea. Seed borne, field and storage fungi produce a large number of toxic metabolites *e.g.* Toxins produce by *A. flavus*, hinder photosynthesis Arya and Monaco (2000).

Arya and Chauhan (1995) studied 6 different varieties of gram seeds. According to their study in 2 years old seeds a total of 12 species were identified.

Some of these included *Aspergillus* sp., *Curvularia* sp., *Rhizopus* sp. *Fusarium* sp. and a few others.

In the present study 9 different fungi were isolated. Fungi were more in untreated seeds. Four different types of Aspergilli were present. The percentage of *Fusarium oxysporum* did not changed after treatment while colonies of *A. niger* reduced and *Cladosporium herbarum* could not found in treated seeds (Table 3). Unidentified sterile mycelium was also encountered in both cases.

The seeds treated with *Trichoderma viride* germinated into healthier plants giving rise to more number of pods although the number of germinated plants did not vary much. This can be due to the production of growth hormones and biocontrol activity of *Trichoderma*.

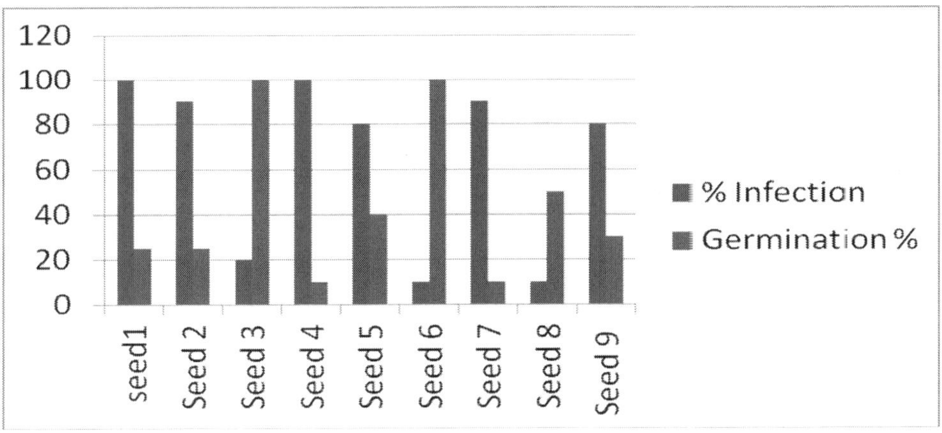

Fig. 1. Percentage germination and % infection in chick pea seeds in different petriplates

Table 2. Percentage germination and seedling growth of chickpea

Treatment	% Germination of seeds	Average no. of leaves**	Average height after 8 days of sowing {cm}
Control 1	100	7	9.2
Control 2	100	5	5.1
Fusarium 1	20	5	7 15
Fusarium 2	10	7	8.6
Trichoderma 1	100	7	9
Trichoderma 2	100	8	10.5
Trichoderma+ Fusarium 1	100	7	11.3
Trichoderma+ Fusarium 2	100	7	9.3

**average of 3 replicates

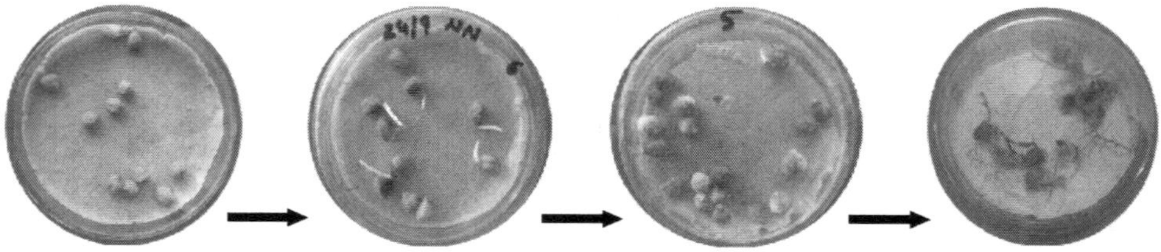

1st day — Germinating seeds on blotter paper — Developing mycoflora on the *Cicer arietinum* seeds — Infected seeds after 10 days on the blotter paper

A. Seeds grown in control 1

B. Seeds grown with *Fusarium* 1

C. Seeds grown with *Trichoderma* 1

Table 3. Percentage occurrence of fungi on treated and untreated seeds of chickpea

Fungi isolated	**Percentage Occurrence**	
	Treated	**Untreated**
Phycomycetes		
Mucor sp.	10	15
Rhizopus stolonifer (Ehrenb.: Fr.) Vuill.	08	08
Ascomycetes		
Aspergillus fumigatus Fres.	12	14
A. niger van Teigh.	10	20
Aspergillus sp.1	10	10
Aspergillus sp.2	09	--
Penicillium citrinum Thom	14	08
Deuteromycetes		
Fusarium oxysporum Schlecht. emend. Snyder & Hansen	07	07
Cladosporium herbarum (Pers.) Link ex Gray	--	10
Phoma glomerata (Corda) Wollenw. & Hochapfel	09	04
Sterile mycelium	11	04

Note: Date based on 100 seeds placed in 10 Petriplates each.

It is evident from Table 2 that in pots having *Fusarium* as maize meal medium the germination of seeds was only 10 and 20%. When the pots were inoculated with *Fusarium* and *Trichoderma* both, the germination was 100% and average height of the plant was 9.3 and 11.3 cm. The performance was better than that in control. *Trichoderma* spp. exhibited higher efficiency in reducing sheath blight of rice and increased grain yield (Tewari and Singh, 2005). Srivastava *et al.* (2012) obtained effective control of chickpea wilt complex caused by *Fusarium, Rhizoctonia* and *Sclerotium* by *T. harzianum* alone or in combination with fungicides. *Trichoderma* was used as a biocontrol agent in chickpea (Shukla, 2001).

Quality of seeds was destroyed due to infestation of these fungi that produce mycotoxins causing health hazards in human beings and animals (Hiscocks, 1965). Fusarubin, moniliformin, vomitoxin, zearalenol and zeraralenone mycotoxins are produced by different species of *Fusarium*. These mycotoxins are hazardous to human and animal health and mostly carcinogens, mutagens and teratogens (Marasas *et al.,* 1984).

Acknowledgements

We would like to thank the Head, Botany department, to provide with the laboratory facilities. We would also like to thank the Head, Cell and Molecular Biology Department, to give the required permissions to continue this research work.

REFERENCES

Arya, C. and Arya, A. (2006) Proper storage conditions for maintaining seed viability and germination of certain leguminous tree seeds. Indian Forester.

Arya, A and Chauhan, R. (1995). Seed Mycoflora of chickpea. *Acta Botanica Indica* 23:293-285.

Arya, A. and Monaco, C. (2000) Seed borne diseases: ecofriendly management scientific Pub. India Jodhpur

Arya, A. and Mathew, D.S. (1991) Seed mycoflora of pigeon pea. *Acta Bot Indica* 19: 102-103.

Bello Gustavo D. and Sisterna, M. (2000) Use of plat extracts as natural fungicides in the Management of seedborne diseases. In: management of fungal plant pathogens. (eds) Arya A. and Perello A.E .Pub by Cabi 51-66p.

Christensen, C.M. and Kaufmann H.H. (1965) Deterioration of stored grains by fungi. Ann, Rev. of Phytopathology 3, 69-84.

Haware, M.P., Nene Y.L. and S. B. Mathur, S.B. (1986) Seed Borne Diseases of Chickpea, Technical Bulletin 1, Danish Government Institute of Seed Technology for Developing Countries, Copenhagen. 1, 1-32.

Hiscocks, E.S. (1965). The importance of moulds in the deterioration of tropical food and feedstuffs. pp 15-26. In: Mycotoxins in foodstuffs. G.N. Wogan (Ed.): MIT Press, Cambridge, Massachusetts.

Kronstad, J.W. (2000) Fungal Pathology. Springer Publication

Landinzinsky, G. (1998) Evolution of crop plants, In: Plant Evolution under domestication. Springer Science.

Marasas, W.F.O., P.E. Nelson and T.A. Toussoun. (1984). Toxigenic Fusarium species, identity and mycotoxicology. The Pennsylvania State Univ. Press, University Park and London, 328 pp.

Neergaard, P. (1979) Seed Pathology, vol. I and II Revised ed. McMillan Press London 1191p.

Richardson, M.J. (1979) An annotated list of seed borne diseases. Phytopathological Papers N-23. Common wealth Mycological Institute, Kew, Surrey, England. International Seed Testing Association, Zurich, Switzerland, 320 PP.

Srivastava, R.K., Singh, R.K. and Prasad, R.D. (2012) Relative antagonistic effect of different isolates of *Trichoderma viride* and *Trichoderma harzianum* against *Rhizoctonia solani. Natl. Acad. Sci. Lett.* 35(1) 49-52.

Shukla, P.J. (2001) Role of *Trichoderma* as a biocontrol agent on seed borne disease in chickpea, M.Sc. Biotech dissertation, Rajkot.

Tewari, L. and Singh, R. (2005) Biological control of sheath blight of rice by *Trichoderma harzianum* using different delivery systems. *Ind. Phytopath*. 58: 35-40

Vishnu Mitre, A. (1974) The beginning of agriculture; palaeobotanical evidence .In J.B. Hutchinson (ed.) Evolutionary studies in world crops. Cambridge.

Zaidi, A., Khan, M.S., Ahemad, M. and Oves, M. (1991) Effects of fungicides on mycoflora of chickpea seeds. *Ind Phytopath* 44: 394-395.

■■■■

Chapter - 7

Integrated Management of Bakanae Disease in Basmati Rice

Anuja Gupta and Ravindra Kumar

ICAR-Indian Agricultural Research Institute, Regional Station, Karnal – 132001

ABSTRACT

Bakanae disease of rice caused by *Fusarium fujikuroi* Nirenberg is widespread in all rice growing countries of the world and is of great concern particularly for Basmati paddy cultivation. The incidence of the disease varies with the regions, rice varieties and the environment. Yield losses due to Bakanae disease has been estimated ranging from 3.0 to 95.4%. Higher crop yields can be attained by using improved crop varieties and integrated pest management technologies. Seed treatment with fungicide is the most important management strategy being followed at present. There is a strong need for development of integrated approach for management of this disease to sustain rice production especially in Basmati/scented varieties. Importance, history of bakanae, symptomatology, etiology of bakanae, survival and spread of the pathogen and different management approaches is being discussed in this chapter.

Keywords: Integrated Management, Bakanae Disease, Basmati Rice

7.1 INTRODUCTION : HISTORY

Rice (*Oryza sativa*) is one of the most important staple food for more than half of the world's population. In India, rice (*Oryza sativa* L.) is grown in an area of 43.50 million ha with a production of 105 million tonnes, which contributes 45% of the total food grain production of the country (Shukla *et al.,* 2015). Rice is grown in almost all the states of India and it provides livelihood to about 70 per cent of the population. Rice is attacked by number of diseases. Amongst many devastating diseases which affect the economics of rice, Bakanae is an emerging disease (Yasin *et al.,* 2003, Bagga and Sharma, 2006) especially in Basmati and aromatic rice. The first known report of '*bakanae*' disease is from Japan since long back in 1828. The disease was first described scientifically in 1898 by Shotaro Hori (a Japanese researcher), who demonstrated that the causative agent of this disease was a fungus, *Fusarium heterosporium.* In 1917, Sawada indicated that the abnormal elongation of the rice seedlings might be due to some stimuli derived from bakanae fungus hyphae. Later in 1938, Yabuta and Sumiki isolated a crystalline compound from culture filtrate of *G. fujikuroi* named as gibberellins.

Heavy incidence of bakanae in paddy field during *Kharif*, 2016

The name "Bakanae" comes from a Japanese word, meaning abnormal elongation of the plants. Bakanae or foot rot disease of rice has been recorded in almost all rice growing countries, especially in Asian countries (Desjardins *et al.*, 2000). Earlier, bakanae disease was of minor importance occurring sporadically and remained confined to some areas only, but in recent years the disease has assumed epiphytotic proportions, causing substantial losses to both quality and quantity of paddy. This shift in disease occurrence may be a consequence of changed climatic conditions, cropping systems and enhanced inputs. Bakanae has become a major disease since last decade spreading in Bangladesh, Pakistan, China, Japan, Nepal and Thailand recording 10-70 percent disease incidence in improved rice cultivars. In India, the incidence of Bakanae is increasing with an alarming pace and moderate to severe yield losses ranging from 15-25 per cent have been reported from Eastern Uttar Pradesh, Haryana, Punjab, Assam, Andhra Pradesh and Tamil Nadu (Rathaiah *et al.,* 1991, Sunder *et al.,* 1998, Pannu *et al.* 2012). The disease has been reported to cause yield losses ranging from 3.0 to 95.4 percent varying with regions and cultivars (Pavgi and Singh, 1964; Ou, 1987; Hajra *et al.*, 2012).

7.2 SYMPTOMATOLOGY

The complex of disease symptoms include seedling blight, root rot, crown rot, stunting and the most classical symptom of etiolation, hypertrophy effect or excessive elongation of infected plants, foot rot, seedling rot, grain sterility and discolouration with ultimate effect on seed/grain quality and yield which have been recorded from different regions of the world (Sun and Snyder, 1981; Ou, 1985; Desjardins *et al.*, 2000).

Symptoms of bakanae disease

The disease can occur both at nursery stage and in the main field. Seedlings which survive the attack in nursery may die soon after transplanting. The affected plants wilt from tip downwards, appear pale yellow in colour, thin and abnormally long (several inches taller) or rarely dwarf. These plants may die prematurely or may develop tall tillers and flower earlier than the healthy plants, but the infected plants mostly bear sterile panicles. Highly diseased plants show collar infection and die within 2-6 weeks. There is profuse branching in the roots and adventitious roots may develop from the lower nodes. A pinkish white cottony growth of the fungal mycelium may be observed at the base of the plant. The fungus produces growth stimulating substances known as gibberellins which are responsible for excessive elongation of the plants (Malonek *et al.*, 2005). Besides gibberellins, the fungus also produces a phytotoxin, fusaric acid – a non-specific substance which is toxic to the plants (Desjardins *et al.*, 2000).

Pinkish-white fungal growth at crown region

7.3 CAUSAL AGENT OF THE DISEASE

Initially, the causal organism responsible for Bakanae disease was identified as *Fusarium moniliforme* Sheldon (Nirenberg, 1976; Ou, 1987), later it was re-identified as *Fusarium fujikuroi* Nirenberg (Telomorph: *Gibberella fujikuroi* (Sawada) Wollenweber). The disease is reported to be caused by one or more *Fusarium* species. *F. moniliforme*, *F. proliferatum, F. andiyazi*, *F. verticillioides*, *F. sacchari* and *F. subglutinans* have been found associated with Bakanae disease from various countries (Agarwal *et al.*, 1989; Desjardins *et al.*, 2000; Zainudin *et al.*, 2008; Dal Pra *et al.*, 2010; Bashyal and Aggarwal, 2013). *F. fujikuroi* and *F. proliferatum* are closely related species. *F. fujikuroi*, regarded as the most virulent species causing bakanae disease, produced excessive gibberellin hormone which is responsible for internode elongation in the infected plants (Zainudin *et al.*, 2008).

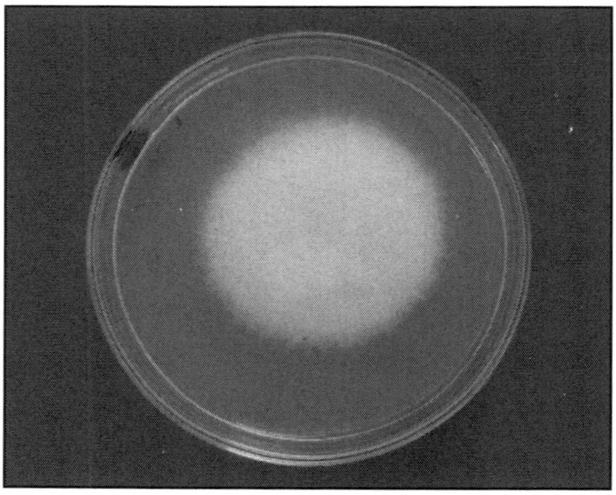

Pure culture of bakanae incitant, *Fusarium fujikuroi*

7.4 SURVIVAL AND DISSEMINATION OF PATHOGENS

The pathogen is both seed and soil-borne which infects the seedlings at the time of germination or at an early growth stage through the injured roots or crown (Ou, 1985; Cumagun *et al.* 2011). Seed borne inoculum is more significant and primary source of inoculum. Soil inoculum gets reduced with passage of time. This fungus survives under adverse conditions in infected seeds in the seed coat and as thick walled hyphae or macroconidia in infested crop residue in the soil (IRRI, 1983; Ou, 1985; Cumagun *et al.,* 2011). Ascospores and conidia adhering to the seed germinate and infect seedlings through roots and crown (Sun, 1975). The production of conidia on diseased or dead stubbles coincides with the flowering stage of the crop resulting in seed infections. The conidia are disseminated by wind and water causing new infections in the rice field. The development of bakanae symptoms is influenced by the fungus strain and the environmental conditions such as temperature and moisture. Soil moisture, soil temperature and nitrogen fertilizer influence the development of the bakanae disease of rice (Nyvall, 1999; Kazempour and Elahinia, 2007). The optimum temperature for pathogen infection is 27-30°C and 35°C for disease development which is also favourable for seedling growth. The disease expression is favoured by high temperatures and relative humidity.

7.5 DISEASE MANAGEMENT

During 1989 and 1991, there was tremendous increase in area under highly susceptible scented cultivar particularly Taraori Basmati which led to increase in the disease incidence and build up of inoculum in seed and soil which resulted in epiphytotics during *Kharif,* 1992. In few fields where farmers used their own infected and untreated seed, the disease incidence was very high (70-80%) leading to ploughing up and re-transplanting of paddy crop. During *Kharif,* 1994 to 2004, the incidence of Bakanae disease declined significantly. However, the disease intensity increased after *Kharif,* 2008, which might be attributed to a significant increase in area under cultivation of highly susceptible scented varieties particularly, Pusa Basmati 1121 and prevalence of favourable environmental conditions for disease development (Sunder *et al.,* 2014).

Seed Treatments

As the disease is primarily reported to be seed-transmitted (Webster and Gunnell 1992), use of clean disease free seed and seed dressing represents the first and most important way to control the spread of the disease. Salt water can be used to separate light weight seeds in a seed lot (Cother and Lanoiselet, 2002). In case of organic cultivation, seed can be soaked in hot water at 60°C for 10 min before sowing to reduce seed infections (Miyasaka *et al.*, 2000). Gergon *et al.* (2007) suggested that seed treatments of rice with NaOCl or benomyl either as seed protectant or eradicant can effectively manage Bakanae disease of rice.

Currently, seed treatment with chemicals is the most common management practice for bakanae disease in India (Gupta *et al.*, 2014) and widely practiced in most of East Asia. Soaking of seeds in 0.3% sodium hypochlorite solution for two hours followed by soaking in fresh water very efficiently reduces inoculum load by 80 to 90% (Webster and Gunnell, 1992). Seed treatment with fungicides like benomyl, thiram, benomyl+thiram, carbendazim+thiram, carboxin+thiram, fludioxonil, mancozeb, iprodione+triticonazole, prochloraz, thiophanate methyl and ipconazole was found effective against the disease (Ou, 1987; Tateishi *et al.*, 1998; Bagga *et al.*, 2007; Ghazanafar *et al.*, 2009; Karov *et al.*, 2009; Ora *et al.*, 2011). Triflumizole, propiconazole, ipconazole and prochloraz were found effective against strains that showed resistance against benomyl and the combination of thiram and benomyl (Tateishi *et al.*, 1998; Karov *et al.*, 2009). Application of carbendazim both as dry seed treatment and seed soaking in fungicidal suspension reduced bakanae incidence significantly (Sunder *et al.*, 2014). Our experimental results also revealed that seed treatment with carbendazim @ 0.25% was very effective in control of the bakanae pathogen (Gupta *et al.,* 2015). Seed/seedling treatment followed by soil drenching of benzimidazole fungicide derosol (0.2%) was found highly effective in controlling the disease (Bhalli *et al.*, 2001). Seed dip in bavistin 50 WP (0.05%) plus streptocycline (0.01%) solution for 12 h and smearing the seeds with talc formulation of *Trichoderma harzianum* (15g/Kg seed) immediately before nursery sowing and seedling root dip for 6 hrs in *T. harziamum* biopowder (15g/liter water) was found most effective against bakanae disease (Pannu *et al.*, 2009). Hongfu and Zhiyuan (2007) invented a fungicide consists of aluminium sulphate as the active ingredient which gave 85% control of bakanae disease (www.patent.ipexl.com/C2N/200510123045). Two seed-borne rice seedling diseases, bakanae disease caused by the fungal pathogen *Fusarium fujikuroi*, and bacterial seedling blight caused by *Burkholderia plantarii*, were successfully suppressed by irradiation of infected rice seeds with atmospheric plasma (Ochi *et al.,* 2016).

Disease management at nursery stage

The control of seed-borne rice seedling diseases e.g. bakanae disease in the seed beds is very crucial to avoid epidemics in rice nurseries as well as in main fields after nursery transplantation otherwise it may lead to severe yield losses. Avoiding the damage on root or crown region (root injury) at the time of nursery uprooting and transplanting is helpful to prevent the potential entry of the pathogen which ultimately reduces the incidence of bakanae disease in the main field. The disease incidence reduced considerably in plots where nursery was uprooted in standing water as against *vattar* (dry) condition which facilitates entry of pathogen through injured roots. Besides seed treatment, nursery application of carbendazim as sand mix broadcast @1g/m^2 seven days before uprooting and as seedling dip in 0.1% carbendazim solution for three hours proved quite effective in reducing the incidence of Bakanae disease (Sunder *et al.*, 2014). Seed soaking, nursery application and seedling dip in carbendazim solution reduced bakanae incidence from 42.4 to 4.7-11.3% along with significant increase in grain yield of paddy cv. Pusa Basmati 1121. Nursery dip treatment with benlate was found highly effective for the control of bakanae disease followed by topsin-M and carbendazim (Javed *et al.*, 1996).

Biological control

Biological control using antagonistic fungi and bacteria is an environment friendly option, and can be a potential alternative to agrochemicals for the management of plant diseases. During the last few decades, many species of potential biological control agents including fungi, bacteria, nematodes etc. have been used for controlling many plant diseases with certain success, but unfortunately only a few have been developed for the practical use at the commercial level. For the management of bakanae, the effect of bacterization of naturally infected seeds on bakanae disease incidence in seedbox and seedbed tests has been studied by Rosales and Mew (1997). They reported reduced disease incidence and effective control of Bakanae disease of rice in IR 58 variety using 10 different strains of bacteria during their study in Philippines. Kazempour and Elahinia (2007) reported that the *P. fluorescens* and *B. cereus* isolates have an excellent potential to be used as biocontrol agents against *F. fujikuroi* in rice at the field conditions. In their studies on combined application of fungicide rovral TS with F15 antagonistic isolate of *P. fluorescens* for seed coating, soil drenching and seed coating + foliar spray they recorded significant decrease in disease incidence by 6.5, 6.75 and 5.5% respectively. *Bacillus* spp. viz., *B. subtilis* and *B. megaterium* have also been effective against bakanae disease (Luo *et al.*, 2005). The most common biocontrol agent *Trichoderma strictipilis, T. atroviride* and *T. neokoningii* reduced the vegetative growth of bakanae pathogen (Bhramaramba and Nagamani, 2013). In India, *T. viride* and *P. fluroscenes* alone and in combination were found effective for the management of bakanae disease of rice. Lower disease incidence was reported in fields treated with FYM 10t/ha + *Trichoderma* + *Pseudomonas* (Wyawahare *et .al.*, 2012). In their study Kumakura *et al.* (2003) reported that soaking of rice seeds in the suspension of strain SKT-1 of *Trichoderma* sp. gave high control of Bakanae disease. This strain has been developed and commercialized with the trade name of "Echohope" against four economically important seed-borne diseases of rice including Bakanae disease (Nagayama *et al.*, 2007).

Use of Plant Extracts for Disease Control

Antifungal activities of various plant extracts have been reported by several workers against *Fusarium moniliforme.* The inhibition in conidia production, conidial germination and mycelial growth of *F. moniliforme* by root bark extract of *Plumbago zeylanica* was reported by (Bambode and Shukla 1974). The complete inhibition in the growth of *F. moniliforme* using oil of *Hedychium spicatum* and *Acorus calamus* was achieved by Mishra *et al.* (2003). They concluded that these oils exhibited both fungistatic and fungicidal properties depending upon the concentration. Yasin *et al.* (2003) found maximum inhibition of mycelial growth of *F. moniliforme* (60.65%) by the leaf extract of *Lawsonia inermis* followed by root extract of *Asparagus racemosus* (50-59%). Antifungal properties of leaf extract of *Andrographis paniculata* and *Lagerstroemia speciosa* and bark extract of *Eucalyptus citriodora* against bakanae disease has also been reported. Essential oils (EO_s) from *Cymbopogon citrates*, *Ocimum gratissimum* and *Thymus vulgaris* were found effective against *F. moniliforme* in controlling the seed infection by 95-100% (Nguefack *et al.*, 2007). Mohana *et al.* (2011) reported that methanol extract of *Decalepis hamiltonii* caused maximum inhibition (89.90%) in mycelial growth of *Fusarium moniliforme*.

Use of compost, vermicompost extract, organic amendments for disease control

Reduction in populations of soil borne phytopathogens and reductions in level of infection through organic amendment is already reported by several workers in different crops (Bonanomi *et al.*, 2007, Marzano, 2012, Kumar *et al.*, 2014). *Fusarium moniliforme* causing foot rot/bakanae disease of rice was controlled biologically using fermented products of composts and vermicompost by Manandhar and Yami (2008). Panneerselvam and Saravanamuthu (1996) reported suppression of pathogen in soil amended with neem cake.

Development of resistant varieties

Management of disease through resistant varieties is still the best option. Breeding for resistance against Bakanae disease is becoming a promising strategy to manage this emerging disease. Several screening methods have been tested and verified till date against bakanae disease in different countries. It has been established that Basmati/scented germplasm and cultivars are more susceptible to bakanae disease as compared to non-scented rice cultivars (Sunder *et al.*, 1998; Bashyal *et al.*, 2012; Pannu *et al.*, 2012; Ghazanfar *et al.*, 2013; Gupta *et al.*, 2014). Hence there is a scarcity of resistant germplasm under Basmati rice category. IARI, New Delhi has recently taken up the challenge to manage this major problem of Basmati paddy. Researchers from IARI, New Delhi used a population of 168 F_{14} recombinant inbred lines (RILs) derived from two *indica* rice parents Pusa 1342, a highly resistant variety and Pusa Basmati 1121, a highly susceptible variety to map quantitative trait loci (QTLs) governing resistance against Bakanae disease. As a resultant the novel QTLs identified during the study are useful resources for efficiently breeding rice cultivars resistant to Bakanae disease (Fiyaz *et al.,* 2016).

Agronomic interventions

Agronomic practices or interventions are 'key' measures to control bakanae disease on farmers' fields. These interventions should be validated and promoted as these are low cost requiring, farmers friendly and have no hazardous effects on environment. Among the various agronomic practices proved effective against bakanae disease few are use of clean and disease free seeds, crop rotation, destruction of crop residues/debris, alteration of planting time, balanced use of fertilizers etc. These are very simple and effective means to manage this disease. Since the pathogen of bakanae disease is mainly seed borne, use of clean, healthy disease free seed is of great help in reducing the disease incidence. Destruction or burning of crop residues can limit the amount of inoculum that may be carried over to the next crop (Gupta *et al.*, 2004). Proper selection of geographic area, use of organic matter, crop rotation, time and mode of planting, selection of resistant varieties, balanced fertilization and irrigation help in minimizing the disease (Burgess *et al.*, 1993; 1996). Bagga *et al.* (2007) found minimum disease incidence in late transplanted rice crop by end of July which may be due to low temperature during the infection period. Higher levels of nitrogen and potassium have also been found to affect the survival and population of fungus. Utmost care and handling should also be taken at the time of harvesting and threshing, proper drying and seed cleaning to prevent the spread of the disease (Saremi *et al.*, 2008).

7.6 INTEGRATION OF SEED TREATMENT, SOIL AMENDMENTS AND SEEDLING DIP TREATMENTS FOR BAKANAE MANAGEMENT IN PB 1121-A CASE STUDY

Paddy variety Pusa Basmati 1121 which was released and notified in 2005 has exceptionally high cooking quality and has become very popular among the farmers and rice exporters as it has extra long kernels (about 8 mm in size) with elongation ratio of more than 2.5 times on cooking, but is highly susceptible to bakanae disease (NCIPM 2008-09). So, there was a dire need of comprehensive studies on the management of this disease. There are many factors responsible in development of the disease, and for control of any disease a full package of practice needs to be followed. Seed treatments, seedling dip treatments before transplanting and soil treatments were integrated in an experimental trial at this station using cultural, chemical and biological approach for the effective management of bakanae disease in paddy cv. Pusa Basmati 1121.

The incidence of the disease in carbendazim and *Tv* seed treatments was 10.3% and 14.3%, respectively as against 15.4% in untreated seeds transplanted in soil without any amendment (control). In FYM + *Tv* amended soils the disease incidence in plants of untreated seed was lower (10.3%) than those raised on *Tv* amended soil (13.9%) and maximum (15.4%) being on untreated soil (Fig. 1). The disease incidence in carbendazim treated seeds was almost similar in all soil types (10.3%, 10.1% and 10.1% in untreated, FYM+*Tv* and *Tv* amended soil, respectively), but the incidence in *Tv* treated seeds sown on untreated soil (14.3%) was higher than those sown on both types of amended soils i.e., FYM+*Tv* (10.2%) and *Tv* (10.5%) amended soils.

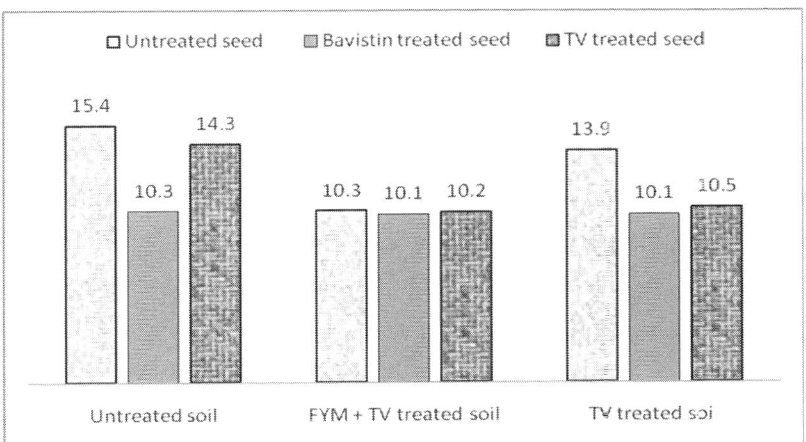

Fig. 1. Effect of seed treatments on per cent incidence of bakanae disease in amended fields

The disease incidence was minimum (5.1%) in plants raised from bavistin seedling dip and sown on soil amended with FYM+*Tv*. The highest disease incidence (27.3%) was recorded in non-amended plots (control) transplanted from nursery of untreated seeds and the seedlings dipped in plain water at the time of transplanting (Table 1). Thus, seedling dip treatment with 0.2% carbendazim supplemented with FYM+*Tv* amended soil was found most effective in reducing the per cent incidence of bakanae disease, irrespective of seed treatments. Bagga and Sharma (2006) also reported that seedling treatment with carbendazim or benlate (0.1%) for 6 and 8 h, significantly reduced the disease incidence up to 92%. There was substantial increase in the grain yield in all the treatments, which were found statistically at par but superior to control.

Table 1. Effect of seed treatments, seedling dips and soil amendments on incidence of bakanae disease and grain yield in the field

Treatments	**%Disease Incidence**				**Yield (q/ha)**			
	Untreated field	**FYM+ *Tv* treated field**	***Tv* treated field**	**Average**	**Untreated field**	**FYM+ *Tv* treated field**	***Tv* treated field**	**Average**
Untreated seed nursery- Plain water dip	27.3a	16.5a	21.4a	**21.7a**	16 b	15 c	13 b	**14 b**
Untreated seed nursery- Bavistin dip	9.5cd	5.1d	8.1cd	**7.6de**	29 a	28 ab	34 a	**30 a**
Untreated seed nursery- TV dip	9.3cd	9.1bcd	12.2bc	**10.2cd**	28 a	26 ab	25 a	**26 a**
Bavistin treated nursery- Plain water dip	13.5bc	14.1ab	13.8b	**13.8b**	26 a	22 bc	28 a	**25 a**
Bavistin treated nursery-Bavistin dip	7d	6.7cd	6.3d	**6.7e**	26 a	33 a	31 a	**30 a**
TV treated nursery- Plain water dip	15.2b	11.2bc	11.2bcd	**12.5bc**	26 a	23 bc	26 a	**25 a**
TV treated nursery- TV dip	13.3bc	9.2bcd	9.7bcd	**10.7c**	28 a	34 a	28 a	**27 a**
LSD 0.05				**3*****				**5*****

Means followed by same alphabet are non-significant at 5%

The time of application of *Trichoderma viride* in the soil also plays an important role in minimizing the disease incidence. It was observed that soil amended with *Tv* 8 days after transplanting (DAT) was more effective in reducing the disease incidence than soils amended with *Tv* 15 DAT (Table 2).

Table 2. Disease incidence in different soil amended at different time

Soil Amendment	% Disease incidence
No *Tv*	3.4
Tv added 8 DAT	1.6
Tv added 15 DAT	2.5

Foliar application of fungicides

Out of different combinations of foliar application, two foliar applications of carbendazim @ 0.25% or *Trichoderme viride* @ 0.4% and a combination of carbendazim and *Trichoderma viride* 30 and 45 days after transplanting effectively controlled Bakanae disease in the field as against control (Fig. 3).

Sunder *et al.* (2014) also reported significant reduction in grain infection from 73.9 to 35.0% by foliar spray of carbendazim at flowering stage. Hajra *et al.* (1994) and Sasaki (1987) have reported that foliar application of benomyl and carbendazim reduced seed borne inoculum of *F. moniliforme* and the disease spread.

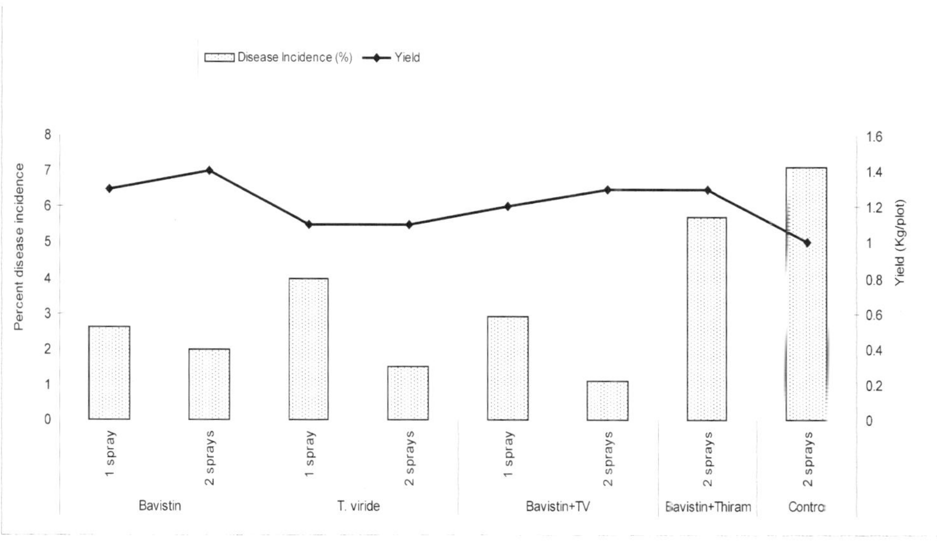

Fig. 2. Effect of foliar sprays on bakanae disease in paddy cv. PB 1121

7.7 CONCLUSION

Bakanae caused by *Fusarium fujikuroi* (Nirenberg) is an emerging disease and a serious threat especially to Basmati/scented rice production around the world. High incidences of bakanae disease result in greater yield losses (sometimes > 90%) across the rice growing countries of the world. The pathogen primarily survives in seed but is also known to survive in soil. Among several species of *Fusarium* genus, reported as causative agent of bakanae disease *F. fujikuroi* has been found frequently associated with the disease as most virulent species all over the world. Varietal screening of rice germplasm/varieties by different workers has revealed that Basmati and aromatic rice germplasm/cultivars are more susceptible to bakanae disease as compared to coarse grain non-scented cultivars. Some strains of *Trichoderma, Pseudomonas* and *Bacillus* have been found effective against the bakanae disease. Currently, seed treatment with fungicides is the most common management practice for control of bakanae disease in India. Integration of seed treatments with seedling dip treatments and soil amendments (chemical/biological) can be effective management strategy for the control of bakanae disease in rice. Therefore, chemical and biological control measures can be integrated with various other appropriate strategies for successful management of bakanae disease for sustaining rice production.

REFERENCES

Agarwal PC, CN Mortensen, and CB Mathur (1989). Bakanae idisease and foot-rot *Fusarium moniliforme* In: Seed-Borne Diseases and Seed Health Testing on Rice. Technical Bulletin No. 3, CAB International Mycological Institute. UK. pp. 31-36.

Bagga PS and Sharma VK. 2006. Evaluation of fungicides as seedling treatment for controlling Bakanae/Foot-rot (*Fusarium moniliforme*) disease in Basmati rice. *Indian Phytopathology* 59(3): 305-308.

Bagga PS, Sharma VK, Pannu PPS. 2007. Effect of transplanting dates and chemical seed treatments on foot rot disease of basmati rice caused by *Fusarium moniliforme. Pl Dis Res* 22: 60-62.

Bambode RS, Shukla VN. 1974. Studies on fungitoxic properties of *Plumbago zeylanica* L. extract. *Punjabrao Krishi Vidyapeeth Res J* 2(2): 110-12.

Bashyal BM and R. Aggarwal (2013) Molecular identification of *Fusarium* species associated with bakanae disease of rice (*Oryza sativa*) in India *Ind. J. Agri. Sci.* 83:72-77.

Bashyal BM, R Aggarwal, S. Banerjee, S. Gupta and S. Sharma (2014). Pathgenicity, ecology and genetic diversity of the *Fusarium* spp. associated with an emerging bakanae disease of rice (*Oryza sativa L.*) in India In: Microbial diversity and biotechnology in food security, (In: Kharwar *et al.* Eds.) Springer 307-314 p.

Bhalli JA, Aurangzeb M and Ilyas MB. 2001. Chemical control of Bakanae Disease of rice caused by *Fusarium moniliforme. Journal of Biological Sciences* 1(6):483-484.

Bhramaramba S and A. Nagamani (2013). Antagonistic *Trichoderma* isolates in control backanae pathogen of rice *Agric. Sci. Digest* 33: 104-108.

Bonanomi G, Antignani V, Pane C and Scala F. 2007. Suppression of soilborne fungal diseases with organic amendments. *Journal of Plant Pathology* 89 (3): 311-324.

Burgess, L. W., D. Backhouse, B.A. Summerell, A.B. Pattison, T.A. Klein, R.J. Esdaile, and G. Ticehutst (1993). Long-term effects of stubble management on the incidence of infection of wheat by *Fusarium graminearum* Schw. Group 1. *Aust. J. of Experi. Agric.* 33: 451-456.

Cother, E. and V. Lanoiselet (2002). Crop nutrition. In: Production of Quality Rice in South-Eastern Australia. (In: Keaky, L.M. and W.S. Clampett, Eds.) (RIRDC: Kingston, ACT, Australia). Plant Health Australia, Rice Industry Biosecurity Plan, 2009. Contingency Plan: Bakanae. 10 p.

Cumagun CJR, Arcillas E and Gergon E. 2011 UP-PCR Analysis of the seedborne pathogen *Fusarium fujikuroi* causing Bakanae Disease in rice. *International Journal of Agriculture & Biology* 13(6):1029–1032.

Dal Pra, M.S. Tonti, D. Pancaldi, P. Nipoti, and I. Alberti (2010). First report of *Fusarium andiyazi* associated with rice bakanae in Italy. Pl. Dis. 94: 1070.

Desjardins AE, Manandhar HK, Plattner RD, Manandhar GG, Poling SM and Maragos CM. 2000. *Fusarium* species from Nepalese rice and production of mycotoxins and Gibberellic Acid by selected species. *Applied and Environmental Microbiology* 66: 1020-1025.

Fiyaz, A.R., Yadav, A. K., Gopala Krishnan, S., Ellur, R. K., Bashyal, B. M., Grover, N., Bhowmick, P.K., Nagarajan, M., Vinod, K. K., Singh, N. K., Prabhu K.V. and Singh, A.K. (2016). Mapping quantitative trait loci responsible for resistance to Bakanae disease in rice. *Rice.* DOI: 10.1186/s12284-016-0117-2.

Gergon EB, Angeles AT and Pautin LR. 2007. Uprising bakanae disease of rice; concerns and options. *Philippine Journal of Crop Science* 32(1):84.

Ghazanfar, M.U., W. Wakil, M. Iqbal, and A. Ahmad (2009). Impact of various fungicides against bakanae disease of rice under the field conditions. 5th International Conference on Plant Pathology in the Globalized Era, Nov. 10-13, 2009, New Delhi, (India).

Gupta AK, Singh Y, Jain AK and Singh D. 2014. Prevalence and Incidence of Bakanae disease of Rice in Northern India. Journal of AgriSearch 1 (4): 233-237.

Gupta Anuja, Ravindra Kumar and V.K. Maheshwari (2015). Integration of Seed Treatments, Seedling Dip Treatments and Soil Amendments for the Management of Bakanae Disease in Paddy Variety Pusa Basmati 1121. *Plant Pathology Journal.* 14 (4): 207-211.

Hajra, K.K., L.K. Ganguly, and D.C. Khatua (1994). Bakanae disease of rice in the West Bengal. J. Mycopath. Res. 32(2): 95-99.

Hori S. 1898. Researches on bakanae disease of rice plant. *Nojishikenjyo Seiseki* 12: 110-19.

International Rice Research Institute (IRRI). 1983. Field problems of tropical rice. Rev. ed. Manila (Philippines): IRRI. 172 p.

Javed, M.S., M. Ayoub, A. Mahmood, and M. Ahmad (1996). Chemical control of bakanae disease of rice by nursery dip treatment method. Pakistan J. Sci. 48: 90-91.

Karov, I.K., K. Mitrev, Sasa, and D.K. Emilija (2009). *Gibberella fujikuroi* (Sawada) Wollenweber- the new parasitical fungus on rice in the republic of Macedonia. Proc. Nat. Sci. Matica Srpska Novi Sad. 116: 175-182.

Kazempour MN and Elahinia SA. 2007. Biological control of *Fusarium fujikuroi*, the causal agent of Bakanae disease by rice associated antagonistic bacteria. *Bulgarian Journal of Agricultural Science* 13: 393-408.

Kishore N, Dubey NK, Tripathi RD, Singh SK. 1982. Fungitoxicity of the leaf extracts of some higher plants against *Fusarium moniliforme*. *Natn Acad Sci Letters.* 5(2): 43-45.

Kumakura K, Watanabe S, Toyoshima J, Makino T, Ichikawa T, Iyozumi H. and Nagayama K. 2003. Effect of *Trichoderma* sp. SKT-1 on suppression of six different seedborne diseases of rice (Oryza sativa). *Annals of the Phytopathological Society of Japan* 69(4): 393-402.

Kumar Ravindra, Asha Sinha, Seweta Srivastava and Saurabh Singh (2014). Evaluation of substrates for mass multiplication of green manure associated fungi for biological control of soil borne phytopathogens. *Indian Phytopathology*. 67(4): 396-401.

Luo, J., X. Guan-lin, L. Bin, Y. Luo, Z. Li-han, W. Xiao, L. Bo, and L. Wen (2005). Gram positive bacteria associated with rice in China and their antagonists against the pathogens of sheath blight and bakanae disease in rice. Rice Sci. 12: 213-218.

Malonek S, Bomke C, Bornberg-Bauer E, Rojas MC, Hedden P, Hopkins P and Tudzynski B. 2005. Distribution of gibberellin biosynthetic genes and gibberellin production in the *Gibberella fujikuroi* species complex. *Phytochemistry* 66 (11): 1296–1311.

Manandhar T and Yami KD. 2008. Biological control of foot rot disease of rice using fermented products of compost and vermicompost. *Scientific World* 6(6):52 – 57.

Marzano SY. 2012. Assessment of disease suppression in organic transitional cropping systems. *Ph.D.* Thesis. University of Illinois, Urbana.

Mishra D, Samuel CO, Tripathi SC. 2003. Evaluation of some essential oils against seed-borne pathogens of rice. *Indian Phytopath* 56: 212-13.

Miyasaka, A., S. Ryoichi, and I. Masataka (2000). Control of the bakanae disease of rice by soaking seeds in hot water for the hydroponically raised seedling method in the long-mat type rice cultivation. Proc. Kanto-Tosan Pl. Prot. Soc., 47: 31-33.

Mohana, D.C., Prasad, P., Vijaykumar, V. and Raveesha K.A. (2011). Plant extract effect on seed-borne pathogenic fungi from seeds of paddy grown in southern India. Journal of Plant Protection Research. 51(2): 101-106.

Nagayama K, Watanabe S, Kumakura K, Ichikawa T and Makino T. 2007. Development and commercialization of *Trichoderma asperellum* SKT-1 (Ecohope), a microbial pesticide. *Journal of Pesticides Science* 32(2):141–142.

NCIPM. 2008-09. Annual report of National Centre of Integrated Pest Management, New Delhi, pp 11.

Nguefack, J., S.K. Nguikwie, D. Fotio, B. Dongmo, V. Leth, A.E. Nkengfack, and P.H.A. Zollo (2007). Fungicidal potential of essential oils and fractions from *Cymbopogon citrates, Ocimum gratissimmum* and *Thymus vulgaris* to control *Alternaria padwickii* and *Bipolaris oryzae*, two seed-borne fungi of rice (*Oryza sativa* L.). J. Essential Oil Res. 19: 581-587.

Nirenberg, H.I. (1976). Untersuchungen uber die morphologische und biologische differenzierung in *Fusarium*-Sektion *Liseola*. Mitt. Biol. Bundesansi. Land-Forstwirtsch. Berlin -Dahlem. 169: 1-117.

Nyvall RF. 1999. Field crop diseases. Iowa State University Press. USA. 1021 p.

Nzojiyobiri JB, Tong X, FengMing S and Ying S. 2003. Resistance induced by *Trichoderma harzianum* NF 9 against *Magnaporthe grisea* and *Xanthomonas oryzae* pv. *oryzae* in rice. *Chinese Journal of Biological Control* 19(3): 111-114.

Ochi, A., Konishi, H., Ando, S., Sato, K., Yokoyama, K., Tsushima, S., Yoshida, S., Morikawa, T., Kaneko, T. and Takahashi, H. (2016), Management of bakanae and bacterial seedling blight diseases in nurseries by irradiating rice seeds with atmospheric plasma. *Plant Pathol.* doi:10.1111/ppa.12555.

Ora, N., A.N., Faruq, M.T. Islam, N. Akhtar, and M.M. Rahman (2011). Detection and identification of seed borne pathogens from some cultivated hybrid rice varieties in Bangladesh. Middle-East J. Scient. Res. 10: 482-488.

Ou, S.H. (1987). Rice Diseases. CAB International, Commonwealth Mycological Institute, Kew, Surrey, England, UK. pp.256.

Ou, SH. 1985. Rice diseases. 2nd ed. Commonwealth Mycological Institute. Great Britain (UK). 380 p.

Panneerselvam, A. and R. Saravanamuthu (1996). Studies on the saprophytic survival of *Fusarium moniliforme* Sheld. in soil under treatment of oil cakes. Ind. J. Agric. Res. 30(1): 12-16.

Pannu, P.P.S., J. Kaur, G. Singh, and J. Kaur (2012). Survival of *Fusarium moniliforme* causing foot rot of rice and its virulence on different genotypes of rice and basmati rice. (Abstr.). Ind. Phytopath. 65S: 149-209.

Pannu, P.P.S., N. Singh, H.S. Rewal, H.S. Sabhiki, and S. Raheja (2009). Integrated management of foot rot of basmati rice(Abstr.). National Symposium of INSOPP on "Plant Pathology in the Challenging Global Scenario", Feb 27-28, 2009, NBPGR, New Delhi, pp.13/32.

Pavgi, M.S. and J. Singh (1964). Bakanae and foot rot of rice in Uttar Pradesh, India. Pl. Dis. Reptr. 48: 340-342.

Rathaiah, Y., G.R. Das, and K.H.U. Singh (1991). Estimation of yield loss and chemical control of bakanae disease of rice. Oryza. 28: 509-512.

Rosales AM and Mew TW. 1997. Suppression of *Fusarium moniliforme* in rice by rice associated antagonistic bacteria. *Plant Disease* 81:49-52.

Saremi, H., A. Ammarellou, A. Marefat, and S.M. Okhovvat (2008). Binam a rice cultivar, resistant for root rot disease on rice caused by *Fusarium moniliforme* in Northwest, Iran. Int. J. Bot. 4: 383-389.

Sasaki, T. (1987). Epidemiology and control of rice bakanae disease. 74 Bull Tohoku Natn Agric. Expt. Stn. 47 p.

Sawada K. 1917. Beitrage uber Formosas-Pilze no. 14. *Trans Natural History Soc Formosa* 31: 131-33.

Shukla, U.N., Srivastava, V.K., Singh, S., Ram, U.S. and Pandey, A.K. (2015). Effect of seedlings age, cultivars and weed management on weed dynamics, nutrient removal and yield of rice *(Oryza sativa)* under system of rice intensification (SRI). *Indian Journal of Agricultural Sciences.* 85(10): 1260–8.

Sun, S.K. (1975). The diseases cycle of rice bakanae disease in Taiwan, Proc. Natl. Sci. Counc. Repub. China. 8: 245-256.

Sun, S.K. and W.C. Snyder (1981). The bakanae disease of the rice plant. In: *Fusarium*: Diseases, Biology and Taxonomy, (Nelson, P.E., T.A. Toussoun and R.J. Cook, Eds). The Pennsylvania State University Press, University Park, PA. pp: 104-113.

Sunder, S., R. Singh, and D.S. Dodan (2014). Management of bakanae disease of rice caused by *Fusarium moniliforme.* Ind. J. Agril. Sci. 84(2): 224–228.

Sunder, S., Satyavir, and A. Singh (1998). Screening of rice genotypes for resistance to bakanae disease. Ind. Phytopath. 51: 299-300.

Tateishi, H., T. Saishoji, T. Suzuki, and T. Chida (1998). Antifungal properties of the seed disinfectant ipconazole and its protection against bakanae and other diseases of rice. Ann. Phytopathol. Soc. Japan. 64: 443-450.

Webster RK and Gunnell PS. 1992. *Compendium of Rice Diseases.* APS Press, The American Phytopathological Society, St. Paul, Minnesota, USA, pp. 62.

Wyawahare, K.K., O.P. Sharma, and Yaspal (2012). Field evaluation of *Trichoderma* and Pseudomonas against bakanae disease of rice caused by *Fusarium fujikuroi.* (Abstr.) Ind. Phytopath. 65S: 56.

Yabuta, T. and Y. Sumiki (1938). On the crystal of gibberellin, a substance to promote plant growth. J. Agric. Chem. Soc. Japan. 14: 1526.

Yasin, S., Khan, T.Z., Akhtar, K.M., Anwer, M. and Ahmad, M. (2003). Economic evaluation of bakanae disease of rice. *Mycopathology* 1(2): 115-117.

Zainudin, NAIM, Razak, A.A. and Salleh, B. (2008). Bakanae disease of rice in Malaysia and Indonesia: etiology of the causal agent Based on morphological, physiological and pathogenicity characteristics. *Journal of Plant Protection Research.* 48(4): 475-485.

Chapter - 8

Response of selected endophytic fungi to different abiotic stresses isolated from *Syzygium cumini* of eastern Uttar Pradesh, India

Sharma VK[1], Kumar J[1], Singh DK[1], Mishra A[1], Gond SK[2], Verma SK[1], Kumar A[3], Singh K[1] and Kharwar RN[1]

[1]Mycopathology and Microbial Technology Laboratory, Centre of Advanced Study in Botany, Banaras Hindu University, Varanasi, India-221005;
[2]Botany Section, MMV, BHU, Varanasi, India;
[3]Dept of Botany, Budhha PG College, Kushinagar, India
E-mail: rnkharwar@gmail.com

ABSTRACT

Four endophytic fungi *i.e. Fusarium* sp., *Trichoderma* sp., *Phomopsis* sp. and *Chaetomium* sp. were isolated from healthy leaves of *Syzygium cumini* and evaluated for their tolerance against different abiotic stresses. Tolerance capacity of the fungi was assessed in terms of their growth at various pH and at different concentrations of heavy metal and salt. At deciding factor of 50% growth as compared to the highest radial growth after 168 h of incubation, the suitable growing pH range was observed 4-8 for *Trichoderma* sp., 4-8 for *Fusarium* sp., and 4-7 for *Chaetomium* sp. and *Phomopsis* sp. respectively. Except *Phomopsis* sp., rest isolates achieved maximum growth without having any salt in the media. *Phomopsis* sp. was found to be tolerant up to 6 % of NaCl concentration in terms of radial growth. All the endophytic fungi had lesser growth on PDA amended with 1 mM of metal salts except the *Phomopsis* sp., with few metals. In general highest growth index was exhibited by *Phomopsis* sp. followed by *Chaetomium* sp., *Trichoderma* sp and lowest by *Fusarium* sp. The high growth index exhibited by some of the fungi signal their suitability for the bioleaching and biosorption applications.

Keywords: Endophytic fungi, salt stress, pH stress, heavy metal stress, stress tolerance, growth index

8.1 INTRODUCTION

Salinity, pH and heavy metals in soil are some major abiotic stresses that limit the growth and productivity of plants. By means of physiological and biochemical adaptation, plants enable itself to survive under these environmental stresses and endophytes play a vital role in augmenting the adaptation of host plant by various ways. Endophytes are microorganisms that reside inside the plant tissues without causing any harmful effects. In the grass-endophyte association, the fungi colonize the intercellular spaces of plant tissue and provide some benefits, such as enhanced drought tolerance, photosynthetic rate, and growth (Clay and Schardl 2002). *Neotyphodium*, a fungal endophyte is considered beneficial because endophyte-infested grasses produce more dry matter, utilize soil nitrogen more efficiently, contributes to water stress tolerance and deter insects (Bayat *et al.* 2009). Microorganisms can convert complex organic substances to simple forms and have ability to produce phyto-hormones, and solubilize the insoluble phosphate as well. Endophytic fungi have also been shown to impart plants with tolerance to salt, drought, heat and diseases (Waller *et al.* 2005; Khan *et al.* 2011a). In a recent study, pretreatment with fungus significantly enhanced germination of soybean seeds and seedling growth. Soybean plants exposed to endophytic fungi exhibited significant enhancement of plant growth under salt stress (Radhakrishnan *et al.* 2013). High salt concentrations in soil and presence of excess ions in the rhizosphere causes heavy damage to plant metabolism, which leads to stunted growth and reduced yield (Shannon 1997). Plants develop some defense mechanism to alleviate this condition by associating with endophytes. Endophytic fungi confer salt tolerance to such plants and decrease yield losses (Rodriguez *et al.* 2004). For example, *Piriformospora indica,* a root inhabiting Basidiomycete, increased plant growth and attenuated the NaCl-induced lipid peroxidation, metabolic heat efflux and fatty acid desaturation in leaves of the salt-sensitive barley cultivar by increasing the activity of antioxidant and amount of ascorbic acid (Baltruschat *et al.* 2008).

Trichoderma is a ubiquitous fungus that is known for its biocontrol capabilities against many economically important plant pathogens. The pH is an important factor affecting the mycoparasitic activity of *Trichoderma* strains. It can grow in a wide range of pH from 2.0 to 6.0 with maximal growth rates at 4.0 and the optimum range being 4.6 to 6.8 (Singh *et al.* 2014).

The filamentous fungi including endophytic belonging to the genera *Aspergillus, Penicillium, Fusarium, Alternaria* and *Geotrichum* have been isolated from heavy metal contaminated sites, of which majority of isolates were tolerant to Pb, Cr, Cu and Zn, whereas only the *Penicillium* sp., was able to grow in Cd stress (Ezzouhri *et al.* 2009). Valix and Loon (2003) conducted an experiment on heavy metals tolerance development in *Aspergillus niger, Penicillium simplicissimum, Aspergillus foetidus* and *Aspergillus carbonarius* strains in the presence of Ni, Co, Fe, Mg and Mn, up to concentrations of 2000 ppm. A high growth followed by a weaker death rate appears to lead to tolerance development of strains in increasing metal concentration, often exceeding the control.

Present study was undertaken to assess the tolerance level of four endophytic fungi, *viz. Phomopsis, Trichoderma, Fusarium* and *Chaetomium*, against different concentrations of salt, pH and heavy metals. The significance of the study was to screen some stress tolerant endophytic fungi of *Syzygium cumini* (Jamun) under different stresses like different concentrations of salt, varied pH ranged from acidic to basic and high level of heavy metals such as copper, cadmium, cobalt, iron, magnesium, manganese, chromium and zinc.

8.2. MATERIALS AND METHODS

8.2.1 Isolation of the endophytic fungi

Mature, green and healthy leaves were collected randomly from the branches from *Syzygium cumini* tree, growing in the botanical garden of Department of Botany, Banaras Hindu University. Collected leaves were brought to the laboratory and processed on the same day. The adhered debris was removed by placing the samples under running tap water for 1 hour, and a final rinse done with double distilled water. The surface sterilization treatment was done adopting the methodology given by Petrini *et al.* (1992). Epiphytic mycelia were removed by immersing the tissues in 70% ethanol for ≅ 2 min and in aq. solution of sodium hypochlorite (2% available chlorine) for ≅2.5 min followed by washing with 70% ethanol for ≅10 sec. The tissues were then rinsed in sterile distilled water and allowed to surface dry in sterile conditions. Leaves were cut into small pieces of $5 \times 5 mm^2$ with a sterile blade and the pieces were placed to Petri plates containing potato dextrose agar (PDA) medium supplemented with streptomycin (200 mg/L) and incubated for 21days at 26±2°C in BOD cum humidity incubator (Calton, NSW, New Delhi). Tissues were observed for the emergence of endophytic fungi at 2 days intervals for 20 days. Immerging hyphae from plant tissues were sub-cultured on PDA Petri plates for identification and enumeration. The efficacy of surface sterilization was checked by the leaf imprint method (Schulze *et al.* 1993).

8.2.2 Test fungi and growth media

Four endophytic genera i.e. *Fusarium* sp., *Trichoderma* sp., *Phomopsis* sp. and *Chaetomium* sp., respectively, were selected for the study. All the four endophytic fungi were tested for their tolerance and growth on PDA medium. The effect of pH on the growth was observed by culturing the endophytic fungi on PDA Petri plates with the pH of the media adjusted from 4 to 10 by HCl and NaOH. The effect of salt stress on the growth was observed by culturing the endophytic fungi on PDA amended with 1%, 3%, 6%, 9%, 12% and 15% NaCl. The effect of metals on the growth was observed by culturing the endophytic fungi on PDA amended with 1mM of metal by using copper(II) chloride dehydrate, cadmium chloride, cobalt(II) chloride, ferrous sulfate heptahydrate, magnesium sulfate heptahydrate, manganese(II) sulfate monohydrate, potassium dichromate and zinc sulfate heptahydrate. The control was kept for every experiment conducted in present study.

8.2.3 Growth test and growth index

The growth test was conducted by sub-culturing the endophytes on the modified PDA Petri plates. The plates were incubated at 26±2ºC in BOD cum humidity incubator. The radial growth of the fungi was monitored at 24hr interval and the diameter of the growing colony was recorded. All the experiments were done in triplicates. The growth index was calculated by dividing the mean radial growth of the fungi in the each respective treatment plate by the mean radial growth of the fungi in control (PDA with pH 5.6) plate.

8.3. RESULTS

In the current experiment, four endophytic fungi i.e. *Fusarium* sp., *Trichoderma* sp., *Phomopsis* sp. and *Chaetomium* sp., residents of *Syzygium cumini* were studied for their growth behavior against different abiotic stresses. The detailed account of the results of the different treatment is given below.

8.3.1 Effect of pH

In general, all the endophytic fungi had shown luxuriant growth at slightly acidic pH (5 to 6) (Figure 1-4). The radial growth of *Trichoderma* sp. was best at pH 5 followed by 5.6 (Control) and 6 while least radial growth was observed at pH 10. *Trichoderma* sp. attained maximum possible radial growth (90±0mm) on PDA plate after 168 h of incubation at pH 5.6, 5 and 6 while lowest radial growth observed after 168 h of incubation at pH 10 (18±1mm) (Figure 1). The radial growth of *Fusarium* sp. was best at pH 5.6 (Control) followed by pH 5 and 6 while least radial growth was observed at pH 10. After 168 h of incubation the radial growth of *Fusarium* sp. was maximum (76.33±0.58mm) at pH 5.6, followed by 63.33±0.58mm at pH 5 and 58.33±0.58mm at pH 6 (Figure 2). The radial growth of *Chaetomium* sp. was best at pH 5.6 (Control) followed by pH 6 and 5 while least radial growth was observed at pH 10. After 168 h of incubation the radial growth of *Chaetomium* sp. was maximum (75.33±0.58mm) at pH 5.6, followed by 62.67±0.58mm at pH 6 and 56.67±0.58mm at pH 5 (Figure 3). The radial growth of *Phomopsis* sp. at pH 5 and 4 was initially (48 to 96 h) more as compared to pH 5.6 (Control) but after 168 h of incubation the radial growth of *Phomopsis* sp. attained maximum possible growth (90±0mm) on PDA plate at pH 5.6, followed by 88.33±0.58mm at pH 5, 81.67±0.58mm at pH 4 and 75±1mm at pH 6 (Figure 4). About 50% of the growth was considered decisive factor as compared to the highest radial growth after 168 h of incubation at suitable growing pH range for *Trichoderma* sp., *Fusarium* sp., *Chaetomium* sp. and *Phomopsis* sp. was recorded to be 4 to 8, 4 to 9, 4 to 7 and 4 to 7 respectively. To get the better comparative result of the growth of all studied endophytic fungi at different pH range, the growth index was calculated as the mean growth in the respective treatment divided by the mean growth of the fungi in control (PDA with pH 5.6) at 120 h of incubation. The highest growth index was found to be 1.02 for *Phomopsis* sp. followed by 1.01 for *Trichoderma* sp. while lowest 0.88 for *Fusarium* sp. at pH 5.0.The lowest growth index for all fungi was observed at pH 10.0. In all the fungi studied, the similar trend of growth index for pH 5.0 and 10.0 was recorded (Figure 5).

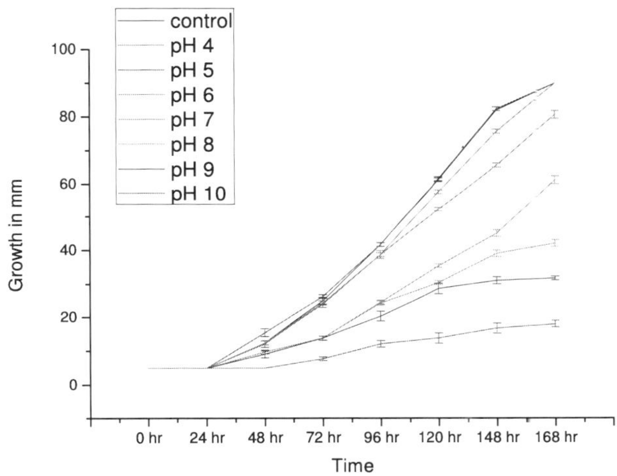

Figure 1. Effect of pH on radial growth (mm) of endophytic *Trichoderma* sp. (Control pH 5.6)

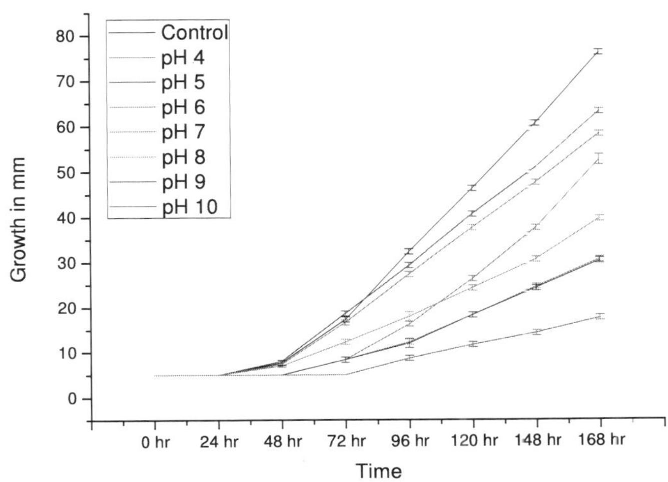

Figure 2. Effect of pH on radial growth (mm) of endophytic *Fusarium* sp. (Control pH 5.6)

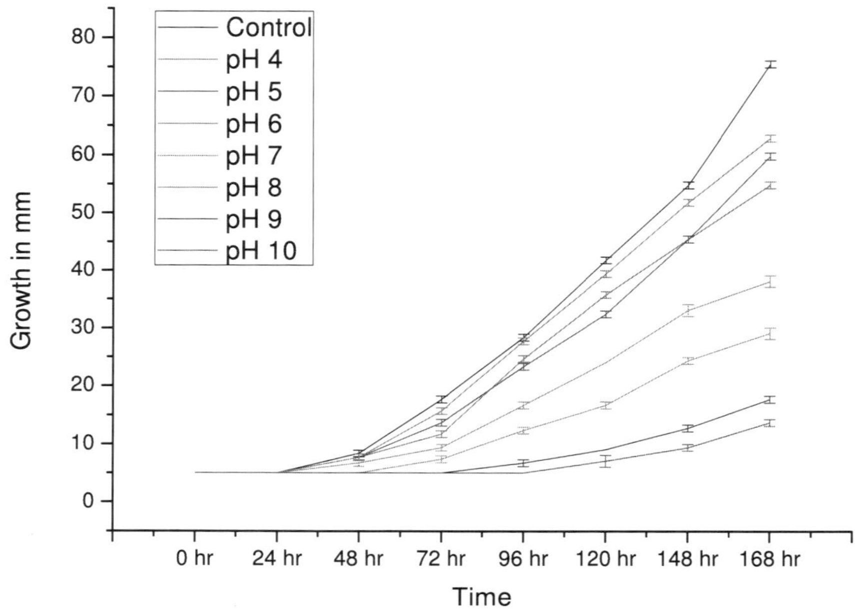

Figure 3. Effect of pH on radial growth (mm) of endophytic *Chaetomium* sp. (Control pH 5.6)

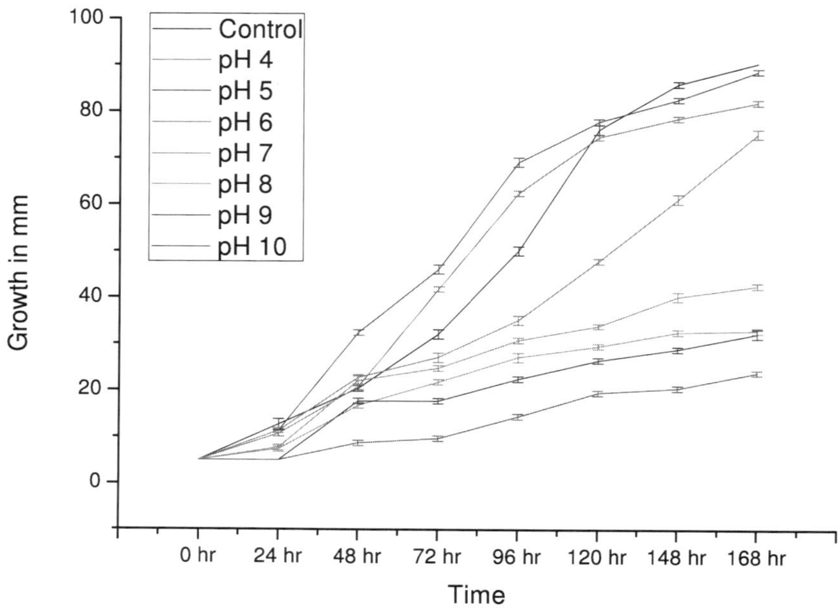

Figure 4. Effect of pH on radial growth (mm) of endophytic *Phomopsis* sp. (Control pH 5.6)

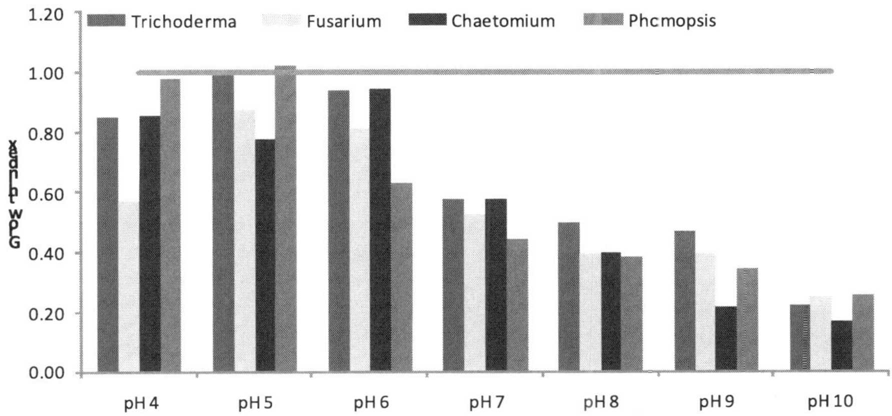

Figure 5. Growth index of various endophytic fungi at different pH (the value above the blue line indicates favorable effect on the growth while values below the blue line indicates decrease growth and pH less tolerant)

8.3.2 Effect of salt

In general, all the endophytic fungi had luxuriant growth under no salt conditions except *Phomopsis* sp. (Figure 6-9). The radial growth of *Trichoderma* sp. was best in control but decreased with the increasing concentration of NaCl while no growth was observed at 12% and 15% NaCl in the growth media. After 168 h of incubation the radial growth of the *Trichoderma* sp. was best in the control (90mm) followed by 1% NaCl (90±0mm), 3% NaCl (85.33±0.58mm), 6% NaCl (51±1mm) and 9% NaCl (32±0.58mm) in the growth medium (Figure 6). The radial growth of the *Fusarium* sp. also decreased with the increasing concentration of NaCl in the growth medium. After 168 h of incubation radial growth of the *Fusarium* sp. was maximum in control (76.33 ± 0.58mm) followed by 1% NaCl (58.33 ± 0.58mm), 3% NaCl (50.67 ± 0.58mm), 6% NaCl (50.67 ± 0.58mm), 9% NaCl (30.67 ± 0.58mm) and 12% NaCl (22.33 ± 0.58mm) in the growth medium (Figure 7). *Chaetomium* sp. showed similar growth pattern as that by *Trichoderma* sp. After 168 h of incubation the radial growth of the *Chaetomium* sp. was best in the control (75.33 ± 0.58mm) followed by 1% NaCl (71.67 ± 0.58mm), 3% NaCl (70.33 ± 0.58mm), 6% NaCl (34.33 ± 0.58mm) and 9% NaCl (19.33 ± 0.58mm) in the growth medium (Figure 8). *Phomopsis* sp. showed similar growth pattern as that by *Fusarium* sp. After 168 h of incubation the radial growth of the *Phomopsis* sp. was best in the control (90 ± 0mm) followed by 1% NaCl (90 ± 0mm), 3% NaCl (90 ± 0 mm), 6% NaCl (90 ± 0mm), 9% NaCl (81.33 ± 0.58mm), 12% NaCl (17.67 ± 0.58mm) and 15% NaCl (13 ± 0mm) in the growth medium (Figure 9). Overall, *Phomopsis* sp. and *Tichoderma* sp. were observed to have more wide range of salt tolerance as compared to other endophytic fungi studied in this experiment. To get a better comparative insight of the growth of all the studied endophytic fungi at different concentration of NaCl in growth media Growth index was calculated as the mean growth of the fungi in the respective treatment divided by the mean growth of the fungi in control (without added NaCl) at 120 h of incubation. The highest growth index was found to be 1.04 of *Phomopsis* sp. at 1% NaCl followed 0.94 by *Trichoderma* sp. at 1% NaCl, while lowest growth index recorded was 0.08 by *Chaetomium* sp. at 12 and 15% NaCl (Figure 10).

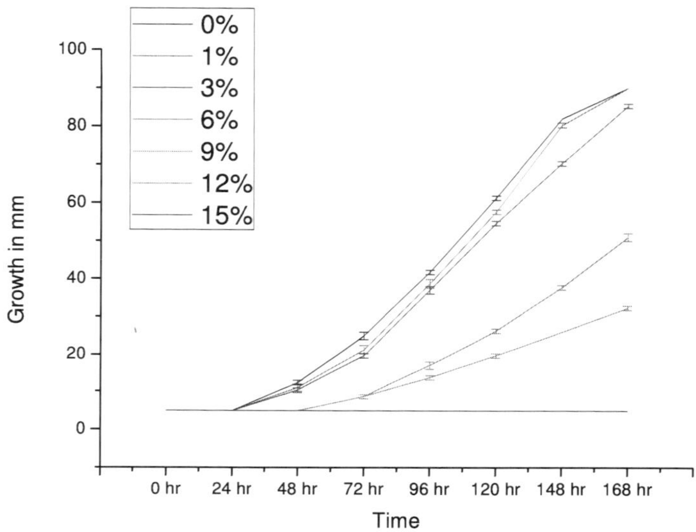

Figure 6. Effect of NaCl on radial growth (mm) of endophytic *Trichoderma* sp.

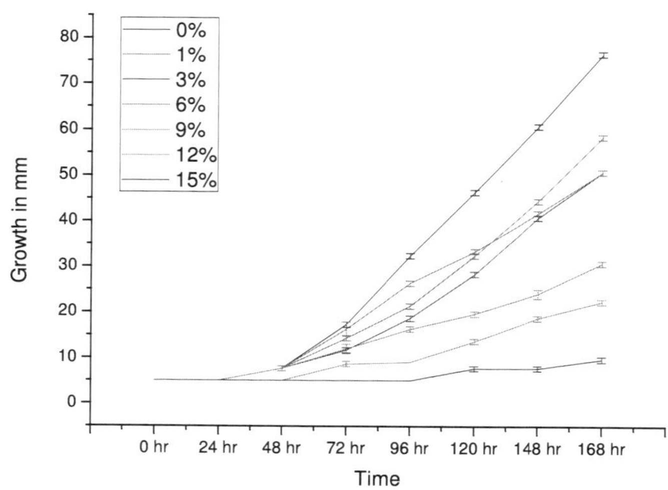

Figure 7. Effect of NaCl on radial growth (mm) of endophytic *Fusarium* sp.

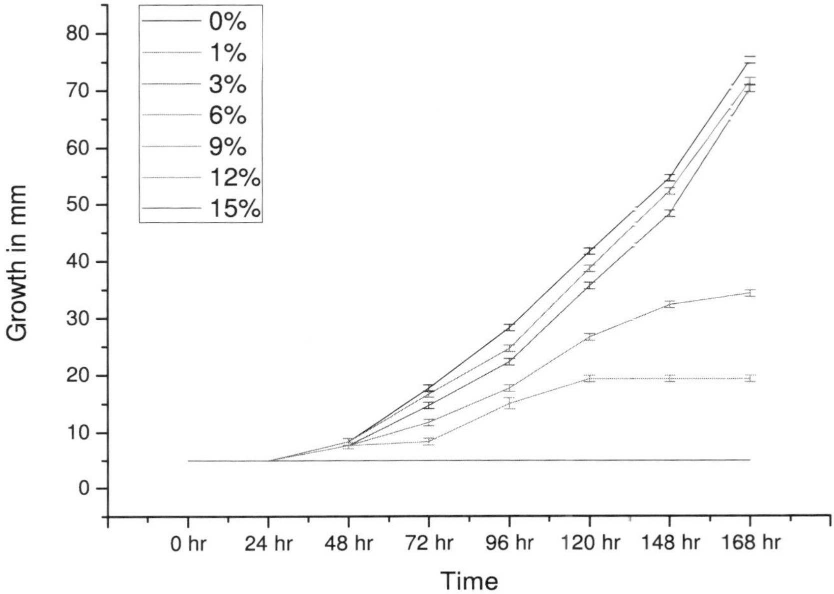

Figure 8. Effect of NaCl on radial growth (mm) of endophytic *Chaetomium* sp.

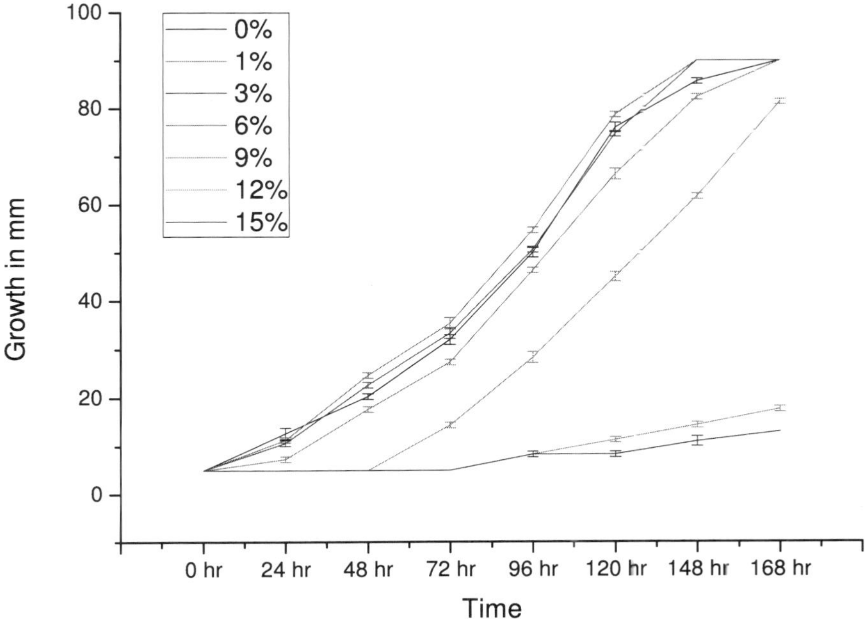

Figure 9. Effect of NaCl on radial growth (mm) of endophytic *Phomopsis* sp.

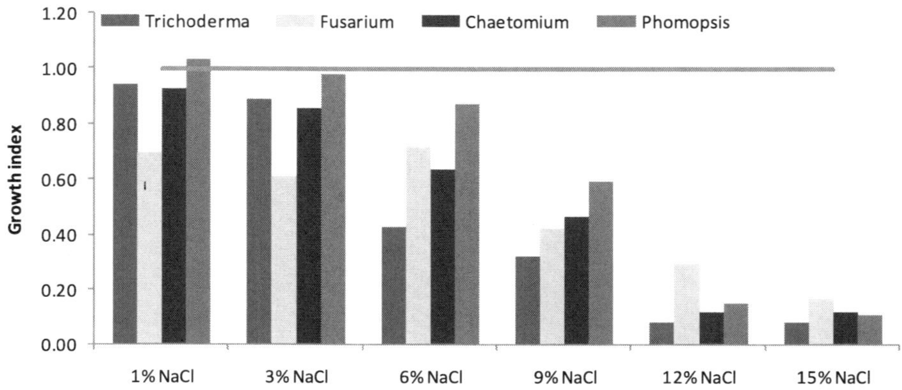

Figure 10. Growth index of various endophytic fungi at different concentrations of NaCl (the value above the blue line indicates favorable effect on the growth while values below the blue line indicates decrease growth and less tolerance

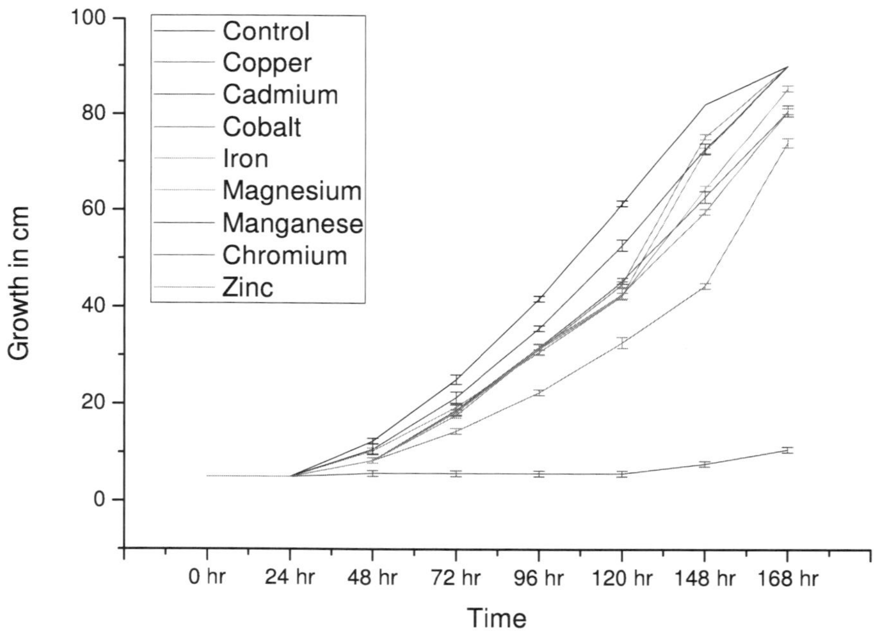

Figure 11. Effect of different metals (1mM) on radial growth (mm) of endophytic *Trichoderma* sp.

8.3.3 Effect of heavy metal

In general all the endophytic fungi had less growth in PDA amended with 1 mM of metal salts except *Phomopsis* sp. in few metals (Figure 11-14). After 168 h of incubation in addition to control the radial growth of *Trichoderma* sp. was maximum (90 ± 0mm) in media amended with metal salt of Co, Fe and Mg respectively followed by Mn (85.33 ± 0.58mm), Cr and Zn (80.67 ± 01.15 and 80.67 ± 0.58mm), Cu (74 ± 1mm) and Cd (10.67 ± 0.58mm) (Figure 11). After 168 h of incubation radial growth of the *Fusarium* sp. was maximum in control (76.33 ± 0.58mm) followed by Zn (48.33 ± 1.53mm), Cr (41 ± 1mm), Fe (28.67 ± 0.58mm), Mg (27 ± 1mm), Mn (23.33 ± 0.58mm), Co (20.67 ± 0.58mm), Cu (20 ± 1mm) and Cd (10.67 ± 0.58mm) (Figure 12). After same period of incubation radial growth of the *Chaetomium* sp. was maximum in control (75.33 ± 0.58mm) followed by Mn (66.33 ± 0.58mm), Mg (62.67 ± 0.58mm), Cr (59.67 ± 0.58mm), Fe (56.67 ± 0.58mm), Cd (54.67 ± 0.58mm), Zn (53.67 ± 0.58mm), Co (53.33 ± 1.15 mm) and Cu (45 ± 1mm) (Figure 13). Again same period of incubation in case of *Phomopsis* sp. showed maximum growth (90 ± 0 mm) in media amended with metal salt of Cu Co, Fe, Mn, Mg, Cr and Zn respectively followed by Cd (74.33 ± 0.58 mm) (Figure 14).

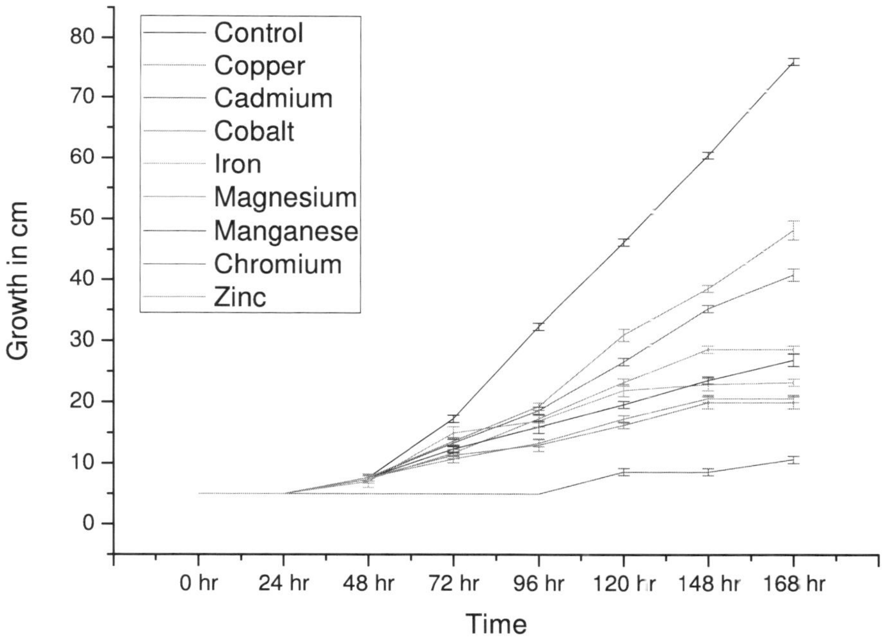

Figure 12. Effect of different metals (1mM) on radial growth (mm) of endophytic *Fusarium* sp.

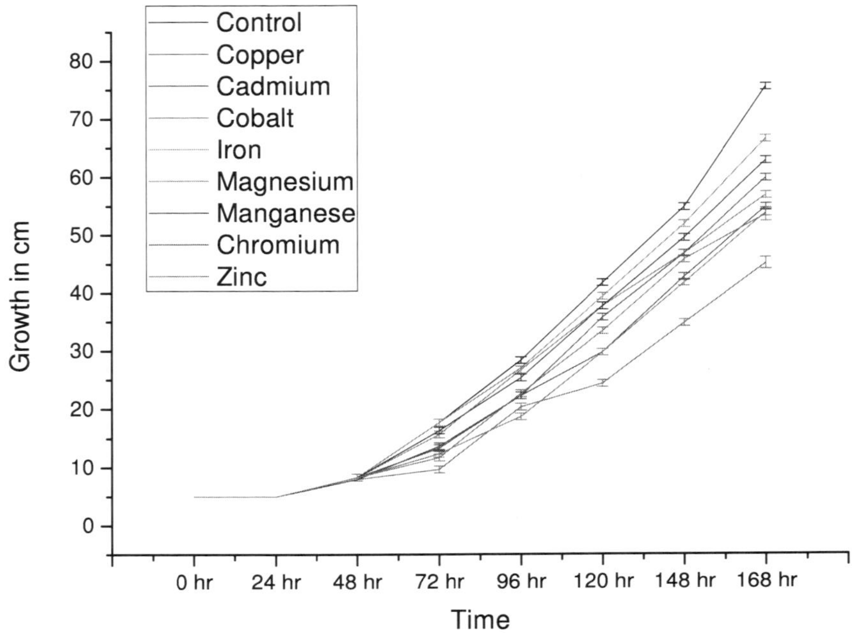

Figure 13. Effect of different metals (1mM) on radial growth (mm) of endophytic *Chaetomium* sp.

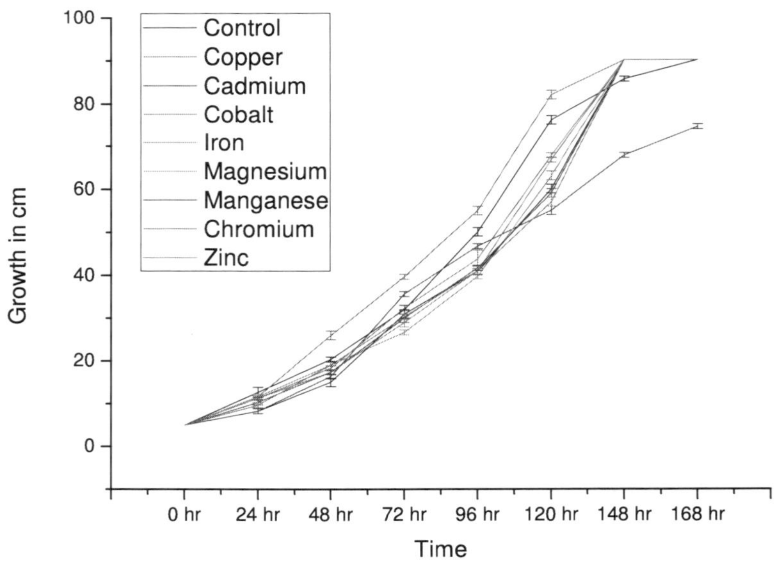

Figure 14. Effect of different metals (1mM) on radial growth (mm) of endophytic *Phomopsis* sp.

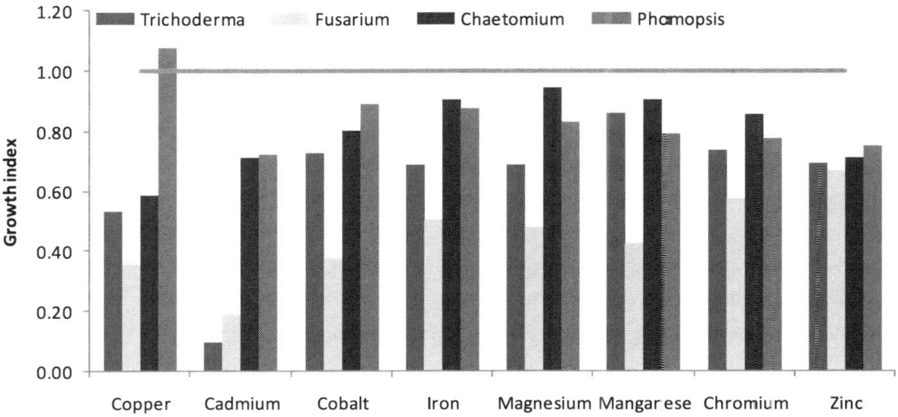

Figure 15. Growth index of various endophytic fungi under 1mM metal salts (the value above the blue line indicates favorable effect on the growth while values below the blue line indicates decrease growth and less tolerant)

In order to get a better comparative insight of metal tolerance to all the studied endophytic fungi at 1mM concentration of metal salts in growth media, Growth index was calculated as the mean growth of the fungi in the respective treatment divided by the mean growth of the fungi in control (without metal salt amendment) at 120 h of incubation. The highest growth index was found to be 1.08 to *Phomopsis* sp. at 1mM Cu followed by 0.94 to *Chaetomium* sp. at 1mM Mn, 0.90 to *Chaetomium* sp. at 1mM Fe and Mg respectively, while lowest growth index recorded was 0.09 for *Trichoderma* sp. at 1mM Cd (Figure 15). The highest growth index (0.86) of *Trichoderma* sp. was observed at 1mM Mg while lowest (0.09) at same concentration of Cd. The highest growth index of *Fusarium* sp. was 0.67 at 1mM Zn and lowest (0.19) at 1mM Cd. The highest growth index of *Chaetomium* sp. was 0.94 and lowest (0.71) treated with 1mM Mn, Zn and Cd, respectively. For *Phomopsis* sp. highest growth index was recorded 1.08 and lowest (0.72) at 1mM of Cu and Cd. At 1mM Cu the growth index of *Phomopsis* sp. was highest (1.08) while growth index of *Fusarium* sp. was lowest (0.35). At 1mM Cd the growth index of *Phomopsis* sp. was highest (0.72) while growth index of *Trichoderma* sp. was lowest (0.09). At 1mM Co the growth index of *Phomopsis* sp. was highest (0.89) while growth index of *Fusarium* sp. was lowest (0.37). At 1mM Fe the growth index of *Chaetomium* sp. was highest (0.9) while growth index of *Fusarium* sp. was lowest (0.5). At 1mM Mn the growth index of *Chaetomium* sp. was highest (0.94) while growth index of *Fusarium* sp. was lowest (0.47). At 1mM Mg the growth index of *Chaetomium* sp. was highest (0.9) while growth index of *Fusarium* sp. was lowest (0.42). At 1mM Cr the growth index of *Chaetomium* sp. was highest (0.86) while growth index of *Fusarium* sp. was lowest (0.58). At 1mM Zn the growth index of *Phomopsis* sp. was highest (0.75) while growth index of *Fusarium* sp. was lowest

(0.67). Overall, *Phomopsis* sp. and *Chaetomium* sp. were observed to be more tolerant to heavy metals than other endophytic fungi studied (Figure 15).

8.4. DISCUSSION

A number of studies have demonstrated the heavy metal tolerance capabilities of fungi isolated from contaminated sites (Valix *et al.* 2001; Valix and Loon, 2003; Zafar *et al.* 2007; Anahid *et al.* 2011; Iram *et al.* 2012). This study was focused on the response of the endophytic fungi isolated from *Syzygium cumini* to abiotic stresses pH, salts and metals. The results of the present study demonstrated that different endophytic fungal isolates from *S. cumini* have different pattern of response to different stresses. Isolates showed medium to high tolerance to the stresses.

Different fungi are known to have different growth, sporulation and metabolite production pattern under different conditions. The growth and interaction between the glasshouse pathogens, *Rhizoctonia solani* Kühn, *Fusarium oxysporum* Schlect., *Pyrenochaeta lycopersici* Schneider & Gerlach, *Phomopsis sclerotioides* van Kesteren, *Sclerotinia sclerotiorum* (Lib.) de Bary and *Botrytis cinerea* Pers. and the antagonistic fungi, *Trichoderma harzianum* Rifai, *Trichoderma viride* Pers., *Coniothyrium minitans* Campbell, *Gliocladium roseum* (Link.) Bain and *Pythium oligandrum* Dreschler were examined on tap water agar, soil extract agar and potato dextrose agar. The effect of media was found significant on the growth rates of the fungi as well as the production of, and response to, volatile and non-volatile antibiotic compounds and hyphal interactions in fungal plant pathogens such as *Rhizoctonia solani*, *Fusarium oxysporum*, *Pyrenochaeta lycopersici*, *Phomopsis sclerotioides*, *Sclerotinia sclerotiorum* and *Botrytis cinerea*and the antagonistic fungi, *Trichoderma harzianum*, *Trichoderma viride*, *Coniothyrium minitans, Gliocladium roseum* and *Pythium oligandrum* (Whipps 1987). According to him media specific responses appeared to be related both to inherent properties of the fungi and to their natural ecological behavior.

Composition of culture medium and cultivation temperature modify the macroscopic features of the fungus. On the different range of temperature and media for the growth of *Phomopsis diachenii,* optimum temperature for the growth and sporulation was found in the range of 22°C to 28°C, Czapek-Dox and PDA are found to be the most appropriate for the cultivation for diagnostic purpose while Czapek-Dox and PDA with an addition of carnation leaves was found best for the stimulation of sporulation (Machowicz-Stefaniak *et al.* 2012). The seven *Trichoderma* species isolated from the rhizospheric soils of selected legumes of different areas of Uttar Pradesh in India were tested *in vitro* at different pH, temperatures and varying agitation speed and it was observed that there was a significant difference in the biomass production among the species at different pH levels and temperature. The most favorable pH range was recorded between 5.5 and 7.5 while favorable temperature range was between 25°C to 30°C and highest biomass was recorded at 150 rpm (Singh *et al.* 2014).

Chaetomium globosum was observed to grow on PDA at different pH range (4.3 to 9.4) but optimal growth and chaetoglobosin C production was highest at neutral pH while sporulation was favored in an acidic environment (Fogle *et al.* 2007). In another study *Chaetomium globosum*, was evaluated on four different media: oatmeal agar, potato dextrose agar, corn meal agar and malt extract agar and observed that colonies were largest with most production of spores and the highest amounts of chaetoglobosins A and C production on oatmeal agar (Fogle *et al.* 2008). *Fusarium* spp. associated with *Aquillaria* spp. were able to grow on 7 different kinds of growth media but the growth rate varied with isolate and growing media (Nurbaya *et al.* 2014). The linear growth of *Trichoderma harzianum*, and its fresh weight and dry weight were found highest on PDA and lowest with water agar (Jahan *et al.* 2013). Effect of pH on the growth of various soil and plant associated fungi have also been studied by various workers. The growing pH range of *Fusarium* spp. was reported from 5.5 to 9 (Marin *et al.* 1995; Sharma *et al.* 2005; Khilare and Ahmed 2012; Rathore *et al.* 2015). The optimum growing pH range of *Trichoderma* spp. was reported from 2 to 7.5 (Bandyopadhyay *et al.* 2003; Begoude *et al.* 2007; Singh *et al.* 2014). The *Chaetomium* sp. was capable of growth on potato dextrose agar ranging in pH from 4.3 to 9.4 while growth and mycotoxin production was favored at neutral pH (Fogle *et al.* 2008). The optimum growing pH range of *Phomopsis* spp. was 6 to 7 in previous reports (Girish and Bhat 2011). While fungi are reported to grow over wide range of pH, in our study slightly acidic pH favored more growth. However, species of *Trichoderma*, *Fusarium*, *Chaetomium* and *Phomopsis* were found to tolerate and grow in pH range of 4 to 8, 4 to 9, 4 to 7 and 4 to 7 respectively.

Most plants do not have adaptive capability to reduce the impacts of stress (Alpert 2000), thus depends on the benefits conferred by fungal endophytes. For stress mitigation at least some plants depend on inter genomic epigenetic processes offered by symbiotic fungi (Rodriguez and Redman 2008). Recent studies had confirmed the role of fungal symbionts in reducing stress related to drought, heat, salt, metal, disease and herbivory to the host plants (Redman *et al.* 2002; Márquez *et al.* 2007; Rodriguez *et al.* 2008; Sherameti *et al.* 2008; Waller *et al.* 2005; Fomina *et al.* 2005; Arnold *et al.* 2003; Saikkonen *et al.* 2013). Tolerance is believed to be habitat-adapted phenomenonand without habitat-adapted fungal endophytes, the plants are unable to survive in their native habitats (Rodriguez and Redman 2008).

Chen (1964) reported that some species of *Aspergillus* and *Penicillium* isolated from soil are able to grow on high salt concentration up to 20 to 30% of NaCl. In this study most of the fungi were not tolerant above 6% NaCl concentration. *Phomopsis* sp. showed highest growth index of 1.04 at 1% NaCl. *Phomopsis occulta* a salt-tolerant fungus when grown in liquid medium at various concentrations of NaCl, i.e. 0, 1, 2 or 3 M, the highest growth rate was found at 1M ($\cong$5.844%) NaCl (Wu *et al.* 2014). In a recent study out of 58 identified species of endophytic fungi associated with *Pinus thunbergii* in coastal shelterbelts of Korea, and 31.03% of them which were mostly *Penicillium* and *Trichoderma* species were found to show salt resistance (Min *et al.,* 2014). The ability to resist high salinity stress is crucial for saline habitats, at the same time salt tolerant endophytes are considered to have potential of enhancing the growth and survival of their host (Min *et al.*, 2014). In endophyte-soybean plant interaction, *Penicillium minioluteum* association significantly promoted growth characteristics (shoot length, shoot fresh and dry biomasses, chlorophyll content, and leaf area) and nitrogen assimilation, by increasing the daidzein and genistein contents in the soybean plants, under salt stress (Khan *et al.*, 2011b). Arbuscular mycorrhizal (AM) also have similar effects of salt tolerance. Mycorrhizal colonization of *Glomus mosseae* fungus in maize plants keep up higher root and shoot dry weights under saline conditions by maintaining higher electrolyte concentrations in roots and lower electrolyte leakage from roots than non-mycorrhizal plants (Feng *et al.*, 2002).

Relative toxicity of heavy metals measured in terms of tolerance index varies with time and type of strain with no particular order that fits all strains. *Penicillium funiculosum* and *Aspergillus foetidus* were found to be the most tolerant to the heavy metals when exposed to heavy metal concentration of 2000 ppm (Valix *et al.*, 2001). In another study heavy metal tolerance of *Aspergillus niger*, *Penicillium simplicissimum*, *Aspergillus foetidus* and *Aspergillus carbonarius* strains in the presence of Ni, Co, Fe, Mg and Mn was reported up to 2000 ppm concentration (Valix and Loon 2003).

Two fungi one closely related to *Cephalosporium* group and other identified as *Acontium velatum* were able to grow in highly acidic pH of 0.3 to 1 and high concentrations copper sulphate (Starkey and Waksman 1943). The Ni was recorded as the most toxic metals for strains of *Aspergillus* and *Penicillium*, *Aspergillus foetidus* was the least tolerant, in particular for Ni, Co and Zn while *Aspergillus niger*, *Aspergillus foetidus* and *Penicillium simplicissimum* showed higher tolerance towards Mo and V (Anahid *et al.* 2011). *Aspergillus flavus*, *Aspergillus niger*, *Aspergillus versicolor*, *Scopulariopsis* sp., *Curvularia* sp., *Helminthosporium* sp., *Humicola grisea* sp., *Nannizzia* sp., and *Fusarium* sp. isolated from contaminated peri-urban agricultural soils of Faisalabad were investigated towards heavy metals tolerance and reported that growth rate of fungal isolates decreased with increase in concentration of metal while most of the isolates were tolerant to metals but some were moderately tolerant and some were sensitive (Iram *et al.* 2012). *Aspergillus, Penicillium, Alternaria, Geotrichum, Fusarium, Rhizopus, Monilia* and *Trichoderma* isolated from wastewater-treated soil had minimum inhibitory concentrations (MIC) of 0.2 to 5 mg ml^{-1} for Cd, 0.1 to 4 mg ml^{-1} for Ni, 0.3 to 7 mg ml^{-1} for Cr, 0.6–9 mg ml^{-1} for Cu and 0.1–5 mg ml^{-1} for Co respectively (Zafar *et al.* 2007). In our study growth index of more than 0.8 were shown by *Trichoderma* sp. for Mn; *Chaetomium* sp. for Fe, Mg, Mn and Cr; *Phomopsis* sp. for Cu, Co, Fe and Mg while growth index of *Fusarium* sp. was below 0.7 for all metals.

The toxicity of heavy metals varies from one fungi to another fungi. Initial adaptation of fungi to heavy metals may reveal the future behavior of tolerance development in the strains with increasing metal concentration, for example high growth followed by a weaker death rate appear to generate into tolerance development of strains in increasing metal concentration often exceeding the control, whereas strong growth followed by equally high death rate appear to result into poorer tolerance (Valix and Loon 2003).

8.5. CONCLUSION

Our preliminary findings indicate that endophytic fungi isolated from *Syzygium cumini* are able to tolerate salt and heavy metal stresses and able to grow on a wide range of pH. Although, there was a particular trend of stress tolerance conferred by different fungi. The high growth index exhibited by some of the fungi provide signal that these fungi are suitable for the bioleaching and biosorption studies. Bioremediation techniques based on the living fungi, is cheaper and environment friendly as compared to the conventional methods of metal removal. Selection is generally based on the tolerance potential exhibited by the fungal strains. Further investigations are needed to optimize and study the bioremediation potential of the studied fungi for the heavy metal.

Acknowledgements

Authors are thankful to the Head and Coordinator, CAS and FIST, Dept. of Botany, BHU, Varanasi, for providing necessary facilities. Financial assistance of CSIR and UGC, New Delhi, is gratefully acknowledged for providing the JRF and SRF. Department of Science and Technology, New Delhi is acknowledged for providing financial assistance to RNK in form of project [SB/EMEQ-121/2014].

REFERENCES

Alpert P (2000). The discovery, scope, and puzzle of desiccation tolerance in plants. Plant Ecology, 151: 5-17.

Anahid S, Yaghmaei, S, Ghobadinejad Z (2011). Heavy metal tolerance of fungi. Scientia Iranica, 18: 502-508.

Arnold AE, Mejía LC, Kyllo D, Rojas EI, Maynard Z, Robbins N, Herre EA (2003). Fungal endophytes limit pathogen damage in a tropical tree. PNAS. 100: 15649-15654.

Baltruschat H, Fodor J, Harrach BD, Niemczyk E, Barna B, Gullner G, Janeczko A, Kogel KH, Schäfer P, Schwarczinger I, Zuccaro A, Skoczowski A. (2008). Salt tolerance of barley induced by the root endophyte *Piriformospora indica* is associated with a strong increase in antioxidants. New Phytol. 180: 501-510.

Bandyopadhyay S, Jash S, Dutta S (2003). Effect of different pH and temperature levels on growth and sporulation of *Trichoderma*. Environment and Ecology. 21: 770-773.

Bayat F, Mirlohi A, Khodambashi M (2009). Effects of Endophytic Fungi on Some Drought Tolerance Mechanisms of Tall Fescue in a Hydroponics Culture. Russian Journal of Plant Physiology. 56: 510–516

Begoude BAD, Lahlali R, Friel D, Tondje PR, Jijakli MH (2007). Response surface methodology study of the combined effects of temperature, pH, and aw on the growth rate of *Trichoderma asperellum*. Journal of Applied Microbiology. 103: 845-854.

Chen AW (1964). Soil fungi with high salt tolerance. Transactions of the Kansas Academy of Science 640: 36-40.

Clay K, Schardl C (2002). Evolutionary origins and ecological consequences of endophyte symbiosis with grasses. Amer Natura. 160: S99-S127.

Ezzouhri L, Castro E, Moya M, Espinola F, Lairini K (2009). Heavy metal tolerance of filamentous fungi isolated from polluted sites in Tangier, Morocco. Af J Microbiol Res. 3: 035-048.

Feng G, Zhang F, Li X, Tian C, Tang C, Rengel Z (2002). Improved tolerance of maize plants to salt stress by arbuscular mycorrhiza is related to higher accumulation of soluble sugars in roots. Mycorrhiza. 12: 185-190.

Fogle MR, Douglas DR, Jumper CA, Straus DC (2008). Growth and mycotoxin production by Chaetomium globosum is favored in a neutral pH. Int J Mol Sci. 9: 2357-2365.

Fogle MR, Douglas DR, Jumper CA, Straus DC (2007). Growth and mycotoxin production by *Chaetomium globosum*. Mycopathologia. 164: 49-56.

Fomina MA, Alexander IJ, Colpaert JV, Gadd GM (2005). Solubilization of toxic metal minerals and metal tolerance of mycorrhizal fungi. Soil Biol Biochem. 37: 851-866.

Girish K, Bhat SS (2011). Physiological variability among isolates of *Phomopsis azadirachtae* from Tamil Nadu. J.Yeast Fungal Res. 2: 65-74.

Iram S, Parveen K, Usman J, Nasir K, Akhtar N, Arouj S, Ahmad I (2012). Heavy metal tolerance of filamentous fungal strains isolated from soil irrigated with industrial wastewater. Biologija. 58: 107-116.

Jahan N, Sultana S, Adhikary SK, Rahman S, Yasmin S (2013). Evaluation of the growth performance of *Trichoderma harzianaum* (Rifai.) on different culture media. J Agri Vet Sci. 3: 44-50.

Khan AL, Hamayun M, Kim YH, Kang SM, Lee IJ (2011a). Ameliorative symbiosis of endophyte (*Penicillium funiculosum* (LHL06) under salt stress elevated plant growth of *Glycine max* L. Plant Physiol Biochem. 49: 852-861.

Khan AL, Hamayun M, Ahmad N, Hussain J, Kang SM, Kim YH, Adnan M, Tang DS, Waqas M, Radhakrishnan R, Hwang YH, Lee IJ (2011b). Salinity stress resistance offered by endophytic fungal interaction between *Penicillium minioluteum* LHL09 and *Glycine max*. L. J Microbiol Biotechnol. 21: 893-902.

Khilare VC, Ahmed R (2012). Effect of different media, pH and temperature on the growth of *Fusarium oxysporum* f. sp. *ciceri* causing chickpea wilt. Int J Adv. Boil. Res. 2: 99-102.

Machowicz-Stefaniak Z, Zalewska E, Król E, Kowalik B (2012). Growth and development of *Phomopsis diachenii* Sacc. in different culture conditions. Acta Sci Pol Hortorum Cultus. 11: 69-80.

Marin S, Sanchis V, Magan N (1995). Water activity, temperature, and pH effects on growth of *Fusarium moniliforme* and *Fusarium proliferatum* isolates from maize. Can J Microbiol. 41: 1063-1070.

Márquez LM, Redman RS, Rodriguez RJ, Roossinck MJ (2007). A virus in a fungus in a plant: three-way symbiosis required for thermal tolerance. Science. 315: 513-515.

Min YJ, Park M, Fong J, Quan Y, Jung S, Lim Y (2014). Diversity and saline resistance of endophytic fungi associated with *Pinus thunbergii* in coastal shelterbelts of Korea. J Microbiol Biotechnol. 24: 324-333.

Nurbaya, Kuswinanti T, Rosmana A, Baharuddin, Millang S (2014). Growth rate and identification of *Fusarium* spp. assosiated with *Aquillaria* spp. from Nunukan regency, North Kalimantan. Int. J. Curr. Res. Aca. Rev. 2: 33-40.

Petrini O, Sieber TN, Toti L, Viret O (1992). Ecology, metabolite production and substrate utilization in endophytic fungi. Nat Tox. 1: 185-196.

Radhakrishnan R, Khan AL, Lee IJ (2013). Endophytic Fungal Pre-treatments of Seeds Alleviates Salinity Stress effects in Soybean Plants. J Microbiol. 51: 850-857.

Rathore SS, Saxena SN, Shrma YK, Mishra BK, Singh B (2015). Effect of pH and salt levels on growth of *Fusarium oxysporum* f. sp. *cumini* isolate from cumin. Int J Seed Spices. 5: 100-101.

Redman RS, Sheehan, KB, Stout RG, Rodriguez RJ, Henson JM (2002). Thermotolerance generated by plant/fungal symbiosis. Science, 298: 1581-1581.

Rodriguez RJ, Redman RS, Henson J (2004). The role of fungal symbioses in the adaptation of plants to high stress environments. Mit Adapt Strat Glob Change. 9: 261-272.

Rodriguez RJ, Henson J, Van Volkenburgh E, Hoy M, Wright L, Beckwith F, Kim YO, Redman RS (2008). Stress tolerance in plants via habitat-adapted symbiosis. ISME J. 2: 404-416.

Rodriguez R, Redman R (2008). More than 400 million years of evolution and some plants still can't make it on their own: plant stress tolerance via fungal symbiosis. J Exp Bot. 59: 1109-1114.

Saikkonen K, Gundel PE, Helander M (2013). Chemical ecology mediated by fungal endophytes in grasses. J Chem Ecol. 39: 962-968.

Schulz B, Wanke U, Draeger S, Aust HJ (1993). Endophytes from herbaceous plants and shrubs: effectiveness of surface sterilization methods. Mycol Res. 97:1447-1450.

Shannon MC (1997). Adaptation of plants to salinity. Advan Agron. 60: 75-120.

Sharma RL, Singh BP, Thakur MP, Thapak SK (2005). Effect of Media, Temperature, pH and Light on the Growth and Sporulation of *Fusarium oxysporum* f. sp. Lini (Bolley) Snyder and Hensan. Ann Plant Prot Sci. 13: 172-174.

Sherameti I, Tripathi S, Varma A, Oelmüller R (2008). The root-colonizing endophyte *Pirifomospora indica* confers drought tolerance in *Arabidopsis* by stimulating the expression of drought stress-related genes in leaves. Mol Plant-Microbe Interact. 21: 799-807.

Singh A, Shahid M, Srivastava M, Pandey S, Sharma A, Kumar V (2014). Optimal Physical Parameters for Growth of *Trichoderma* Species at Varying pH, Temperature and Agitation. Virol Mycol 3: 1-7.

Starkey RL, Waksman SA (1943). Fungi tolerant to extreme acidity and high concentrations of copper sulfate. J Bacteriol. 45: 509-519.

Valix M and Loon LO (2003). Adaptive tolerance behavior of fungi in heavy metals. Min Engin 16: 193-198

Valix M, Tang JY, Malik R (2001). Heavy metal tolerance of fungi. Minerals Engineering, 14: 499-505.

Waller F, Achatz B, Baltruschat H, Fodor J, Becker K, Fischer M, Heier T, Huckelhoven R, Neumann C, Von-Wettstein D (2005). The endophytic fungus *Piriformospora indica* reprograms barley to salt-stress tolerance, disease resistance, and higher yield. Proc Nat Acad Sci USA. 102:13386-13391.

Whipps, J. M. (1987). Effect of Media on Growth and Interactions between a Range of Soil-Borne Glasshouse Pathogens and Antagonistic Fungi. New Phytologist, 107: 127-142.

Wu H, Zhang F, Williamson N, Jian J, Zhang L, Liang Z, Wang J, An L, Tunnacliffe A, Zheng Y. (2014). Effects of Secondary Metabolite Extract from *Phomopsis occulta* on β-Amyloid Aggregation. PloS one. 29: e109438.

Zafar S, Aqil F, Ahmad I (2007). Metal tolerance and biosorption potential of filamentous fungi isolated from metal contaminated agricultural soil. Bioresource Technol. 98: 2557-2561.

Chapter - 9

Status of Aeromycoflora and Aeroallergens in the Vadodara City

Bablu Prasad and Arun Arya

Department of Environmental Studies, Faculty of Science,
The Maharaja Sayajirao University of Baroda, Vadodara 390002
E-mail: aryaaruarya@rediffmail.com

ABSTRACT

Environment is cause of concern today at one place there are problems like Global warming, increase of green house gases, acid rains, ozone hole and on the other hand there are problems of air pollution due to vehicles, dust pollution due to anthropogenic activities. There are reports that these pollutants are resulting into more stressed human life and at the same time microbes and pathogens may cause allergy and respiratory problems. Epidemiological studies report the prevalence of respiration allergy from 15 to 30%. Poor air quality has negative effects on the environment in which we live.

Quantitative analysis of aeromycoflora was tried in 7 out door and 5 indoor places in Vadodara. Air sampling was done in Railway station, Fatehgunj area, SSG hospital, Mandvi gate, Khanderao Market, ONGC Circle and Genda Circle and indoor air of university library, canteen, classroom, CPCB office and residence at Makarpura was analyzed during 2008-2009. Of the outdoor samples stations air of Railway station, SSG hospital and Khanderao market harboured more number of fungal colonies, while indoor air of canteen was more polluted. Presence of *Aspergillus*, *Rhizopus*, *Penicillium* and *Cladosporium* were more at various places than other fungal species. These four species are common aeroall-ergens too. This investigation focuses on the identification of microorganisms prevailing in air, causing biodeterioration at different places Laheripura gate in Vadodara city.

Keywords: Aeromyco flora, indoor air, aeroallergens, black crust, pollutants, plaster, Conservation.

9.1 INTRODUCTION

Urbanization has caused air, water, soil and noise pollution, which is increasing in magnitude and intensity in all major cities of the country. The industries hurl millions of tonnes of toxic effluents, noxious fumes, corrosive gases, dirt and smokes into urban environment. Presence of sulphur dioxide and nitrogen oxides in atmosphere results in acid rains. The automobiles affect human life in several ways, while they have made living easy and convenient they have also complicated human life by increasing the deadly diseases, accidents and fatalities. As a means of quick and convenient transport it is "necessary" in the modern world. It conquers any distance, saves time but toxic fumes emitted by engines may cause break in our lives. In 1988 the death toll was 49,000 in USA. In country like India where the traffic is more chaotic the figure may be many times higher.

The developing nations although have fewer automobiles than the rich developed nations yet they produce far more pollution per vehicle because they have very little technology for emission control. Pollution at ground level affects the children most. Petrol vehicle pollute more than the diesel, and the two and three wheeler are worst pollutant as because they mix mobil oil with petrol. Although diesel vehicles emits lower amount of carbon monoxide and hydrocarbons than petrol vehicles, they emit 30 to 50 times more particulate matter. These are carried by air current and settle in the lower respiratory tract of human lungs when inhaled. The pollution is further more in disuse city areas where traffic is slow. The co and hydrocarbon are highest, while it is minimum at uniform speed. CPCB has recorded increase in SPM levels in all major cities.

There are more than 48 different kinds of motorized vehicles (car, buses, motorcycles etc.) and ten types of non-motorized vehicles including cycle and tri-rickshaws. India has world's highest number of two wheelers in India. In most of the developed nations there are multi-axle trucks while in India there are two-axle trucks. Overloading and adulteration of fuel makes the situation grim. The Supreme Court regulation of use of CNG in Delhi has changed the scenario.

The black crust coating the surfaces of building materials located in urban (polluted) environment may contain all kinds of organic compounds present in aerosols and particulate matter. Wet and dry deposition processes combined with plaster component result in dirty, grey-to-black crust formation, in which aerosols, spores, pollen, dust and every class of particulate matter are entrapped in the mineral matrix. Analysis of the organic compounds extracted from black crusts demonstrate them to be mainly composed of molecular markers that are characteristic of petroleum derivatives. The composition of each crust is governed by the composition of the particular airborne pollutants in the area. Secondly stone and plaster surfaces are usually colonized by bacteria, algae and fungi which accelerate their deterioration. The mechanisms of deterioration, the main microbial genera involved and factors which may affect the degree of colonization and attack are discussed. The chief factor determining microbial growth on constructional materials is moisture. Thus it is important for conservators to consider critical points for the better conservation and preservation. Damp surfaces are readily colonised by microbial cells settling from the air. This leads to the formation of a biofilm, which can trap dust and other particulate materials, increasing its disfiguring effect. In addition, the biofilm can act as a reservoir for potentially dangerous microorganisms. Materials may be protected against microbial growth by the use of biocides.

The four historical gates Laheripura gate, Panigate, Gandi gate, Champanar gate are situated in eastern, western, northern, and southern direction of Mandvi gate, respectively. Mandvi gate is in heart of Vadodara city. These four gates are closely related with the history and culture of the city. Mehmud Begda the sultan of Gujarat died in 15th century A.D. and the Sultanate passed to his son Khalikhan who assumed the name Muzaffar and constructed a fort to the East of ancient vadapadraka which later on evolved into Vadodara. The fort was known are Kila-e-Daulatabad and present day Laheripura gate in part of the western ramparts of the fort. Under its august façade the British and the Arab soldiers of the Gaekwad fought a bloody battle in 1802. Coppersmiths residing in that area where known as Laheris. The word Laheripura is a variant of the original Lahari. Similarly other three gates also have their own significance of transformation of names. All these four gates are constructed with sand stone up to certain height. The upper part constructed with bricks overlaid by plaster. Wooden and iron beams, doors, windows, ceiling are also present which needs conservation work. The stone areas of these gates were subjected to lime coats and OBD (Oil Bond Distemper) coats in past. These coats are not only concealing the architectural details but also disturbing aesthetic view of the gates, besides causing damage to the stone surface.

The microorganisms present on external stone walls of building are fungi, bacteria (including actinomycetes), algae (including cyanobacteria), lichens and protozoa (Warscheid and Braams, 2000; Zanardini *et al.,* 2000; Tomaselli *et al.,* 2000). Microorganisms are able to obtain different elements for their metabolism, e.g., calcium, aluminum, silicon, iron and potassium, by biosolubilization from the stone matrix. Such microbial biosolubilization involves the production of organic and inorganic acids by the metabolic activity of algae, lichens, fungi and bacteria. This acidic deterioration is one of the best known biogeochemical mechanisms of rock decay (Herrera *et al.,* 2004).

9.2 OCCURRENCE OF FUNGI ON MONUMENTS

Four historical gates are surrounded by heavy vehicular traffic. The road passing through these gates are linking old city with new city of Vadodara, hence, these gates are prone to heavy vehicular pollution, besides pollution caused due to nearby by hotels, shopping and residential complexes. The complex physical and chemical interactions of atmospheric pollutants with the mineral material and microorganism have accelerated the decay of stones (Warscheid and Braams, 2000; Zanardini *et al.,* 2000; Tecer and Cerit, 2002). Microbiological deterioration, caused by microorganisms growing on the material surfaces, follows the chemical deterioration (Zanardini *et al.,* 2000). Biodeterioration has usually been considered to be a degradation process following the initial deterio-rating effects of inorganic agents. The pollutants were thought to condition stone surfaces for microbial contamination due to structural changes and the enrichment of inorganic and organic nutrient substrates.

Fig: Laheripura Gate (Historical Gate) at Vadodara city of Gujrat state : front view,

9.3 SAMPLING AND ISOLATION OF FUNGI

Totally 05 Samples were collected from various places of the Lehripura gate of Vadodara city of Gujrat state and brought to the laboratory under aseptic conditions. The isolation of fungi was done by culturing the samples and by direct incubation of samples in moist chamber. Two different agar media were taken for the selection of basal media. Media employed were Czapeck-Dox and Potato dextrose agar. Out of two media Czapeck's-Dox medium was selected as the basal medium for subsequent studies, because this medium supported good mycelial growth and excellent sporulation for all the best organisms and its composition is simple due to which modifications and substitution of various ingredients were possible. The purified fungal cultures were identified by using mycological techniques and were compared with the available authentic literature, reviews and mycological manuals (Gilman *et al.*, 1995; Sharma, 1974)

Percentage of frequency

Frequency occurrence was calculated as follows:

$$\%\ \text{Frequency} = \frac{\text{No. of samples in which specific organism occurred}}{\text{Total number of samples examined}} \times 100$$

Based on the frequency occurrence the fungi were grouped as Rare (0-25% frequency), Occasional (26-50% frequency), Frequent (51-75% frequency), and Common (76-100% frequency) species.

Table 1. Occurrence, percentage frequency and frequency class of different Fungal species in Lehripura gate, Vadodara city of Gujrat state

S.N.	Name of Fungi isolated	LG 1	LG 2	LG 3	LG 4	LG 5	Frequency %	Frequency class
1	*Trichoderma viride* Pes.	-	+	+	+	-	60	F
2	*Rhizopus stolonifer* (Ehrenb ex Fr.) Linder.	-	+	+	+	-	60	F
3	*Aspergillus niger* Van Tiegh	-	+	+	+	-	60	F
4	*Curvularia clavata* Walker	-	+	-	+	-	40	O
5	*Aspergillus terreus* Thom.	-	-	-	+	-	20	R
6	*Alternaria alternata* (Fr.) Keissler	-	-	-	+	-	20	R

LG= Laheripura gate; (+) = presence of species; (-) = absence of species; C = common; F = frequent; O= occasional

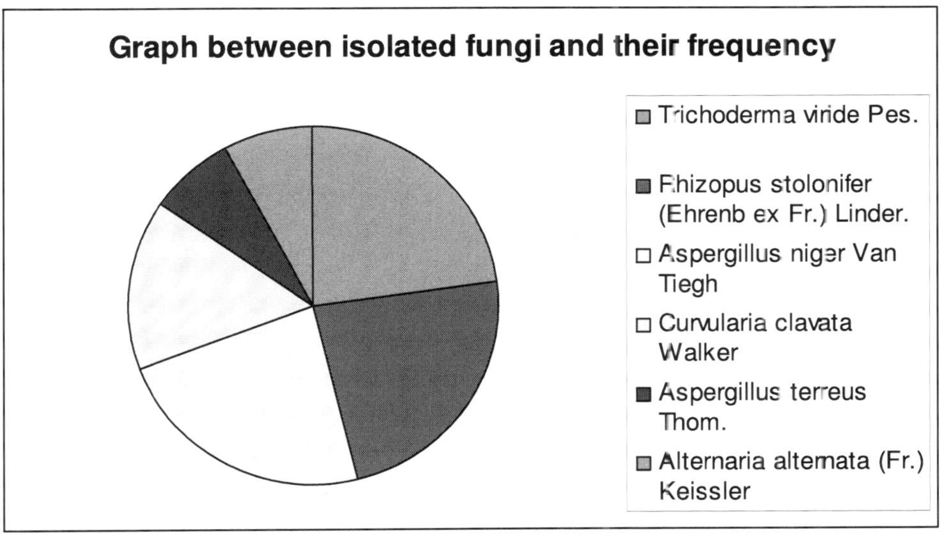

Graph: showing isolated fungi and their frequency

9.4 POLLUTANT AEROMYCOFLORA IN VADODARA

Air pollution due to domestic heating and transportation has increased the atmospheric concentration of inorganic and organic compounds in the form of gases, aerosols or particulate matter. The concentrations of air pollutants such as SO_2, NOx and particles released into the atmosphere as a result of the burning of the fossil fuels. Sulfur dioxide is one of the major gaseous components of polluted atmospheres in urban areas. Because oxidation of sulfur dioxide results in the formation of sulfuric acid, there is a strong correlation between the sulfaction of Calcium compounds present in the sandstone and the atmospheric concentration of sulfur dioxide, resulting the formation of gypsum. During the crystallization of gypsum, airborne organic pollutants, carbonaceous particles, dust, etc. are accumulated at the surface of buildings and trapped in the mineral matrix. This results in the formation of a hard, grey to black crust. This crust enriches the substratum and anthropogenic compounds may influence, to a great extent, the colonization and growth pattern of microorganisms in stones located in polluted environments (Saiz-Jimenez, 1997).

Microorganisms are able to obtain different elements for their metabolism, e.g., calcium, aluminum, silicon, iron and potassium, by biosolubilization of materials. Such microbial biosolubilization involves the production of organic and inorganic acids by the metabolic activity of algae, lichens, fungi and bacteria. This acidic deterioration is one of the best known biogeochemical mechanisms of rock decay (Herrera *et al.*, 2004).

The recent researches demonstrated that numerous funguses present on the stonework can utilize, for their growth, the airborne organic compounds and inorganic particulate matter settled on the stone surface. They mainly arise from the incomplete combustion of fossil fuels and dust (Saiz-Jimenez, 1997; Zanardini *et al.,* 2000). This evidence indicates that heterotrophic microorganisms can act as first colonizers in the areas with a high level of organic pollutants (Zanardini *et al.,* 2000).

9.5 CONCLUSION

The deterioration of stone monuments in tropical regions due to the action of biodeteriogens has long been recognized, only this research brings up that the microorganisms can play a significant role on biodeterioration of stone monuments under pollution. The evaluated and the results are compared with the other researches, it is seen that there are a few aerobic heterotrophic bacteria as colony-forming units in the other sides (researches) but the other main microbial flora (fungi and actinomycetes) are similar with moderate climatic regions (Praderio *et al.,* 1993 and Zanardini *et al.,* 2000). However, Vadodara is a hot city with average temperatures and that microbial metabolism of the dwelling microorganisms of stone surfaces can be restricted by extreme weather. The small amount of aerobic heterotrophic bacteria may be due to the extreme hot weather, but the other microbial flora such as fungi and actinomycetes may develop like a normal population in moderate climatic regions.

Acknowledgements

The authors wish to thank Dr. Arun Raj, T., Superintending Archaeologist, at the Archaeological Survey of India, Raipur Circle, Dr. K.S. Rana, Director (Science) at the Archaeological Survey of India, Dehradun for their guidance and, we would like to thank Head, Deptt. of Botany, The M.S. University of Baroda for helping us during the isolation of microorganism.

REFERENCES

Dukes, W.H., (1972) *Conservation of Stone: Chemical Treatments,* The Architects' Journal Information Library, 156, (34): 433-438.

Gilman, J. C. (1995), *A Manual of Soil Fungi.* Print Well Publication, Jaipur (India).

Herrera L.H, Arroyave C, Guiamet P, de Saravia SG and Videla H. (2004) Biodeterioration of peridotite and other constructional materials in a building of the Colombian Cultural Heritage. Int Biodeterior Biodegrad 53:135–41.

Praderio G, Schiraldi A, Sorlini C, Stassi A and Zanardini E. (1993) Microbiological and calorimetric investigation on degraded marbles from the Ca' D'Oro facade (Venice). Thermochim Acta 227:205–13.

Saiz-Jimenez C. (1997) Biodeterioration vs biodeterioration: the role of microorganisms in the removal of pollutants deposited onto historic buildings. Int Biodeterior Biodegrad 40: 225– 32.

Shah, N.R. and Shah, A. (2000) *Studies on some fungal biodeterioration.* Pub. Bhartiya Kala Prakashan, Delhi, pp.208.

Sharma, K.R. (1974), Colonisation of saprophytic microfungi and bacteria on the aerial parts of Sesamum orientale and Gossypium hirsutum. Ph.D. Thesis Univ. of Delhi (INDIA).

Tecer L, and Cerit O.(2002) The effects of air pollution on carbonate stone monuments in urban areas (Sivas, Turkey). Fresenius Environ Bull; 11: 505–9.

Tomaselli L, Lementi G, Bosco M and Tiano P. (2000) *Biodiversity of photosynthetic micro-organisms dwelling on stone monuments.* Int Biodeterior Biodegrad ;6:251 – 8.

Warscheid Th and Braams J. (2000) *Biodeteriration of stone: a review.* Int Biodeterior Biodegrad 46: 343– 63.

Zanardini E, Abbruscato P, Ghedini N, Realini M and Sorlini C. (2000) *Influence of atmospheric pollutants on biodeterioration of stone.* Int Biodeterior Biodegrad 45: 35– 42.

■■■■

B: Climate Change and Sustainable Development

प्रकृति के प्रहरी

पेड़ों का हिलना
पत्तों का चारो और फैलना
फूलों का खिलखिलाना
डालियों का झुक झुक कर स्वागत करना
लम्बी बेलों का मिलकर
नत मस्तक होना
प्रकृति की सृजनात्मकता एवं
शक्ति का अनूठा प्रदर्शन है।
मानव जीवन के वृक्ष हैं आधार
सभी ऋतुओं को जीवांत बनाने में,
वन हैं वनस्पतियों को संवारे हुये
कार्बन का संग्रहण करते और
स्वच्छ वायु देते
प्रकृति के इन प्रहरियों को हमारा शत्–शत् नमन।

Chapter - 10

Effect of Fluorspar mining waste on germination and Plant Growth of *Zea mays* and *Vigna radiata*

Disha Thakkar and Arun Arya

Department of Environment Studies, Faculty of Science,
The Maharaja Sayajirao University of Baroda, Vadodara- 390002.
Email: Sarojarun10arya@rediffmail.com

ABSTRACT

Largest deposits of Fluorspar ore are available in the Ambadungar area of Kadipani in Vadodara district. Fluorspar is extensively used in iron industry. Reclamation of Waste Mining land is a major problem. Due to opencast mining the high amount of harmful substances present inhibit the normal plant growth and developing a healthy ecosystem. Plantation is the oldest technology available for the restoration of lands degraded by human activity. The study included growing of maize and cowpea in soil containing different percentage of mining soil wastes. Biomass of seedlings and chlorophyll content of raised plants will provide a clue whether plantation in mining land is possible or not. Physico-chemical properties and microbiological characteristics of soil were analyzed.

Keywords: Fluorspar, Wild plants, *Zea mays* L., *Vigna radiata* L., Mining, Reclamation, Heavy metals and Chlorophyll.

10.1. INTRODUCTION

Fluorspar forms a sizeable and economically exploitable mineral deposit in Kadipani, Vadodara, India. The Gujarat Mineral Development Corporation (GMDC) is mining the mineral F by open cast mining method. Flourite occurs in association with carbonatite suite of alkaline rocks belonging to Deccan- trap sub province of lower Narmada Valley. The quality of fluorite is both acid-grade and metallurgical-grade (less than 96% CaF_2 content) (Merh, 1995).

Due to open cast mining a large area is disturbed and requires reclamation. To suggest suitable plantation a preliminary study was undertaken. Heavy metal stress, contributed by elements like Cd or Mn is one of the crucial factors that limit the distribution and Productivity of major health crop plants poses severe health hazards to human.

One possible approach is to make use of the hyper-accumulators plants and the phytoremediation techniques. Furthermore, a general solution to this problem is chelation, which is generally understood as carbon binding to a compound resulting in a neutrally charged complex that can move more freely through a variety of substrates. Several chelators are known to perform this function in soil plants.

Restoration has been attempted on an experimental scale in various parts of the country and its implementation for the whole area has also started at some locations. In some cases the plantation of exotic and horticulture species has also been considered as restoration, while in other cases, e.g. Neyveli, the degraded land has been made productive and changed to agricultural land.

10.2 METHODOLOGY

A Survey has been conducted at the mining site of Kadipani, District Vadodara at an interval of a month to collect soil and plant samples. Soil and plant samples were collected from mined, nearby unmined and dumping areas. Random samplings from different mining areas have been conducted. The samples were brought to the laboratory in polythene bags, and then study of soil samples and plants have been done.

a. Physico- Chemical properties of the mining soil

Various parameters of the mining soil sample were checked and analyzed in the laboratory. The pH, Electrical conductivity, Moisture Content, Water Holding Capacity, Metallic content (EDXRF-Spectrometer), Organic Matter content, Organic Carbon, Available Phosphorus, Available Nitrogen, Exchangeable Sodium and Potassium.

All analysis was carried out according to the procedures suggested in **"Handbook of Methods in Environmental Studies Air, Noise, Soil and Overburden Analysis, S.K. Maiti (Vol-2) 1994".**

b. Growing of maize and moong in mining soil samples

In this method, 3 Replicates of maize and moong were planted in

1. Normal garden soil
2. 50:50 % Normal + mining soil
3. 75:25 % Normal + mining soil
4. 100 % mining soil.

10 seeds of maize and moong were added in each replicates. After that it was kept in botanical Garden where proper light, temperature and water was

provided. The seed germination of 2, 4, 6, 10, 20, 30 days was recorded. The plant biomass and plant height was recorded in the period of 7, 14,21,28,56 days.

The **Percentage Germination rate, Chlorophyll Content, Biomass, Leaf Size, Plant Height and Ascorbic Acid Content** were measured.

Chlorophyll a, b and total chlorophyll were determined by following the method of Arnon (1949). Biomass is regarded as the natural mass of organisms (fresh weight) in situ, just as they are or it can be measured in terms of the dried organic mass (dry weight). Ascorbic acid content was estimated following the method of Keller (1997). The leaf area/size is measured through CI-203 Laser Leaf area meter.

10.3 RESULTS AND DISCUSSION

Pot experiments were carried out in different percentages of mining soils and garden soils. The survey revealed that total chlorophyll (Table 4) content was poor in maize as compared to moong leaves. It is evident from Table 5, thus, the biomass and chlorophyll content of maize and moong revealed that the mixing of mining soil with normal soil, plantation and healthy crops can be grown in the waste mined land.

The results are indicative of more growth in both the cases i.e maize and moong. After one month it is observed that the biomass and chlorophyll content in maize and moong plants were healthy and fully grown in 50-50% (normal garden soil + mining soil). It is evident from Table 3, that the vegetation prevailing around mining area indicates 10 species of trees, 9 species of herbs and 6 species of shrubs. The comparison of leaf size samples of Kadipani mining site with Vadodara was carried out. It was found that leaf of charel and bhoi ringadi was greater in size at Vadodara.

Table 1. Percentage Germination in corn

S.N.	**% Germination in *Zea mays* L.**								**Soil sample**
	2 days	**4 days**	**6 days**	**10 days**	**20 days**	**30 days**	**40 days**	**50 days**	
1.	5	6	10	14	22	35	65	92	Soil 1
2.	9	10	12	15	24	38	75	95	Soil 2
3.	7	9	11	14	21	28	68	84	Soil 3
4.	4	6	9	13	27	35	62	80	Soil 4
5.	6	7	10	12	23	30	67	90	Soil 5
6.	3	6	9	11	20	27	60	88	Soil 6

Table 2. Shows the Plant Diversity of Kadipani Mining Site

S.N.	Scientific Name	Common Name
1	*Aerva lanata* (L.) Juss.	Ambaahaldhar
2	*Ailanthus excela* Roxb.	Arduso/Wild Neem
3	*Annona squamosa* L.	Sitafal
4	*Alangium salvifolium* (L.f.) Wang	Ankul
5	*Blumea lacera* L.	Kakronda
6	*Cassia glauca* Linn.	Bhootalan
7	*Cassia tora* L.	chakod
8	*Diospyros melanyolon* Roxb.	Timru
9	*Echinops echinatus* Roxb.	Utkanto
10	*Ficus hispida* (L.f.)	Dedhumado
11	*Holoptelia integrifolia* (Roxb.) Planch	Charel/ popti
12	*Indigo feracordifolia* B. Heyne ex Roth.	Gokhru
13	*Liandenbergia muraria* (Roxb.)	Patharchatti
14	*Prosopis juliflora* (Sw.) Dc	Gandobaval
15	*Pongamia pinnata* (L.) Pierre.	Indian beech/ Karanj
16	*Salvia santolinifolia* Boiss.	-
17	*Solanum xanthocarpum* Schrad.	Bhoiringdi
18	*Tamarindus indicus* L.	Tamarind / Amli
19	*Tectona grandis* L.f.	Teak / Sag
20	*Woodfordia fruticosa* (L.) Kurz	Dhavti

Table 3. Indicates the comparison of leaf size taken from Kadipani mining site and Vadodara

No	Scientific name	Common name	At mine site [leaf size/ leaflet] in cm		At Vadodara [leaf size/ leaflet] in cm		% Reduction in leaf length	
			Length	Width	Length	Width	Length	Width
1.	*Tectona grandis* L.f.	Teak	35	24	38.5	25	9.09	4
2.	*Woodfordia fruticosa* (L.) Kurz	Dhavti	7	2.5	7	3	-	16.66
3.	*Solanum xanthocarpum* Schrad.	Bhoi ringdi	5.5	1.5	9	6	38.8	75

No	Scientific name	Common name	At mine site [leaf size/ leaflet] in cm		At Vadodara [leaf size/ leaflet] in cm		% Reduction in leaf length	
			Length	Width	Length	Width	Length	Width
4.	*Alangium salvifolium* (L.f.) Wang	Ankul	12	2.8	13	2.9	7.69	3.44
5.	*Ficus hispida* (L.f.)	Dedhumado	22	10	37	14	45.95	28.5
6.	*Annona squamosa* L.	Sitafal	11	6.5	17	6.7	35.29	2.98
7.	*Holoptelea integrifolia* Planch.	Charel (popti)	3	1.8	9.5	6	68.42	70
8.	*Diospyros melanaxylon* Roxb.	Timru	11.5	5	26	11.4	55.76	56
9.	*Cassia glauca* L.	Bhootalan	12	6.8	15	7	20	2.8
10.	*Aerva lanata* L.	Ambaa haldhar	2.5	1	4	2	37.5	50
11.	*Prosopis juliflora* (Sw.) DC	Gandobaval	3	1.5	3.1	1.7	3.22	11.7
12.	*Ailanthus excelsa* Desf.	Araduso	11.5	5	14	5	17.85	-

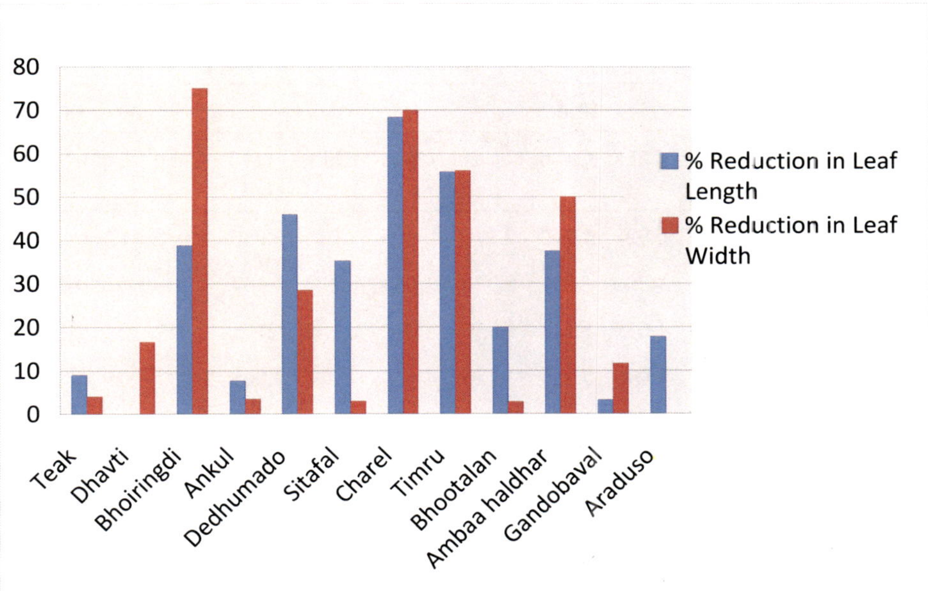

Graph 1. % Reduction in leaf Length & leaf Width of Kadipani mining site and Vadodara

Figure 1: Maize and Moong cultivation in the botanical garden.

Table 4. Chlorophyll content of *Zea mays* L. and *Vigna radiata* L.

Sr. no	Type of soil sample	Plant	Chlorophyll content (mg/g)								
			10 days			20 days			30 days		
			Chl a	Chl b	Tot. chl	Chl a	Chl b	Tot. chl	Chl a	Chl b	Tot. chl
1.	**Soil 1**	*Zea mays* L.	0.997	1.947	2.178	0.993	1.078	1.880	1.070	1.166	1.930
2.	**Soil 2**	*Zea mays* L.	1.264	1.082	1.376	1.064	0.574	1.147	0.786	0.940	1.605
3.	**Soil 3**	*Zea mays* L.	1.055	1.148	1.356	1.032	1.74	2.38	0.917	0.979	1.902
4.	**Soil 4**	*Zea mays* L.	1.096	0.985	1.656	1.198	1.95	2.065	0.731	0.926	1.659
5.	**Soil 1**	*Vigna radiata* L.	1.122	1.310	2.358	1.349	1.041	2.072	1.124	1.539	2.665
6.	**Soil 2**	*Vigna radiata* L.	0.837	0.561	0.926	1.379	1.119	2.520	1.360	1.170	2.559
7.	**Soil 3**	*Vigna radiata* L.	1.357	1.119	2.520	1.192	1.043	2.265	1.085	1.452	2.540
8.	**Soil 4**	*Vigna radiata* L.	1.192	1.065	2.268	1.280	0.584	1.769	1.204	1.109	2.318

Figure 2: Mining Site, Kadipani Gujarat

Table 5. Biomass of *Zea mays* L. and *Vigna radiata* L.

No	Type of soil sample	Plant	Biomass (in g)											
			10 days				20 days				30 days			
			Fresh wt		Dry wt		Fresh wt		Dry wt		Fresh wt		Dry wt	
			S wt.	R wt.	S wt.	R wt.	S wt.	R wt.	S wt.	Rwt.	S wt	R wt	S wt	R wt
1.	Soil 1	Corn	0.784	0.714	0.090	0.150	0.892	0.635	0.102	0.115	1.802	0.280	0.426	0.104
2.	Soil 2	Corn	0.849	0.294	0.093	0.042	1.620	0.488	0.395	0.132	1.837	0.560	0.228	0.084
3.	Soil 3	Corn	1.614	0.430	0.471	0.087	0.894	0.347	0.160	0.098	1.173	0.247	0.209	0.109
4.	Soil 4	Corn	0.710	1.230	0.65	0.140	0.996	0.207	0.223	0.052	0.724	0.307	0.415	0.192
5.	Soil 1	Moong	0.332	0.047	0.048	0.006	0.548	0.085	0.070	0.039	1.710	0.60	1.025	0.26
6.	Soil 2	Moong	0.429	0.066	0.062	0.005	0.968	0.115	0.342	0.047	1.970	0.66	0.790	0.34
7.	Soil 3	Moong	0.680	0.274	0.096	0.062	0.840	0.138	0.198	0.040	1.053	0.305	0.980	0.47
8.	Soil 4	Moong	0.309	0.040	0.055	0.009	0.801	0.062	0.194	0.022	1.621	0.562	0.887	0.32

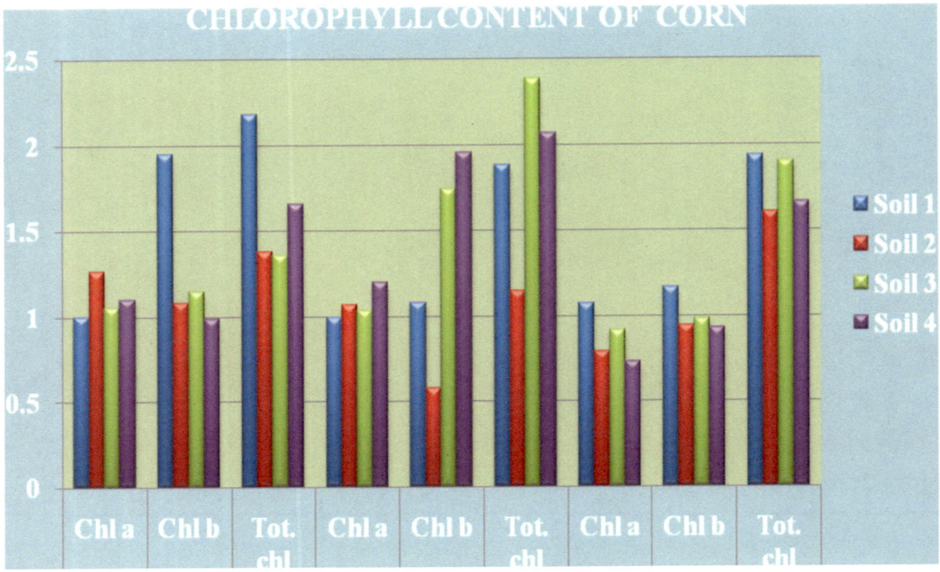

Graph 2: Shows the Chlorophyll content of *Zea mays* L.

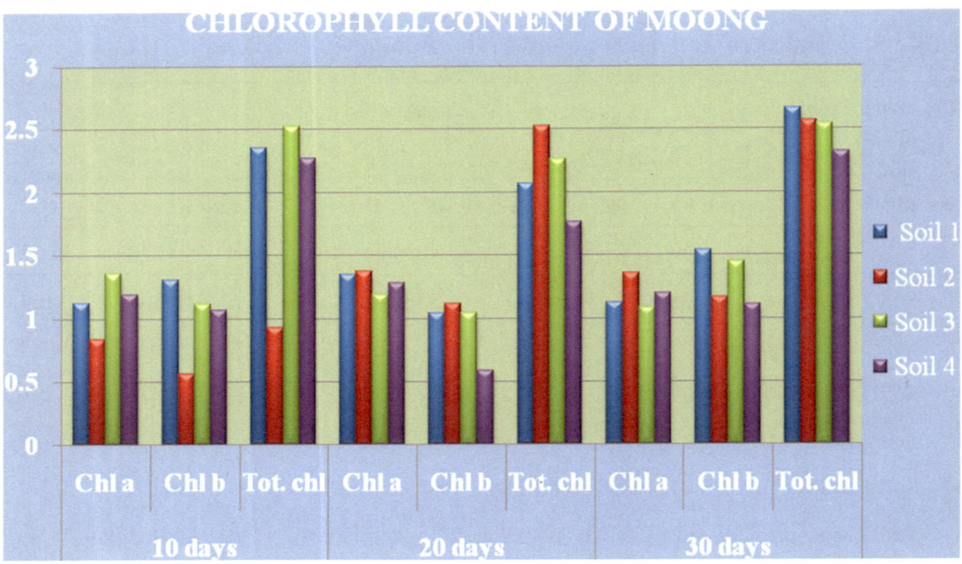

Graph 3: Shows the Chlorophyll content of *Vigna radiata* L.

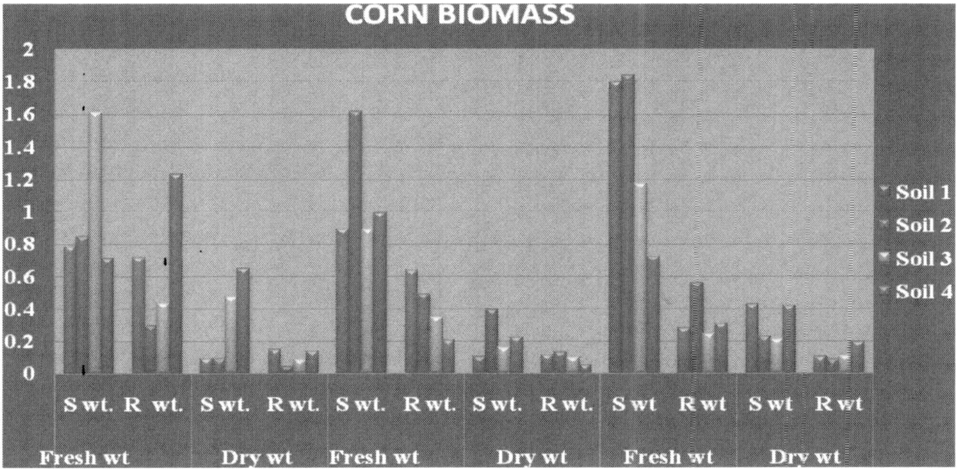

Graph 4. Shows the Biomass of *Zea mays* L.

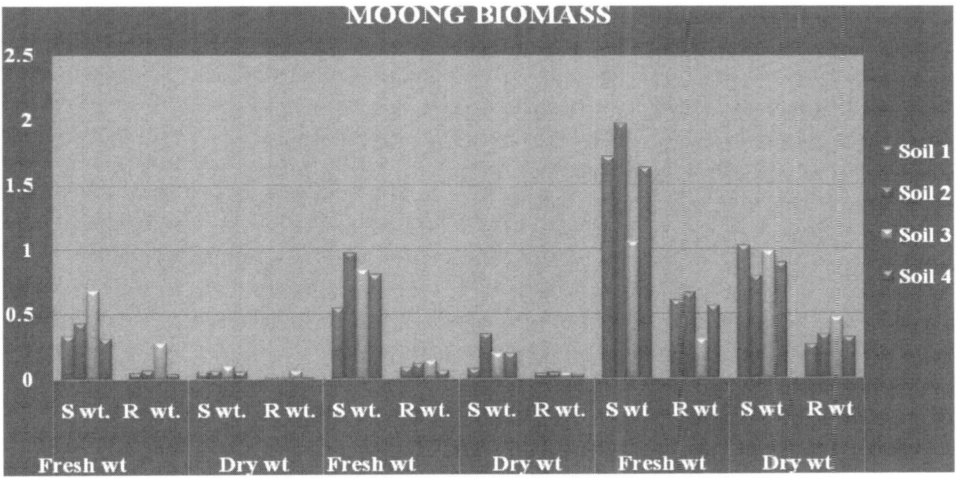

Graph 5: Shows the Biomass of *Vigna radiata* L.

10.4 CONCLUSIONS

- Plantation is the oldest technology available to restore the degraded lands. The physico-chemical analysis was done. The pH of soil is alkaline. The elemental analysis of mined soil and the soil containing maize and moong seedlings was done. It was revealed that sample form mined soil shows higher calcium content.
- Pot experiments were carried out in different percentages of mining soils and garden soils. The survey revealed that total chlorophyll content was poor in maize as compared to moong leaves. The

maximum chlorophyll content was recorded in the 50-50% (normal garden soil + mining soil). In case of moong, the fresh weight and dry weight was less upto 20 days. The fresh weight of maize was higher in 25%-75% (Mining soil + Normal soil). Thus, the biomass and chlorophyll content of maize and moong revealed that the mixing of mining soil with normal soil, plantation and healthy crops can be grown in the waste mined land.

- The results are indicative of more growth in both the cases i.e maize and moong. After one month it is observed that the biomass and chlorophyll content in maize and moong plants were healthy and fully grown in 50-50% (normal garden soil + mining soil).
- By plantation, the degraded mined area can be restored aesthetically and hence the spoiled area can be recovered by growing maize, moong and such other crops by mixing with normal garden soil.

REFERENCES

Allan, R.J. (1995). Impact of mining activities on the terrestrial and aquatic environment with emphasis on mitigation and remedial measures in Heavy metals: problems and solutions. Edited by W. Salomons, U. Forstner, and P. Mader. Springer-Verlag, Berlin. pp. 119–140.

Aneja, K.R. (1996) Experiments in Microbiology, Plant Pathology, Tissue Culture and Mushroom Cultivation. 2nd ed. Pub. Wishwa Prakashan.

Anonymous (1996) Soil Survey Report of Ambadungar Flourspar Project Area (GMDC). Kadipani, Chhota Udepur, Vadodara, Gujarat, Narmada Water Resources and Water Supply Department, Govt. of Gujarat, 1996.

Arya, A. (2006) *Mycorrhiza* - Application of AM fungi for Biomass Production in Neem pp. 259-265.

Badrinath, S.D., Chakraborty, A. and Khan, S. (1994) *The Impact of Mining on the Environment – Problems and Solution*. Proc. Int. Symp., Oxford and IBH, New Delhi, pp. 159–172.

Basu, B.B. (1994). *Minerals and Ecology* (ed. Banerjee, S.P.), Oxford and IBH, New Delhi, pp. 83–93.

Baver, L.D.B. (1940). *Soil Physics*. John Wiley & Sons, Inc., New York.

GES, (1997). *Ecological Restoration of Mining Sites of GMDC*; Report 2 Vol. II – Ambadungar Fluorspar Mines, Kadipani, Gujarat Ecological Society, Vadodara, pp. 85.

Gilman, J.C. (1957). *A Manual of Soil Fungi*. 2nd ed. Iowa State University Press. Iowa. pp. 450.

Jha, A.K. and Singh, J.S. (1992). *Restoration of Degraded Land: Concepts and Strategies* (ed. Singh, J.S.), Rastogi Publication, Meerut, 1992, pp. 212–254.

Jordan, W.R. III, Peters, R.L. and Allen, E.B. (1988). *Environ. Management*, 12, pp.55–72.

Maiti, S.K. and Banerjee S.P (1994). *The Impact of Mining on the Environment Problems and Solution.* Proc. Int. Symp., Oxford and IBH, New Delhi, pp. 369–379.

Merh, S.S. (1995). *Geology of Gujarat*, Geological Soc of India, Bangalore.

Mueller, G.M., Bills, G.F. and Foster, M.S. (2004). *Biodiversity of Fungi*- Inventory and monitoring methods Elsevier.

Ramarao, P. and Ram Reddy, M.A. (1964). *Studies on Soil Fungi*- II. Mycopath. Et Mycol. Appl. 34: 311-318.

Rusell, E.J. (1953). *Soil Conditions and Plant Growth.* Longman green & Co., London.

Satyanayana, T. and Johri B.N. (2012). *Microorganisms in Environmental Management*, Springer pp. 81-113, 719-745

Sharma, P.K. (1989). *Raman Principles and Practice of Mineral Exploration.* Geological Soc. of India, Bangalore.

Shrivastava, K.L. (2009). *Economic Mineralization* Scientific Publishers (India).

Tyagi, S.N., Khan, J.A. and Sharma, M.L. June (2007). Forest Resource Survey Chhotaudepur Forest Division (Vadodara district). Gujarat Forest Department.

Verma, S.K and Verma, M (2012). *Plant Physiology, Biochemistry and Biotechnology*, S. Chand pp.103-117

Wadia, D.N. (1996). *Minerals of India.* National Book Trust India.

Wainwright, M. (1992). *An Introduction to Fungal Biotechnology.* John Wiley and Sons, INC. pp.202

Warcup, J.H (1995). Isolation of Fungi from Hyphae present in Soil. *Nature*, UK. pp.175: 953.

■■■■■

Chapter - 11

Toxicological potentials of heterogeneous chemical mixtures

Kauresh D Vachhrajani and Kiran Morya

Department of Zoology, Faculty of Science, The Maharaja Sayajirao University of Baroda, Vadodara-390002, Gujarat, India.
E-mail: kauresh@gmail.com

ABSTRACT

Toxicology deals with the study of interaction of chemical substances with living system at different doses for the evaluation of safety exposure limits of chemicals like drugs and man-made and natural substances. The toxicological studies are important for improving the health of humans, animals and their environments. Toxicity testing assist to define flow of events including chemical exposures, distribution, metabolism and importantly its interactions with biologically significant macromolecules, resulting into a toxic end point. The information on chemical substances obtained from toxicological research predicts the safe exposure levels. Thus, research provides understanding of mechanisms of toxicant action and then this information can be used in the prophylaxis and therapeutics. With the identification of specific type of chemical or material hazard toxicological studies can provide scope for proper planning and safety measures to prevent exposure. Further, toxicological data are also important in development of legislation aimed at protecting health and environment by assessing possibility that a particular chemical possess a significant risk to life and environment. Therefore, today the science of toxicology has become one of the most interesting interdisciplinary fields. Biologist, chemist, engineers, ecologist and other scientists have worked together for assessment of environmental effects of chemicals within this discipline.

Keywords : Toxicological potentials, Heterogeneous chemical mixtures, Heavy Metals

11.1 POTENTIAL HEALTH HAZARDS OF XENOBIOTICS

There are numerous toxic chemicals present in the environment and over past century humans have introduced a large number of chemical substances. Some of these chemicals are essential for life and are useful in many ways, but many are toxic and have potential to cause hazard to health and environment. Few chemicals have indirect effect on human health by disturbing balance of environment. Some effects of chemicals are reversible while others are not. The toxic effect of a chemical or its metabolites in a living system are produced only when it reaches to appropriate site of action at a particular concentration for sufficient duration. For comprehensive characterization of the potential hazard of a specific chemical agent, we need to know type of effects, exposure time and the dose required to produce that effect. In other words, the toxicity of a substance can be affected by many different factors such as route of exposure, duration and frequency of exposure, physical form of chemical substance, an individual's overall health *etc.* (Fig. 1). The spectrum of undesired effects of chemicals is broad.

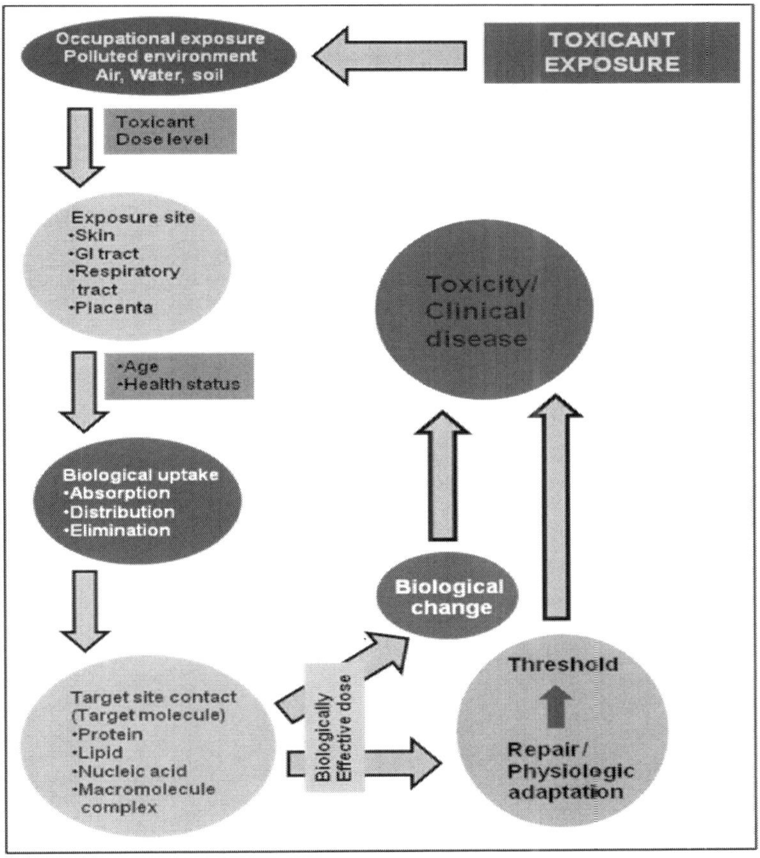

Fig. 1.

Laboratory experiments on known contaminants determine whether true causal relationship exist between potential causes and effects noted in the field. Comprehensive animal studies have been made on the effect of various contaminants present in the environment (Table 1). Much of the literature on the toxicity of individual substances has detailed the toxic responses at relatively higher dosage of exposure than those actually detected as environmental levels.

Many of the substances, such as cadmium, arsenic, chromium, lead, Mercury, nickel, dioxins, polycyclic aromatic hydrocarbon, phenolic compounds, phthalates, and pesticides are considered to be carcinogenic, however, evidences that these substances cause cancer at detected environmental levels are very less (Ames and Gold, 1990; Mishra *et al.,* 2010; Soto and Sonnenschein, 2010). The potential health hazards of the environmental toxicants are detailed in Table 1.

Table 1: Experimental studies on Chemicals Toxicity

Cadmium: Renal tubular dysfunction, Osteomalacia, anemia, Neurodegenerative changes with reduction in learning and memory retention capacity, and necrosis of periportal liver cells.

Chromium: Causes allergic dermatitis, pulmonary carcinogenicity, nephrotoxicity. Induces testicular damages with degeneration of seminiferous epithelium. Induces neoplasm of epithelium of oral cavity and small intestine.

Mercury: Repeated exposures resulted in severe disturbances in the central nervous system, Dementia, loss of memory, excitability, and insomnia. Induce oxidative injury in renal proximal tubule and liver.

Lead: Causes damages in prefrontal cerebral cortex, hippocampus, and cerebellum and leads to neurologic disorders like information processing difficulties, memory problems, reduction in sensory and motor reaction times.

Arsenic: Skin lesions, peripheral neuropathy, encephalopathy, hepatomegaly, anaemia, bone marrow depression, and renal necrosis.

Nickel: Atrophy of the nasal olfactory epithelium, cause lung inflammation, fibrosis, and alveolitis.

Organohalogen compounds: Toxic to neural and glial cell, suppress immune responses, impair reproductive functions, produce hepatotoxicity, hypothyroidism, cardiovascular toxicity, bone, skin, and tooth toxicity and cancer.

polycyclic aromatic hydrocarbon: induce cardiovascular disease and pulmonaly dysfunction

Phenols: Neurotoxic, hematotoxic, hepatotoxic, nephrotoxic, provoke mutagenesis and carcinogenesis.

Phthalates: Reduce testicular hormone production; also affect reproductive functions characterized by reduction in fertility, sperm density and motility, and ovarian and testicular weights.

Pesticides: Affects the tracts in the nervous system and Inhibit the cholinesterases activity. Also affect the immune responses.

Epidemiological studies are very important in the assessment of human health to chemicals exposure. Contamination of our environment with huge amount of chemical wastes because of rapid industrialization and urbanization has raised the concerns about the development of specific disease and potential long term impact on the quality of life and the environment. In spite of our suspicions, we have very few clear data, linking human illness to low or moderate environmental exposures of many potentially toxic substances. Few epidemiological studies over a last few decades have described the health hazard by chemical pollution (Table 2). Most of these epidemiological studies have been the outcome of occupation exposures and exposure to the general public through environmental pollution, chemicals escape from the industry or from accidental spill or leakage. This information would be particularly relevant in situations where one or two specific substances are of concerns but may be less useful for investigating the more general exposures to the toxicants emitted simultaneously in the environment from various industrial processes, which tend to be heterogeneous in nature. Therefore, epidemiology of multiple chemicals exposures is a relatively unexplored field in occupational and environmental health. Combined chemicals exposures are common events, therefore, studies on potential health hazards from such exposures should be encouraged. The data obtained from chemical mixture toxicity studies will be very useful to relate environmental chemical pollution and causes of increasing health complications. It can help to identify and evaluate health risk, mode of action and establish exposure response relationships form which meaningful permissible levels can be derived.

Table 2: Epidemiological studies on toxicity of chemicals

Cadmium: Itai-Itai disease, renal damage

Chromium: highly carcinogenic

Mercury: Minamata disease with characteristic of endoneurial fibrosis and degenerated myelin sheath.

Lead: Anemia, pronounced effect on developing central nervous system of young children

Arsenic: Cause skin cancer, hyperkeratosis and hyperpigmentation.

Nickel: Lung fibrosis with risk of respiratory cancer, cardiovascular and kidney disease

Organohalogen compounds: Erythema followed by acneform eruption, Low birth weight and increase breast cancer incidences, cardiovascular diseases and immunologic disorder

Polycyclic aromatic hydrocarbon: Classified as human carcinogens and cause lung and bladder cancer.

Phenols: Hematotoxic and hepatotoxic, provoked mutagenesis and carcinogenesis.

Phthalates: Induce developmental and reproductive toxicity and alter thyroid, respiratory function.

Pesticides: Organochlorines adversely affect semen quality and caused testicular cancer in males, induce menstrual cycle abnormalities and spontaneous abortions. Organophosphate pesticides caused thyroid toxicity and inhibited cholinesterases activity.

11.2 HETEROGENOUS CHEMICAL MIXTURE

Human exposure to pollutants, ambient and/or occupational, is almost always heterogeneous. The toxicological profile of individual chemical may describe its pharmacological properties and toxicological responses however, does not necessarily explain the actual toxic potentials to humans when the exposure is heterogeneous. The research carried out during past several decades has identified potential hazardous chemicals, their toxic manifestations and permissible limits for release of pollutant into the environment. In past for chemical risk assessment the toxicological studies evaluated the effect of single chemical in biological system.

These chemicals in our bodies are a necessary evil without any toxicological importance, particularly at such low levels and represent the toxicological *Unknown Unknowns* (Mumtaz, 2010). Exposure to persistent chemicals could lead to their accumulation and to increasingly high tissue concentrations (El-Shahawi *et al.,* 2010). The potentials to accumulate, to biodegrade and to excrete the chemicals are variable among individuals and hence the toxic responses are also different. Such chemical body burdens and their possible variations from person to person, together with the unlimited combinations of chemical mixtures that may be inherent in human populations, are difficult to be assessed by routine toxicity testing.

The Department of Health and Human Services Center for Disease Control and Prevention, CDC (2009) sampled large human population for survey of chemical body burden (sample size ranged from a low of 1854 samples for PCB and a high of 8945 for cadmium or lead. It was alarming to note that 219 chemicals were present in the blood and/or urine of the population and 75 of these compounds were recorded for the first time in USA population. These analytical surveys have been repeated during the past decade and the number of detected chemicals is increasing in each survey which suggests that probably due to inadequacy of analytical methods or very low concentrations of the chemicals all the chemicals are not detected in the blood or urine samples. It is suspected that total number of environmental pollutants and other chemicals as body burden in the USA population could be much higher than 219 chemicals. It is stated in the report (2009) that the presence of chemical at very low levels in human blood or urine sample does not necessarily indicate that it is biologically available at toxic levels or it is the causative agent of a particular toxic response. However, mere presence of such several chemicals does magnify the potential risk to the individual. Unfortunately we do not know how many such chemicals are present at very low or undetectable levels in our body. The human health risk remains unassessed for such unknown heterogeneous chemical mixtures. The use of different combination drugs in therapeutics has also increased in past two decades (Atangwho *et al.,* 2010; Roy *et al.,* 2011). An individual, who suffers from more than one ailment simultaneously, like diabetes and hypertension, will be taking combinations of variety of medicines. The therapeutic doses of drugs, if not cleared out from body completely may result into low level body burden with great potentials to interact with already existing toxicant burden (Mumtaz *et al.*, 1998). This is bound to lead to a very challenging situation where the toxicants and drugs form much different mixture of chemicals in the body than one could expect.

For the toxicological evaluation, it is essential to understand the basic concepts of combined action and interaction of chemicals and to be able to distinguish between the whole mixture analysis (top-down approach) and component interaction analysis (bottom-up approach) (Groten *et al.*, 2001). Generally, the mixtures are grouped as simple and complex mixtures depending on their composition (Fig. 2 and 3).

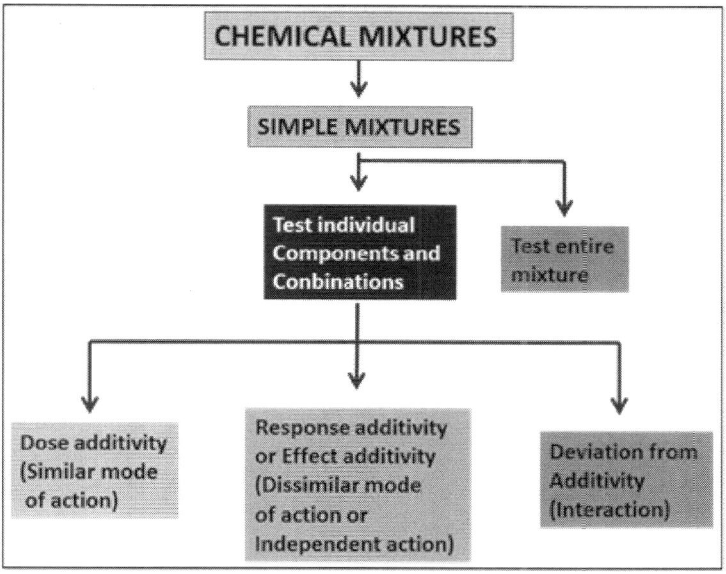

Fig. 2. Diagrammatic representation of possible interaction based study protocol for simple mixtures

As per one of the classification, the simple mixture has much less number of component chemicals (*e. g.* up to 8-10 only), while the complex mixture has many more chemical components (*e. g.* up to 100 or more). The other system refers to the type and combinations of chemicals. Components of the same chemical class, like metals or organochlorine pesticides are simple mixtures where the number of chemical components is not important. The complex mixtures, therefore, would be composed of components of different chemical classes; like a mixture of hazardous waste site or that of effluent arising from different industries. If the qualitative and quantitative composition of any complex mixture is known it becomes easier to evaluate the mixture toxicity or at least it is possible to categorize it, however, if the compositions largely unknown or undeterminable it is difficult to speculate the chemical interactions and thus the ultimate toxicity. The toxicological evaluation would, however, describe relevant toxicity; these data can be used to characterize the hazard following exposure to the entire mixture and for the risk assessment. Although, information on dose-response relationship can be generated, in such cases it is difficult to determine the interactions of component chemicals and to define the mode of action of the mixture.

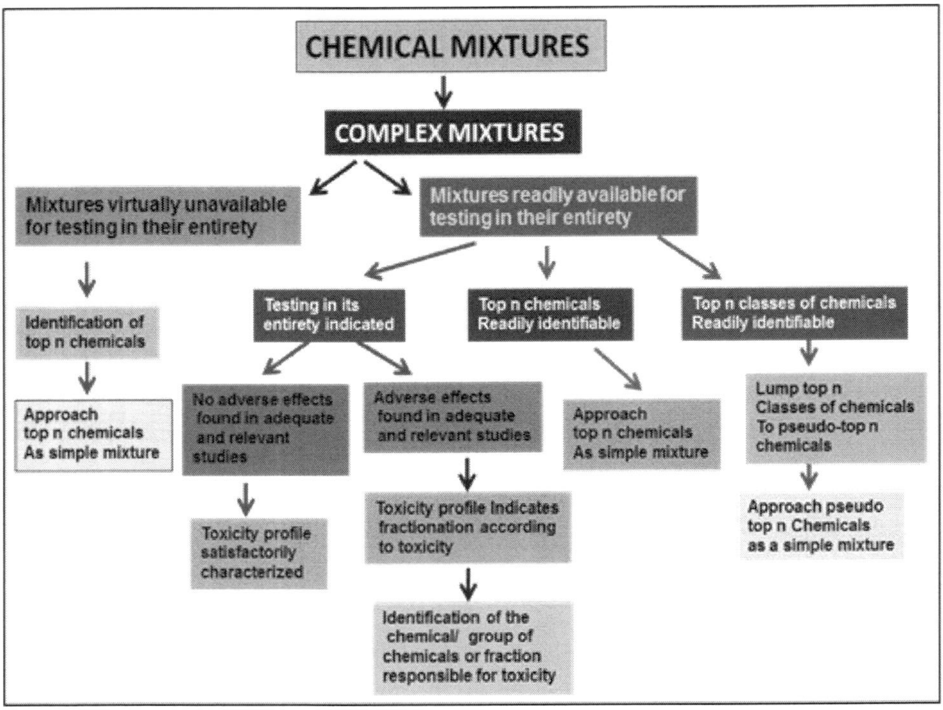

Fig. 3 Diagrammatic representation of possible interaction based study protocol for complex mixtures

The literature is piled up with the reports on experimental and epidemiological evidences of adverse effects of chemicals on different systems (Tables 1 and 2). Although, toxicity studies of single substances are important for obtaining basic toxicological information like LD_{50} values, threshold doses, acceptable daily intake, No observed adverse effect levels etc. however, there is strong evidence that chemicals in a mixture belonging to same or different category can produce effects greater than the effects of each component when tested singly. This inference has been investigated experimentally by combining chemicals at level below or equal to no observable adverse effect level (Walter *et al.,* 2002; Kortenkamp, 2008). The papers dealing in the toxicology of mixtures are gradually increasing but several of these studies, tested two or more compounds at relatively higher doses. Initial studies were conducted with plankton where they work on the planktonic micro fauna and showed a decline in the plankton diversity in the polluted regions (Thomas *et al.,* 1980; Nanda, 2002).

Similarly, studies on fish toxicity were carried out to assess the effects of mixtures of more than two chemicals (Khangarot and Ray, 1990). Recent studies have been documented on effect of chemical mixture on higher organisms and on different cell lines exhibiting the additive and interactive effects of the mixture toxicant (Hotchkiss *et al.,* 2004; Pomati *et al.,* 2008).

Several environmental laws including the Safe Drinking Water Act amendments (SDWAs), the Clean Air Act, the Food Quality Protection Act etc. have acknowledged the significance of potential exposure of chemical mixtures. The scientific literature on mixture toxicity has few publications describing method for conducting additive or interaction studies. The most notable examples are the testing of various petroleum products together and chemicals mixture from drinking water (Chapin *et al.,* 1989; Simmons *et al.,* 1994; Patrick-Iwuanyanwu *et al.,* 2011). Another example of the testing mixtures is the studies of commercial formulations of polychlorinated biphenyls congeners (Kostyniak *et al.,* 2005). Few studies have been conducted on the assessment of combined toxicity of defined chemical mixtures consisting of Metals/ pesticides/ fertilizers (Jadhav *et al.,* 2007; Shalaby and El-Mageed, 2010). Gruener and Lockwood (1980) examined the mutagenicity to hamster embryo cells of chemical mixtures of drinking water. Similarly, genotoxicity of chemical mixtures have also been studied in different cell lines (Lah *et al.,* 2005). Combinations of several nephrotoxicants were also studied in the form of mixture on animals (Jonker *et al.,* 1993). Systemic toxicity studies with a couple of chemicals demonstrated development of cellular and sub cellular lesions (Brzoska *et al.,* 2003). Very low levels of 25 different pollutants in groundwater were reported to behave differently and induce toxicity in experimental animals (Heindel *et al.,* 1995). However there is a lack of information on prolonged, repeated toxicity studies on mixtures at low doses. The study on heterogeneous mixture may alter basic toxicological findings and understanding acquired by single compound exposure. We believe that exposure to heterogeneous toxicant mixture may provide better understanding of the toxic potentials of industrial complexes. With increase in the number of chemicals in a mixture formulation there may be greater chance of chemical interaction.

11.3 SINGLE COMPOUND, SPECIFIC GROUP AND HETEROGENEOUS MIXTURE TOXICITY

Majority of research regarding the potential impacts of contaminants on life is derived from laboratory toxicity studies on single chemicals. With the improvement of technology and methodology, the toxicology gradually evolved from a science of high dose to a science of environmentally relevant doses. It also progressed from insensible conclusions to insightful end point which includes measurement of biochemical and functional changes in different systems. These changes toward low doses and more sensitive end points are encouraging as detailed information on mechanism of action can be obtained. The existing literature indicated abundance of toxicity information on single compounds on several model species. For each study, test animals were exposed to different concentrations of test chemical to obtain the valuable information of both behavioral and physical responses. Toxicological studies conducted earlier during mid-20th century were designated to identify lethal dose (LD_{50}). The most common test was the acute toxicity test, in which animals were given a single dose of chemical. As part of many toxicology studies, lethality studies are still conducted, but as a first step towards providing some insight into the toxic potentials of the new chemical.

Now a day's investigations on structure- activity relationships and molecular mechanisms of action are fundamental of single compound toxicity studies. A great deal of effort has already been devoted to elucidate the toxic responses and mechanism of toxicity and there is a growing body of literature which reflects the same (Table 1).

The evidences suggest that carcinogenic action of certain dioxin and related compounds is due to interactions of specific protein receptor (Mandal, 2005; Schwarz and Appel, 2005). In contrast, the study of toxicity mechanism of metals has suggested the generation of reactive electrophilic species which undergoes covalent binding to biological macromolecule ultimately leading to cancer (Flora *et al.,* 2008). Of all the toxicity studies that address the assessment of mixtures, the majority focus on binary mixtures or two chemicals at a time. The assessment was carried out at relatively high chemical concentrations and mostly chemicals belonging to same class were selected (Chukwuand and Okhumale, 2009). The studies were also carried out on combined toxicity (more than two) of defined chemical mixtures consisting of nephrotoxicants, pesticides, endocrine disrupting chemicals or carcinogens (Crofton *et al.,* 2005). In these studies the mixtures were formulated with those compounds whose mode of action and pharmacological properties were similar or comparable. Few mixture studies were concentrated on very complex mixtures such as industrial effluent, diesel fuel mixture or contaminated water samples (Huang *et al.,* 2008). The studies on environmentally relevant concentration of mixture *(i.e.* at low doses) are very less to rule out the potential for interactions amongst low doses of many different chemicals in living organisms.

Our environment (either at waste site, in the aquatic bodies, soil or the air environment) consist of chemical mixtures of several different compounds that are not produced together, but may occur together. For example a waste site might contain several heavy metals together with organophosphate, pesticides and some solvents. Therefore, studies on toxicity of mixture derived from different compounds (in terms of their nature and structure) are essentially required. But unfortunately this type of toxicological database is inadequate for risk assessment of chemical mixture.

Voluminous literature is available to disclose the occupational health hazards due to exposure to chemicals in industries; to determine the mode of action of toxic chemicals; to conduct safety evaluation of chemicals used in industries and everyday life. Literature is also available on remedial or preventive measures to safeguard health and environment from contaminants. Several of Indian researchers have suggested that a variety of accompanying changes in antioxidant defense enzyme is associated with production of reactive oxygen species due to chemical exposures. A review written by Chowdhury (2009) revealed that exposure to different heavy metals caused an irreversible toxic insult to male reproductive system and could interfere with the spermatozoa directly.

11.4 PRESENT STATUS OF ENVIRONMENT IN GUJARAT

Gujarat is one of the top industrialized states in India and consistently, it has maintained a higher pace of industrialization. The industrial belt of Gujarat is known as "Golden Corridor", runs along the main north – south highway. This industrial belt is considered as the robust backbone of Gujarat's economic development. This area includes the large industrial estates in Ahmedabad, Nandesari, Ankleshwar and Vapi. These areas contain thousands of individual industrial units, including dye factories, pesticide, textile, pulp and paper producer, pharmaceutical, engineering and chemical companies *etc.* The chemicals used or produced by these industries are listed in Table 3. This rapid industrial development throughout the state has resulted in generating abundant industrial wastes having both organic and metal contents (Table 4) (Labunska *et al.,* 1999). These industrial wastes are discharged through Asia's first and longest effluent channel into estuarine region of Mahi River after proper effluent treatment for detoxification. It starts from Dhanora (take off point) near Nandesari Industrial area, where it receives treated effluent from industries located in industrial complex in between Vadodara and Nandesari, and other industries at Nandesari through common effluent treatment plant. It receives effluent from oil refinery, fertilizers and chemicals factories at Koyali and chemical industries mostly of pharmaceuticals and dyes from Padra taluka of Vadodara district and Jambusar taluka of Bharuch district. A small river, called Mini River passes in between Nandesari industrial area and nearby industrial complexes. All these industries release various contaminants into this river. The Mini River meets River Mahisagar and ultimately meets the Gulf of Khambhat leading to Arabian Sea. Therefore, in this way the released industrial effluent is diluted into the environment at various points.

Table 3: List of hazardous chemicals identified in waste water and solid waste from the industries of Nandesari area near Vadodara

Hazardous Chemical(s) Used/ Products /By Products
Caustic soda, Ethylene, Propylene butadiene, Toluene, Xylenes, Glycol, Ethylene-oxide
Caustic Sauda, Sodium Hydrochloride, Liquid chlorine, Compressed hydrogen gas, Aluminium chloride
Caprolactam, Methyl ethyl ketoxime, Melamine, Anhydrous ammonia, Nylon 6,

Ammonium sulphate

Petrol, Diesel, Naphtha, Kerosene, Linear alkyl benene, Hydrocarbons, Hydrogen sulfide, Phenols, Benzopyrene, Ammonia

Nitric acid

Dimethyl hydantoin, Halogenated hydantoins, Succinimide, Acetic acid, Acetone cyanohydrine, Ammonia, Bromine, Isopropanol, Methanol,

Hydrochloric acid, Residue of rock phosphate

Ethylene Amine, N-Methylpyrrolidone, ammonia, sodium hydroxide, Methyl methacrylate, Glyceryl monolaurate

Monochloroacetic acid, Isopropyl alchol, Methyl alcohol, Ammonium hydroxide, Cholo acetetes

Carbon Sludge

Chloral hydrate, Hydrochloric acid, Chlorine

Acetyl, Trityl Chloride, Diisopropyl propionitrile, Polyporous prilled ammonium nitrate, Soda amine, Isopropyl bromide, Benzyl chloride, Methanol

Borax anhydrous, Boric acid, Zinc carbonate, Saccharin, Halazone Usp, Etp Sludge, Hdpe Bags,

Monochloroacetic acid, Ascorbic acid, Hydrochloric acid

Metalic sodium, Benzophenone, Chloro acetophenone, Naphthoic acid, Ethyl amines

Vinyl sulphone esters, Hydrochloric acid

Bitumen emulsions

Vinyl sulphone ester, Acetic acid

Chlorinated paraffin, Dicalcium phosphate, Glacial acetic acid, Sodium sulphate, Meat tenderizer, Zinc oxid, Methyl 12 hydroxy stearate, Sulphuric acid, Sodium metabisulphite

Table 4: Chemicals found in sediment, solid waste, industrial waste water and river water sample collected in the vicinity of the industries estate around Vadodara (Labunska *et al.,* 1999)

Organic compound	Heavy metal
Dichlorobenzenes	Cadmium
Trichlorobenzenes	Chromium
Tetrachlorobenzenes	Cobalt
Pentachlorobenzenes	Copper
Chlorinated benzenamines	Lead
Chlorinated toulenes	Manganese

Organic compound	Heavy metal
Chlorinated naphthalenes	Mercury
Polychlorinated butadienes	Nickel
Naphthalene and its derivatives	Zinc
9 H-Fluorene	
Phenanthrene and its derivatives	
Alkyl phenol derivatives	
Biphenyl and its derivatives	
Alkylated benzene derivatives	
Organosulphur compounds	
Linear alkanes and alkanes	

11.5 CHEMICAL POLLUTION: PRESENT AND FUTURE SCENARIO

Gujarat's rivers are bearing the burden of industrial pollution. All the major rivers and streams of Gujarat are in bad state due to effluent discharged by industries as the quality of effluent at final discharge point is always with high organic content and also not meeting with standard discharge norms. Several times, because of the high costs involved in effluent treatment, the chemical factories dump their waste materials in open or in effluent channel without any treatment processes illegally. Not only this has polluted the ground water but it has also spoiled the fertile lands. Moreover, despite the fact that effluent is rich in toxic hazardous wastes, it is regularly used by farmers for irrigating the farms on the sides of channel, leading to toxic and hazardous organic and metal compounds being absorbed in the food chain. Studies carried out in our department earlier, on environmental impacts of the effluent along the effluent channel, have demonstrated the high level of heavy metals in vegetables, fruits, food grains, fish muscle and even in human hairs (Sharma, 1995). The study clearly showed that over the years the soil had accumulated several toxic compounds and as a result of this biomagnification was observed in all edibles items (Table 5).

The nonstop dumping of effluent wastes of diverse kinds into the Mahi and Mini River made their water unreceptive for aquatic life and inappropriate for human consumption. The inhabitants of nearby areas, reported about the contamination of water tanks and wells. These kinds of complaints became a regular feature in this area and local newspapers or television news is reporting same types of evidences on daily basis. In fact, families staying in the region face severe health problems. The local population in all these industrialized areas has complained of skin irritation, perforation of nasal septum, high blood pressure (among youngsters), renal stone, asthma, cancer and infertility. These industrial units also release noxious and poisonous gases other than solid and liquid hazardous wastes. The chemical zone has the strong odour of chemicals and pesticides. Landfills also contribute towards the emission of toxic gases into the air.

Table 5: Accumulation profile of several toxicants in various eatables and other items grown in the farms irrigated with industrial effluent directly from effluent channel (Sharma, 1995)

Metal (mg/kg)	Wheat	Bajra	Drumstick	Brinjal	Potato	Tomato	Grass
Copper	89.8	204	39	29	92	11.8	3.4
Chromium	3.4	63.2	3.1	2	1.6	7.6	2.1
Znic	367.4	427.4	86.8	48.1	22.6	10.8	108.4
Nickel	2.1	3.2	2.2	18.8	2.3	2.3	3.9
Lead	2.4	2.9	9.2	26.3	9.8	91.3	40.8
Cadmium	21.3	10.7	7.9	8.8	8.8	6.2	3.0
Iron	41.3	25.7	12	32.1	10.1	31.2	21.2

The government of Gujarat is going ahead with its various infrastructure development plans, to make the second most urbanized Indian state into a dream destination for business, but ignored the impact of chemical pollution caused by the industrial discharge on public health and the environment. In Nandesari village around 200 hectares of agricultural land has been turned into a chemical industrial estate. Moreover, according to the future planning for economic development, government has targeted Gujarat's 1600 kilometer Arabian Sea coastline for port based industries. Mega cement plants are coming up on the coast in Kachchh and Saurashtra region, while huge refineries of the Reliance and Essar group of industries are under implementation alongside the protected marine national park in the Gulf of Kachchh.

With massive investment in the petrochemical, chemical and pharmaceutical industries, the state is facing problem of pollution in the major industrialized areas. Moreover, its future economic development planning can lead to severe horrified changes in environment and health, if the steps of environment protection measures were neglected or not followed strictly. Therefore, the studies like "impact assessment of industrial chemicals or industrial effluent on systems" and "toxicity report of assessment of industrial complex mixture" will draw the attention of higher authorities to take the necessary action to reduce the level of pollution.

REFERENCES

Ames BN and Gold LS (1990) Chemical carcinogenesis: too many rodent carcinogens. Proc Natl Acad Sci, 87: 7772-7776.

Atangwho IJ, Ani IF, Agiang MA, Effiong GS and Ebong PE (2010) Comparative effect of chlorpropamide and combined leaf extracts of *Azadirachta indica* and *Vernonia amygdalina* on blood glucose and biochemical parameters of alloxanized rats. J Med Med Sci, 1: 248-253

Brzoska MM, Moniuszko-Jakoniuk J, Pilat-Marcinkiewicz B and Sawicki B (2003) Liver and kidney function and histology in rats exposed to cadmium and ethanol. Alcohol and Alcoholism, 38: 2-10.

CDC (2009) (Center for disease control and prevention): Fourth national report on human exposure to environmental chemicals. Available at: http://www.cdc.gov/exposurereport/pdf/FourthReport.pdf

Chapin RE, Phelps JL, Schwetz BA and Yang RSH (1989) Toxicology studies of a chemical mixture of 25 groundwater contaminants: III: Male reproduction study in $B6C3F_1$ mice. Fund Appl Toxicol, 13: 388-398.

Chukwuand LO and Okhumale BO (2009) Mode of joint action response to binary mixtures of three refined petroleum products by Nile tilapia *Oreochromis niloticus* fingerlings. Scientific Res and Essay, 4: 806-811.

Crofton KM, Craft ES, Hedge JM, Gennings C, Simmons JE, Carchman RiA, Carter H and DeVito MJ (2005) Thyroid hormone–disrupting chemicals: evidence for dose-dependent additivity or synergism. Environ Health Perspect, 113:1549–1554.

Chowdhury AR (2009) Recent advances in heavy metals induced effect on male reproductive function—a retrospective. Al Ameen J Med Sci, 2: 37-4 2.

El-Shahawi MS, Hamza A, Bashammakh AS and Al-Saggaf WT (2010) An overview on the accumulation, distribution, transformations, toxicity and analytical methods for the monitoring of persistent organic pollutants. Talanta 80: 1587–1597.

Flora SJS, Mittal M and Mehta A (2008) Heavy metal induced oxidative stress and its possible reversal by chelation therapy. Indian J Med Res, 128: 501-523.

Groten JP, Feron VJ and Sühnel J (2001) Toxicology of simple and complex mixtures. Trends in Pharmacol Sci, 22: 316-322.

Gruener N and Lockwood MP (1980) Mutagenic activity in drinking water. Am J Public Health, 70: 276-278.

Hotchkiss AK, Rider CV, Furra J, Howdeshell KL, Blystone CR, Wilson VS, and Gray Jr LE (2010) In utero exposure to an AR antagonist plus an inhibitor of fetal testosterone synthesis induces cumulative effects on F1 male rats. Reprod Toxicol, 30: 261–270 .

Heindel JJ, Chapin RE, George J, Gulati DK, Fail PA, Barne LH and Yang RSH (1995) Assessment of the reproductive toxicity of a complex mixture of 25 groundwater contaminants in mice and rats. Toxicol Sci, 25: 9-19.

Huang J, Liao J, Liu Y, Lu S, Chou C, Chan W, Chen S and Ueng T (2008) Motorcycle exhaust induces reproductive toxicity and testicular interleukin-6 in male rats. Toxicol Sci, 103: 137–148.

Jadhav SH, Sarkar SN, Patil RD and Tripathi HC (2007) Effects of subchronic exposure via drinking water to a mixture of eight water-contaminating metals: a biochemical and histopathological study in male rats. Arch Environ Contam Toxicol, 53: 667–677.

Jonker D, Jones MA, Bladeren PJ, Woutersen RA, Til HP and Feron VJ (1993) Acute (24 hr) toxicity of a combination of four nephrotoxicants in rats compared with the toxicity of the individual compounds. Food Chem Toxicol, 31: 45-52.

Khangarot BS and Ray PK (1990) Acute toxicity and toxic interaction of chromium and nickel to common guppy *Poecilia reticulata* (Peters). Bull Environ Contam Toxicol, 44: 832-839.

Kortenkamp A (2008) Low dose mixture effects of endocrine disrupters: implications for risk assessment and epidemiology. International Journal of Andrology Volume 31: 233–240.

Kostyniak PJ, Hansen LG, Widholm JJ, Fitzpatrick RD, Olson JR, Helferich JL, Kim KH, Sable HJK, Seegal RF, Pessah IN and Schantz SL (2005) Formulation and characterization of an experimental pcb mixture designed to mimic human exposure from contaminated fish. Toxicol sci, 88: 400–411.

Labunska I, Stephenson A, Brigden K, Santillo D, Stringer R, Johnston PA and Ashton JM (1999) Toxic Hotspots: A Greenpeace Investigation of Gujarat Industrial Estates. Technical Note, No: 5/99, Greenpeace Research Laboratories, Exeter.

Lah B, Zinko B, Narat M and Marinsek-Logar M (2005) Monitoring of genotoxicity in drinking water using in vitro comet assay and ames test. Food Technol Biotechnol, 43: 139–146 .

Mishra S, Dwivedi SP and Singh RB (2010) A review on epigenetic effect of heavy metal carcinogens on human health. The Open Nutraceuticals J, 3: 188-193 .

Mumtaz MM (2010) Principles and Practice of Mixtures Toxicology. Edited by Moiz Mumtaz, WILEY-VCH Verlag GmbH & Co. KGaA, Weinheim ISBN: 978-3-527-31992-3 .

Mumtaz MM, De Rosa CT, Groten J, Feron VJ, Hansen H and Durkin RR (1998) Estimation of toxicity of chemical mixtures through modeling of chemical interactions. Environ Health Persepect, 106:1353-1360.

Mandal PK (2005) Dioxin: a review of its environmental effects and its aryl hydrocarbon receptor biology. J Comp Physiol B, 175: 221-230.

Nanda A (2002) Studies on the impact of pollution on the invertebrate diversity of Mahi River. Ph.D Thesis, The M. S University of Baroda, Vadodara .

Pomati F, Orlandi C, Clerici M, Luciani F and Zuccato E (2008) Effects and Interactions in an Environmentally Relevant Mixture of Pharmaceuticals. Toxicol Sci, 102: 129–137.

Patrick-Iwuanyanwu KC, Onyemaenu CC, Wegwu MO and Ayalogu EO (2011) Hepatotoxic and nephrotoxic effects of kerosene and petrol-contaminated diets in Wistar albino rats. Res J Environ Toxicol, 5: 49-57.

Roy DN, Sen G, Chowdhury KD and Biswas T (2011) Combination therapy with andro Grapholide and d-penicillamine enhanced therapeutic advantage over monotherapy with d-penicillamine in attenuating fibrogenic response and cell death in the periportal zone of liver in rats during copper toxicosis. Toxicol Appl Pharmacol, 250: 54-68 (2011).

Schwarz M and Appel KE (2005) Carcinogenic risks of dioxin: mechanistic considerations. Reg Toxicol Pharmacol, 43: 19–34.

Sharma A (1995) Environmental Impact Assessment along the effluent channel from Baroda to Jambusar and at its confluence with Mahi estuary at the Gulf of Khambhat with special reference to heavy metals. PhD Thesis, The M S University of Baroda, 1-326.

Shalaby SEM and El-Mageed AEM (2010) Biochemical targets affected by sub acute doses of new pesticide mixtures tested on albino rats. J Plant Protec Res, 50: 513-519.

Simmons JE, Yang RS, Svendsgaard DJ, Thompson MB, Seely JC and McDonald A (1994) Toxicology studies of a chemical mixture of 25 groundwater contaminants: hepatic and renal assessment, response to carbon tetrachloride challenge, and influence of treatment-induced water restriction. J Toxicol Environ Health, 43: 305-25.

Soto AM and Sonnenschein C (2010) Environmental causes of cancer: endocrine disruptor as carcinogens. Nat Rev Endocrinol, Doi:10.1038/nrendo.2010.87.

Thomas WH, Hollibaughl JT, Seibertl DLR and Wallace GT (1980) Toxicity of a Mixture of Ten Metals to Phytoplankton. Mar Ecol Prog, 2: 213-220.

Walter H, Consolaro F, Gramantica P, Scholze M and Altenburger R (2002) Mixture toxicity of priority pollutants at No Observed Effect Concentration (NOECs). Ecotoxicol, 11: 299-310.

Chapter - 12

Organic Farming: A new way to increase the Basmati Rice production

Shakuntala E Pillai and Ravinayak Patlavath

Department of Botany, Faculty of Science,
The Maharaja Sayajirao University of Baroda, Vadodara -390002, Gujarat, India.

ABSTRACT

Rice has been one of the most important human food crop on earth. The twelve thousand year old antiquity reveals that no other crop of economic importance has provided food calories to man as much as that of rice. Explosive increase in world population, deterioration of arable land and availability of quality irrigation water are forcing crop production into more and more marginal environments. Rice is a particularly versatile remedy. A dish of rice prepared with clarified butter, meat, sour fruit or any lentil helps to build new tissue and imparts strength and rotundity. Fried rice alleviates vomiting and dysentery; pulverized, it alleviates vomiting, thirst and burning sensation of the skin. Immature or newly harvested rice builds tissue. Old or well-matured rice promotes the healing of fractured bones. Old rice is easy to digest than new rice

Chapter deals with use of bio fertilizer in place of conventional synthetic pesticide use of botanical pesticides is suggested to give a better yield of Basmati rice. In recent years role of PGPR is emphasized, the present paper explains how bio fertilizers will enhance the microbial population and better nutrition and use of bio pesticide will reduce the disease losses.

Keywords: Organic Farming, Yield, Basmati Rice, Biofertilizer

12.1 INTRODUCTION

Rice (*Oryza sativa*) is one of the three most important cereal crops in the world, forms the staple diet of half the world's population; a majority of rice being (95% of the world's rice) produced and consumed in southeast Asia (Bhattacharjee *et al.* 2002). Rice is one of the major agricultural product of India and it earns a good proportion of foreign exchange. Basmati rice variety, being the major rice variety in demand, two third of its production is exported every year from our country (Table 1). Rice varieties of therapeutic value have been known to exist in use for a very long time. Susruta (400 BC), Kashiyapiya-Krishisukti (800-900

AD) and Acharya Bhavamisra (15-16th Century) are some of the records. Red Yeast rice used as dietary supplement helps in reducing blood cholesterol. The *njavara variety* of rice from Kerala is traditionally used for the treatment of rheumatism, arthritis, neurological problems, blood pressure etc. (Patra and Marndi, 2012)

Table 1. India's trade flow for basmati rice during 2010-11 to 2015-16

Year	Export of Basmati (Million tonnes)
2010-11	2.37
2011-12	3.18
2012-13	3.50
2013-14	3.76
2014-15	3.70
2015-16	4.04
2016-17 (April-July2016)	1.55

(Source: nic.gov.in)

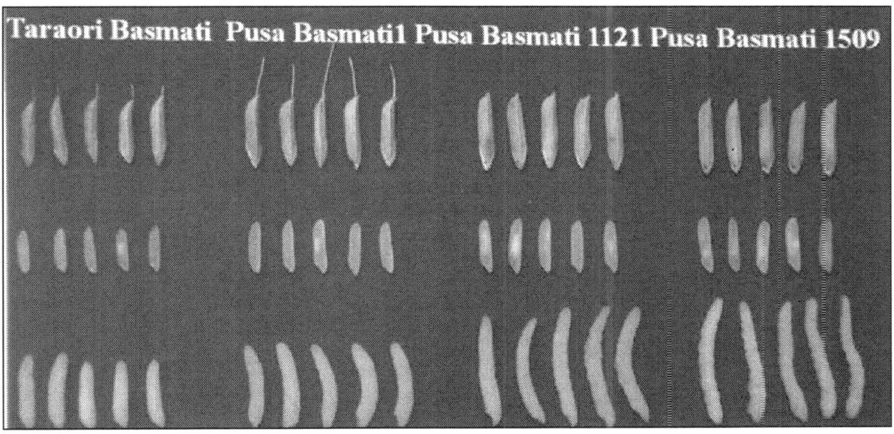

Figure 1: Grain characteristic of different basmati rice varieties.
Source: IARI, New Delhi.

It is a variety of aromatic rice with long, slender-grain. The ideal characteristic of the export quality of Basmati rice concentrates on many properties of the grain like, length, aroma, taste, texture, shape and chemical composition (Table 2) (Siddiq *et al.,* 1997). North Indian states like Punjab, Haryana, Uttar Pradesh, Uttarakhand, Himachal Pradesh, Delhi are major producers of basmati rice and some regions of Jammu, Kashmir, Assam, West

Bengal also cultivate this rice variety. Indian market of basmati rice has competition from other countries like Pakistan and Texas (USA). Non-basmati aromatic rice varieties like Jasmine and Sadri rice from Iran, Thailand, China, Myanmar, Afghanistan etc also compete for Indian basmati rice at the international market. India has exported 40,45,796.25 metric tons(MT) of basmati rice to the world and earned Rs. 22718.44 crores in the year 2015-16. China, Saudi Arabia, Iran, UAE, Iraq and Kuwait were the major importers in this year (http; apeda.gov.in….).

12.2 INDIAN BASMATI RICE

Basmati 370, Basmati 386, Type 3 (Dehradun basmati), Karnal local/Taraori Basmati (HBC-19), Basmati 217 and Ranbir Basmati IET (11348) are the premium quality basmati rice varieties exported from India. Pusa Basmati-1 (IET 10364), Pusa Basmati-1121, Punjab Basmati-1 (Bauni Basmati), Kasturi (IET-8580), Haryana Basmati-1 and Mahi Sugandha are few other varieties that are exported in 2015-2016. These 12 varieties are notified as 'the true basmati rice' varieties under the seed Act 1966. Indian government certifying agency for agricultural products, AGMARK, has laid certain specification of Basmati raw milled rice for export. These include the following characters (http: agmarknet.nic.in…….)

1. The grains should be long slender of white creamy-white or greyish colour and Translucent.
2. The rice : a) Shall be the dried matured kernels of *Oryza sativa* L. and have uniform size, shape, and colour; b) Shall possess in a marked degree the natural fragrance characteristic of Basmati rice both in the raw and cooked state; c) Shall not have been artificially coloured and shall be free from polishing agents; d) May contain up to 3 percent of grains with appreciable amount of bran thereon; e) Shall be free from musty or obnoxious odour and shall carry no sign of mould or containing webs and dead or live weevils. f) Shall have length 6.0 mm. and above and length-breadth ratio 3 and above; g) Shall be in sound merchantable condition.

Table 2: Characteristic of cooked Basmati rice varieties

Quality trait	Basmati 370	Haryana Basmati-1	Pusa Basmati-1
Length (l) in mm	6.76	6.74	6.82
Breadth (b) in mm	1.76	1.65	1,73
l/b ratio	3.84	4.08	3.94
Elongation ratio	1.88	1.7	1.89
Water uptake (g/100g)	221	338	228
Volume expansion ratio	3.92	3.8	3.81

Source: Bhattacharjee *et al.* (2002) Ahuja *et al.* (1995)

Green revolution and Basmati rice cultivation

Green revolution (GR) boosted rice cultivation along with other major staple crops like maize and wheat. The government of many developing countries like India invested a lot of money on crop research, infrastructures, and programs to promote land under cultivation. Plant breeders started developing rice varieties that are easy to grow with increased yield (Oad *et al.* 2006). Considering the economic importance and high market value of basmati rice, many plant breeders took interest in developing new varieties. A major issue with the original basmati variety is that (1) it is a tall variety and (2) flower induction is photosensitive which restricts its cultivation to certain geographical area and climatic conditions. Thus, less photosensitive dwarf varieties of basmati, Pusa Basmati-1, and Haryana Basmati-1 were introduced in the market (Ahuja *et al.* 1995). Pusa Basmati-1 has more yield than the traditional variety HBC-19. Other characters focused during breeding were aromatic properties, grain length, cooking quality, yield, the shelf life of the aromatic compounds etc. (Bhattacharjee *et al.*, 2002). Pusa-1121 has better aromatic property than Pusa-1.

12.3 ADVERSE IMPACT OF GR

Although green revolution initially increased agricultural productivity, it has led to an adverse impact on the environment. During GR, all local rice varieties were replaced by commercially important high yielding rice variety (Oad *et al.*, 2006). GR also lead to the development of rice varieties which can be grown in any season which promoted farmers to grow same crop through-out the year. As a result of which only handful of commercial rice varieties were grown throughout the year (Shiva *et al.*, 2004). The major disadvantage of this would be a threat from an outbreak of new pest or disease attack, which can lead to huge loss. Moreover, the crop diversity was lost which disturbed the balance in pest-predator relation and soil ecology (Shiva *et al.*, 2004). The government, at the onset of GR, gave subsidy to farmers on chemical pesticides, inorganic fertilizers, development of irrigation system, bore wells, canals etc. to boost agricultural yield (Oad *et al.*, 2006). This promoted extensive use of pesticides and inorganic fertilizers leading to the destruction of soil structure in such a way that it can no more support growth without fertilizer (Shiva *et al.*, 2004). The unintended consequence of green revolution involves depletion of ground water level, ecosystem imbalance, reduced crop diversity, deforestation, soil infertility, environmental pollution etc. These have affected the yield of many different crops including basmati rice in India. Many developing countries, including India, are now switching back to organic farming to restore the nature and increase the agricultural productivity.

12.4 GREEN REVOLUTION-2 STARTS WITH ORGANIC FARMING

Organic farming is not new to India, it is the traditional way of farming excluding the use of chemical pesticides and inorganic fertilizer. General awareness about organic farming and the organic food product is growing in the population with which the national and international markets are now focusing on organic food products. The demand for organic basmati rice has already been developed and is marketed by brands like 'India gate basmati rice'. The government of India initiated a body, National program for organic production (NPOP), for inspection and certification of organics products in India. Agricultural and Processed Food Products Export Development Authority (APEDA) will inspect and certify organic basmati rice for export. In 2008-09, around 5630 MT of organic basmati rice was grown in Punjab, Uttar Pradesh, Haryana and exported from India through APEDA (www: rkmp.co.in). The increase in demand for organic rice has motivated many farmers to switch to organic farming. A lot of basmati research is now diverted to increase productivity, nutrient content, pest control by traditional way of farming.

In order to increase basmati rice productivity, strategies of organic farming has to be implied on different aspects of farming focusing soil nutrient management, disease-pest control, and post-harvest storage issues. Following are the traditional ways of farming that has been implied for basmati and non-basmati rice varieties to increase the yield.

1. Monoculture to Multi-culture

Monoculture means growing one crop at a time in both seasons (Rabi and Kharif). During GR, many farmers growing basmati rice, cultivated rice throughout the year or followed rice-wheat crop rotation. With mono-crop, it was easy for farmers to cultivate, irrigate and control pest -weeds. This leads to great loss of biodiversity, climatic changes, accumulation of pest and soil nutrient depletion (Shiva *et al.,* 2004). Traditional farming followed crop rotation, mixed farming or multicultural. In crop rotation, different crops are cultivated on the same land one after the other. Mixed cropping includes growing different crops in the same field at the same time. As not all plants have same nutritive needs of the soil is not leached and pest build-up is controlled. Network Project on Organic Farming (NPOF) have tried different cropping systems (Table 3) for rice which can be used for increasing basmati rice as well.

Table 3. NPOF tested cropping system for rice

Crop	Field Trails Executed
Rice – Wheat/ Potato/ Mustard/ Lentil	Ranchi, Jharkhand
Rice – Durum Wheat/ Berseem	Ludhiana, Punjab
Rice – Potato – Okra	Ludhiana, Punjab
Rice – Garlic	Ludhiana, Punjab
Carrot/ Rice (Pre Kharif) – Rice (Kharif)	Umiam, Meghalaya
Potato/Rice (Pre Kharif) – Rice (Kharif)	Umiam, Meghalaya
Tomato/ Rice (Pre Kharif) – Rice (Kharif)	Umiam, Meghalaya
	Umiam, Meghalaya

Source: NPOF, Principles in organic farming.

In crop rotation or mixed farming, 40% of the cultivated land should be occupied with legume crop at least in one season, as these plants increase soil fertility and replace nitrogen content of the soil (anonymous; Shiva *et al.*, 2004). For mixed farming crops should be selected carefully, high nutrient-demanding crops should be planted with low nutrient-demanding crops like vegetables, flowers etc. Indian agriculture research institute (IARI), carried a field trail of organic farming in 2003-06, where they followed Basmati rice-wheat-green gram crop rotation system with organic manures increased rice yield up to 65% over the controls (Singh *et al.* 2007). Farmers should select alternate crops based on their needs and season compatibility.

2. Chemical Fertilizer to Biofertilizer

Basmati rice grows easily in any type of soil but heavy neutral soil like clay, loam clay with good water holding capacity are excellent. Saline and alkaline soils are not suitable for basmati rice cultivation. Extensive use of chemical pesticides, inorganic fertilizer and mono-culture have led to the destruction of soil structure and fertility. A fertile soil is characterized by its organic content, microorganisms and other living forms like worms, ants, nematodes etc. Organic content of soil ideally should be 0.8-1.5% of total microbial content (bacteria, fungi, actinomycetes) should be above 10^8 CFU/gm of soil. There should be 3-5 earthworms/cu ft. of soil (Anonymous; Shiva *et al.* 2004).

i. Field pretreatment: As mentioned earlier crop rotation involving legume crop in one season increases the soil fertility. Growing legume crop to improve soil quality is known as green manuring. For basmati rice cultivation of *Sesbania aculeata*, cowpea and mung bean prior to rice cultivation improves the soil physio-chemical property (Nayyar *et al.* 2000). Incorporation of *S. aculeata* increased basmati yield by 2.38%, 4.14%, and 10.82% in the first, second and third year of cultivation respectively than the other two (Pooniya *et al.* 2012).

ii. Seed pre-treatment: Basmati seeds are washed with salt water (1kg/10 l) of multiple times followed by a day treatment with Emison (0.5g/l), Agromycin (0.25g/l) or Streptomycin (0.1g/l) and then washed thoroughly next day with water (http: agrifarming.in). Other treatments practiced for paddy crop, in general, involves treatment with either *Azospirillum*, phosphate solubilizing bacteria (PBS), *Azotobacter* + PBS or *Azospirillium* + PBS(21).

iii. Main field Treatment for cultivation: In general, for basmati rice the dry field is ploughed (2-3 weeks) then submerged in water (5-10 cm) for 3-4 days. During this period different combination of organic manure like farm yard manure, vermicompost, paddy straw, sunhemp are incorporated into the field (http: agrifarming.in). Biofertilizers are added just before or after plantation of basmati seedling. Biofertilizers are basically microorganisms that enrich soil nutrient by their natural metabolic process. Following are different biofertilizers used for paddy cultivation (Shiva *et al.,* 2004).

Blue-green algae: These are photosynthetic bacteria capable of living in symbiotic association with other organisms. The fix CO_2 and increase the organic content of soil. These are incorporated in the submerged field as they grow well in this condition.

Azolla: It is an aquatic pteridophyte fern found growing on the surface of paddy field. It fixes atmospheric nitrogen with help of a symbiotic blue-green algae (*Anabaena*) that grows on the lower surface of the fern leaf.

Azotobacter: It is bacteria capable of fixing atmospheric nitrogen into ammonia. Unlike symbiotic bacteria *Rhizobacteria* these can fix nitrogen independently.

Azospirillum: Similar to *Azotobacter* it is a free living nitrogen fixing bacteria.

Different combinations of above biofertilizers are used to cultivate basmati and non-aromatic rice to increase the productivity in organic farming.

- Incorporation of blue-green algae (15kg/ha), *Azolla* (1.0 t/ha), vermi-compost and farm yard manure (5.0 t/ha), to Pusa Basmati-1-wheat-green gram crop system, worked well with 114-116% increase in yield (Singh *et al.*, 2007).
- The combined use of *Anabaena oscillarioides* CR3, *Brevundimonas diminuta* PR7 and *Ochrobactrum anthropi* PR10 PGPRs significantly improved basmati rice yield by 21.2%, as compared to the application of recommended dose of NPK fertilizers in Pusa basmati-2511- wheat cropping system (Rana *et al.* 2015).
- For sandy clay loam soil in Andhra Pradesh, application of farm yard manure, neem leaf manure, sheep manure and poultry manure (60 kg/ha of each) significantly increased the yield of four arnomatic rice varieties (Pusa Basmati 1, Tararoi Basmati, RNR 2465 and RNR 18833) (Rao *et al.* 2013).
- The standard protocol used for general any rice variety is incorporation of BGA (10 kg/ha) 10 days after planting; *Azolla* (1 t/ha) after 3 weeks of planting (http: rkmp.co.in).
- Another field study shows with basmati rice-wheat crop system, higher productivity was with FYM+GM (*S. aculeate/ Leuceana*) + BGA combination but net higher return was with GM (*S. aculeata/ Leuceana*) + BGA(21).

An important point to be noted is that first year of organic farming doesn't give high productivity as mentioned above. All of the above results were obtained after 3-4 years of organic farming. Thus the repetitive use of organic farming techniques brings back the normal soil structure.

12.5. PESTICIDES IN DISEASE CONTROL

Disease and pest attack in the field cause a huge loss in yield of basmati rice. The major threats for basmati rice crop are neck blast, bacterial leaf blight, sheath rot, stem borer and white backer plant hopper (Siddiq *et al.,* 1997). According to Srinivasan 1997 records, the annual use of pesticides in agriculture is around 80,000 tons and rice field accounts 23% of it. It has been reported that only 0.1% of pesticide actually reached the target pests and rest go to the non-targeted insects and soil. Thus, indirectly use of pesticide has increased the number of resistant pests in the field. In 1990, there were around 40 insect pests attacking rice crops, in the year 2000 it has increased to 200. The most common pesticides used for rice cultivation are HCH, carbofuran, butachlor, triazophos, quinalphos, chloropyriphos etc.

During GR, research was more concentrated on developing high yielding basmati rice varieties; rather it should have been high yielding and disease resistant varieties. For example, there are no specific chemicals to control bacterial leaf blight, the ideal solution would be to generate a disease resistant variety. A three-gene pyramid, *Xa21, xa13 and xa5,* have been identified to give resistance to bacterial blight and it is introduced in Samba Masuri, a non-aromatic rice variety (Sundaram *et al.*, 2008). Efforts are being made to introduce these genes in commercial basmati rice. Blast resistance gene Pi2-Pi54 and leaf blight resistance genes *Xa21* and *xa5* has been successfully incorporated into Pusa basmati-1121 and Pusa Basmati-6 background. The new resistant strains are promoted for field trails in 2015 as per Indian national basmati trail (Ellur *et al.,* 2016). As mentioned earlier, crop rotation and mixed cropping control the accumulation of pest in the field. Additionally, using biocontrol methods disease or pest attack can be controlled in eco-friendly manner.

Biocontrol methods involve the use of biological agents or preparations to control disease and pest. Following are the different ways by with biocontrol can be used for rice cultivation.

i. Cultural control: Proper sanitization of the field, sufficient nutrition of soil, selection of appropriate crops and destruction of the previous crop prior to cultivation are included under cultural control. These methods ensure that plants are grown healthy and their natural immune system is not compromised (Shiva *et al.*, 2004).

ii. Mechanical control: It involves physical removal of pest and its different stages from the plant. This method is useful only when the field is very small. For large fields installation of bird perches, light traps, sticky colored plates, and pheromone traps are very effective (Shiva *et al.*, 2004; www. rkmp.co.in).

iii. Botanical control:

- Neem Extract: This good pest control agent was found effective against more than 200 species of insects. It is effective against grasshoppers, beetles, green rice leaf hopper, spider mites, moths which are commonly found on rice crops.

- Cow urine diluted in 1:20 ratio is effective pest repellant.
- Dashpharni extract: It is a fermented extract obtained from crushing leaves from neem, *Vitex negundo, Carica papaya, Aristolochia, Tinospora cordifolia, Annona squamosa, Ricinus communis, Nerium indicum, Calotropis procera,* chilli paste, garlic paste, cow dung and cow urine (Anonymous).
- Tobacco extract: The leaves, stalk, and stem are used to control aphids, mites, Rice stem borer (Shiva *et al.,* 2004).
- Garlic extract: Bulbs leaves, flowers, and oils are used to control aphids, bacterial disease, beetle, nematodes and rice fungal blast (Shiva *et al.*, 2004).

iv. Biological control: This includes the use of various natural enemies of the pest like pathogen, parasite or predators.

Fungal biocontrol agents: *Trichoderma viride, T. harazianum* control soil borne diseases. *Beauvaria bassiana, Metarizium anisopliae, Numeria rileyi, Verticillium* sps are pest specific biocontrol agents (Shiva *et al.,* 2004).

Bacterial biocontrol agents: *Bacillus thurengiensis* stenebrionis and *B. thurengiensis* sandigo are effective against insects belonging to class coleopteran (Shiva *et al.,* 2004). *Pseudomonas fluorescence* manage most of the seed borne and soil borne diseases. Endophytic strain, *B. oryzicola* YC7007 has been recently shown to controlbakanae disease ofrice caused by a fungi *Fusarium fujikuroi* (Hossain *et al.,* 2016). Pseudomonad EA105, isolated from rice soil was found to control fungal sheath blight caused by *Magnaportha oryzae* (Spence *et al.,* 2014).

Insect biocontrol agents: *Trichogramma* are minute insect wasps that are endoparasites found growing in insect eggs. *T. japonicum* Ashmead and *T. chilonis* Ishii, are used in fields against Rice Yellow Stem Borer and Leaf Folder for basmati crops (http: agrifarming. in….).

3. Prevention of post-harvest pest attack

Post-harvest losses for paddy crop is 2.76% of the total production (Table 4). The losses at transport level and during harvesting can be controlled by using the appropriate method in each step. Losses during storage is a major issue which is due to pest attack and needs to be controlled even for maintenance of quality of the rice.

Table 4. Estimated total loss of rice at production levels

Operations	**Loss (% of total production)**
Transport from field to threshing floor	0.79
Threshing	0.89
Winnowing	0.48
Transport from threshing floor to storage	0.16
Storage	0.40
Total	**2.71**

Source: Marketable Surplus & post-harvest losses of paddy in India, 2002, Directorate of marketing and inspection, Nagpur

Traditional methods used to prevent pest attack is listed below (Shiva *et al.*, 2004).

- Proper drying of the grain in sun
- Smoking of the storage rooms with neem leaves
- Mixing of wood ash, charcoal or cow dung ash with grain
- Fire oil lamp trap to attract and kill insects
- Organic control for storage pest involves mixing of Leaf powder of *Vitex negundo* with the paddy grains.

12.6. CONCLUSIONS

The productivity of basmati rice is increased by adopting organic farming and maintaining crops health. The health of field is maintained by using organic manure, bio fertilisers, biocontrol agents and diverse crop rotation. The present environmental problems created by our own deeds in past will be solved only by switching to organic farming and focused scientific research in the same direction.

REFERENCES

Anonymous. (2001). Report of Task Force on Organic Farming, Department of Agriculture and Cooperation, Ministry of Agriculture, Government of India, 2001. p. 76.

Ahuja, S.C., Panwar, D.V.S., Ahuja, U. and Gupta, K.R. (1995). Basmati rice: the scented pearl.

Bhattacharjee, P., Singhal, R.S. and Kulkarni, P.R. (2002). Basmati rice: a review. International Journal of Food Science & Technology, 37(1), 1-12.

Ellur, R.K., Khanna, A., Yadav, A., Pathania, S., Rajashekara, H., Singh, V.K. and Prakash, G. (2016). Improvement of Basmati rice varieties for resistance to blast and bacterial blight diseases using marker assisted backcross breeding. Plant Science, 242, 330-341.

Hossain, M.T., Chung, E.J. and Chung, Y.R. (2016). Biological Control of Rice Bakanae by an Endophytic *Bacillus oryzicola* YC7007. The Plant Pathology Journal, 32(3), 228-242.

Nayyar, V.K. and Chhibba, I. M. (2000). Effect of green manuring on micro-nutrient availability in rice-wheat cropping system of Northwest India. Long-term Soil Fertility Experiments in Rice-Wheat Cropping Systems. Rice-Wheat Consortium Paper Series, 6, 68-72.

Oad, G.L., Oad, F.C., Bhand, A.A. and Siddiqui, M.H. (2006). Performance of aromatic rice strains for growth and yield potentials. Asian Journal of Plant Sciences.

Patra B.C. and Marndi B.C. (2002) Exploration, collection, conservation and characterization of medicinal rice germ plasm. Oryza 49(4):305-306

Pingali, P.L. (2012). Green Revolution: Impacts, limits and the path ahead. Proceedings of the National Academy of Sciences, 109(31), 12302-12308.

Pooniya, V. and Shivay, Y.S. (2012). Summer green-manuring crops and zinc fertilization on productivity and economics of basmati rice (*Oryza sativa* L.). Archives of Agronomy and Soil Science, 58(6), 593-616.

Ram, M., Davari, M.R. and Sharma, S.N. (2011). Effect of organic manures and biofertilizers on basmati rice (*Oryza sativa* L.) under organic farming of rice-wheat cropping system. International Journal of Agriculture and Crop Sciences, 3(3), 76-84.

Rana, A., Kabi, S.R., Verma, S., Adak, A., Pal, M., Shivay, Y.S. and Nain, L. (2015). Prospecting plant growth promoting bacteria and cyanobacteria as options for enrichment of macro-and micronutrients in grains in rice–wheat cropping sequence. Cogent Food & Agriculture, 1(1), 1037379.

Rao, K.T., Rao, A.U., Ramu, P.S., Sekhar, D. and Rao, N.V. (2013). Effect of organic manures on performance of scented varieties of rice in high altitude areas of Andhra Pradesh. Int. J. Curr. Microbiol. App. Sci, 2(11), 339-346.

Shiva, V., Pande, P. and Singh, J. (2004). Principles of organic farming: Renewing the Earth's harvest. Navdanya, A - 60, Hauz Khas, New Delhi, India, 186p. 2004.

Siddiq, F.A., Muralidharan, K. and Shobha, R.N. (1997). *Basmati Rice*. Pp. 1–14. New Delhi: Directorate of Rice Research, Indian Council of Agricultural Research.

Singh, Y.V., Singh, B.V., Pabbi, S. and Singh, P.K. (2007). Impact of organic farming on yield and quality of basmati rice and soil properties.

Spence, C., Alff, E., Johnson, C., Ramos, C., Donofrio, N., Sundaresan, V. and Bais, H. (2014). Natural rice rhizospheric microbes suppress rice blast infections. BMC plant biology, 14(1), 1.

Sundaram, R. M., Vishnupriya, M.R., Biradar, S.K., Laha, G.S., Reddy, G.A., Rani, N.S. and Sonti, R.V. (2008). Marker assisted introgression of bacterial blight resistance in Samba Mahsuri, an elite indica rice variety. Euphytica, 160(3), 411-422.

Websites

1. http://apeda.gov.in/apedawebsite/SubHead_Products/Basmati_Rice.htm
2. agricoop.nic.in/imagedefault1/Rice_Apr15.pdf
3. http://agmarknet.nic.in/amrscheme/punerice044.htm
4. www.rkmp.co.in/
5. http://www.agrifarming.in/

Chapter - 13

Intelligent Traffic and Fast Transport System to Reduce Vehicular Pollution in Certain Indian Cities

Arun Arya

Environmental Studies, Faculty of Science,
The Maharaja Sayajirao University of Baroda, Vadodara 390002
E-Mail: sarojarun10arya@rediffmail.com

ABSTRACT

Traffic is the joint consequences of land use and transport capability. Traffic science deals with the scientific understanding and application of scientific methods to solve the problems of traffic jam, parking, including lane changing, timing of signal lights etc. Knowledge of traffic flow and kind of vehicles plying on roads is very essential to frame the rules and monitor it. In most of the modern cities ever increasing number of two and four wheelers has aggravated the traffic problems. The number of accidents have increased and loss of human lives have occurred due to careless driving. With modern designs the speed of vehicles have increased at the same time at certain places accidental fires have resulted in the cars or buses due to their faulty designs.

Keywords: Intelligent Traffic, Transport System, Vehicular Pollution, Vadodara, Bicycle, Palki, Gondola, Tanga, Tram.

13.1 NEED FOR A BETTER TRANSPORT SYSTEM

Long back nomadic people learnt horse riding and were hunting wild animals in the forest to meet their food requirements. Bones and stones were frequently used as their weapons. About 12000 years ago they domesticated other animals like bullock, goat and sheep. Sumerians living in Mesopotamia, what is now called as Iraq, 7,000 years ago were practicing agriculture with the help of wooden plough. They discovered a wheel which was made from solid wood. Later on a wheel with spokes was developed to reduce its weight. You can not imagine a racing car that runs with a speed of 320 km/h, or a supersonic car travelling with 1228.49 km/h and normal car with 150 km/h without the invention of a wheel. Bicycles were invented in late 1700 s. Later on a German inventor Gottlieb

Daimler made motorbike. Now we have an option to travel by local train, tram, diesel or electric driven trains, or by CNG or petrol driven car, taxi or a bus plying on the city roads (Bhatt, 2005). In Mumbai 45 per cent of the work trips are made in buses and trains. Good cement concrete and bitumen roads were built for smooth flow of traffic. Multi lane express ways were designed to accommodate fast traffic. In certain places motor boats served the purpose of transport in waterways. Venice an Italian city is connected by such water ways and tourist enjoy taking a trip on boats called gondola.

13.2 GLIMPSE OF TRAFFIC PRESENT DURING MUGHAL PERIOD

Travel by 'Palkis' or on horses was routine thing in past. Grand Trunk road now known as National Highway No 2 was built by Sher Shah Suri from Agra to Sasaram in 16th century. The 2500 km. long road was extended from Peshawar to Hawra during British period. It had check posts at regular intervals and shade providing trees were on both sides to provide relief to traveler from hot sun rays in summers. Large – scale movement of trading caravans consisting of thousands of bull or horses caused rise of dust in the air. Sir Thomas Roe, first ambassador of East India Company to the Mughal Court, records that he had to wait at a road crossing for three hours while a herd of bulls was passing from Surat to Agra via Ajmer. SPM in the air and presence of soot and smoke due to burning of wood and cow doing was major source of pollution during mughal period.

Mode of transmission is cycle rickshaw in many cities of the country till today. Upto 1996 hand held rickshaw was major source of local transport in Kolkata city besides buses, tram and taxi. ***Jin Riki Sha*** in Japanese, literally means a man-powered vehicle. It was invented in Japan by Samurai Yusuke Tzumi in 1870. Which was introduced in Shimla in 1880, and in Kolkata in 1914. Hard work done by these rickshaw pullers have made our visits easier. Although no pollution is caused by hand held or cycle rickshaw the poor health standards of the rickshaw pullers have been a cause of concern. There is need to adopt better mode of transport. In India 13 distinct types of vehicles ply having a speed of 5kmph for a bullock cart and 120 kmph for a fast moving car. Non- motorized vehicles like use of *Tanga* and *Ekka*, two horse driven vehicles were popular during mughal period in many cities like Lucknow, Agra, Delhi etc. High cost of good horses, their maintenance and competitive fare of auto rickshaw has changed the scenario now and there are not many *Tanga* on roads. We find such mode of travel on special occasions like any marriage or a procession. The horses and elephants are preferred choices for such events. Military and border security forces make use of horses and camels to reach difficult places to keep a vigil on enemy (Plate 1. Camel as transport).

13.3 TRAFFIC IN OLD BARODA CITY

Baroda had good roads during the time of Maharaja Sayajirao Gaekwad III. Present Jail road also known as Indira avenue was full of lush green banyan trees.

The road was darker due to shade during day time and ladies prefer to travel via Raopura road. The Maharaja has instructed to sprinkle water on the roads every evening during summer to prevent problem of dust. A bullock cart having a tank full of water used to visit and sprinkle water in Kirti Stambh and Dandiya bazar roads. Maharaja tried to built more than 75 gardens to provide recreation centers and reduction in heat during summer. Sayaji baug is known for its beauty throughout the country. It is interesting to note that Sayajibaug earlier called Kamati baug and Laxmi Vilas Palace gardens were developed by Sir John Henery Director of Kew Botanical gardens U.K.

There is a need to generate more awareness among citizens of Baroda. Roads are not clean, there is encroachment at several places by temporary shops, larries or places of worship. Such obstacles create traffic problems on busy roads. The major roads in the area of Raopura and Dandiya bazar are intercepted by numerous narrow lanes, a fast driven vehicle from these lanes may colloid with a vehicle passing on the main road. No cautions are displayed on main roads. Zebra crossing on various roads are missing. Fly overs are being built but few of these lack any good scientific study of traffic load. To reduce the vehicular pollution no serious effort is being made. In some roads plantation is visible on road dividers but frequently bigger trees are uprooted to widen the roads. It is advisable to plant more trees on both the sides of roads. A clean and green city will keep the air clean and maintain the atmospheric temperature (Plate 2 Showing bike as popular means of transport).

13.4. INTERNATIONAL STATUS OF ROADS AND TRAFFIC MANAGEMENT

We can take few lessons from our neighboring country China. In 1981 a nationwide voluntary tree planting campaign was started, in which every able bodied person of 13 years and above is supposed to plant least 3 trees annually; 18th March is designated as tree planting day. Fine is charged from defaulters. Villagers are allowed to take five times of the cost of a cut or damaged tree. This afforestation drive has generated several good green patches. Many of the species planted in Shanghai and Beijing area are of *Cedrus deodara*, *Rubinia pseudoacacia*, Poplar, *Platanus orientalis*, maple, walnut, *Magnolia grandiflora*, camphor and *Nerium* plants. In Beijing bicycle is still the principal mode of transport and one can find living fossil tree *Ginkgo biloba* planted in fairly large numbers as an avenue tree. This plant has been introduced in few Indian gardens as well. The common Vad is now replaced by Ashopalao, on both sides of roads; of various trees Ashopalav has the highest numbers of trees in Vadodara.

I was fascinated to see the clean roads and disciplined traffic on the narrow roads of Torino city in Italy. The roads were having foot path on both sides and were having uniform tree plantation. The same road was used for the movement of buses and trams. Electric signals and video camera were helping the smooth flow of traffic, while absence of police deployment on crossing was more amusing to me. The people were aware of the traffic rules and pedestrians were given priority in crossing the roads. The mechanized cleaning of roads was common in Marrakech city of Morocco, which I visited in 2008. The big palm trees were having porous pipe irrigation system to prevent the water loss. The city was comparatively hotter in the day and besides water scarcity one can see good plantation at various places. Mozzarella garden full of xerophytic plants was a place of great attraction to tourists in the city. We seldom come across such beautiful roads and controlled traffic in any Indian cities.

Plate 1. A. Camel ride, **B.** Two seater Bike with a complete family on ride, **C.** Traffic chaos in a modern city, **D.** Flyover in modern city

13.5. VEHICULAR POLLUTION IN DIFFERENT CITIES

Vadodara faces higher pollution levels of RSPM, $PM_{2.5}$ and SPM particularly during busy evening hours in Kala Ghoda and Nyay Mandir. Similar observations were recorded for Delhi Agra and Nagpur and automobile exhaust was the major region. When studies were done in Shillong the total suspended particulates concentration was higher on the roads and reduced to lower level 600 meter away from road. The burning of fossil fuel and TSP were major sources of pollutants in Indian roads.

Plate 2. A. Traffic jam in Delhi which prompted odd even formula in the city. **B.** Fast moving bikes, giving speed and polluted air both.

Like fast communication we need bigger, broader and clean roads for smooth traffic flow. The traffic on roads is heterogeneous, pedestrians, bicycle and different vehicles move with different speed on narrow roads and these may colloid during over crossing. Stray cows are a menace and work as speed breakers. The problem has been more crucial due to tenfold increase in two and four wheelers in past 25 years in the Vadodara. In 1985 there were 85.000 two wheelers now their number is 8 lakhs. Number of cars has increased from 10300 in 1985 to 1 lakh presently. The city of Vadodara lacks a good public transport system, like Delhi Metro there is a need to develop a fast life time, which will reduce the pollution and ever increasing vehicle numbers. Euro III norms of fuel were implemented in 2010 and now Euro IV norms are in process of implementation (Plate 2 Heavy Car Traffic on a Delhi Road).

The blue sky is a rare seen in many major cities of the country. During past decade we have witnessed huge fog in Delhi which hampered road, air and rail transport. Our capital city was one of the ten most polluted cities of the world, when Supreme Court banned the use of diesel driven buses and auto in the city in April 2001. Particulate matter pollution is higher in 126 *cities* of the world. A study conducted in 1986 in Barcelona and Helenski and Paris and London during 1982-1992 revealed significant increase in number of Asthma cases. A World Bank study indicates that one out of ten children in the age group of 5-16 years suffers from bronchial asthma. Proper transportation plans are needed for our cities along with efficient enforcement of traffic laws. F5 simplifies the network architecture and provides flexibility to manage network resources efficiently and cost-effectively, deploy new services quickly, and manage changes more gracefully—all in a solution that delivers the highest, carrier-grade performance and reliability (Anonymous 2012). The latest available data show that, nationally, congestion levels at certain times of the day improved between 1998 and 2003. Average traffic speeds in the morning peak and to a lesser extent in the off-peak period, for example, rose over the period. They were still lower, however, than they were in 1995. Average traffic speeds, for example, fell by up to six per cent (four miles per hour) depending on the time of day between 1995 and 2003. Speeds have fallen as the volume of traffic has continued to grow – up by 14 per cent on all roads between 1995 and 2002. Around seven per cent of the network suffers heavy congestion at peak times, while a further 13 per cent of the network suffers heavy congestion on at least half the days of the year (NAO, 2004).

The Transport Research Laboratory has estimated that weight of traffic accounts for some 65 per cent of congestion, while accidents and incidents account for 25 per cent, and road works account for 10 per cent. The government is seeking to tackle congestion through a combination of measures intended to increase the capacity, and improve the management, of the network. Measures include new road building and improvement schemes. The Department recently published a report1 on the feasibility of charging road users for general access to the road network to reduce congestion. Yet, these schemes will take time to deliver. Action is needed in the near term to deal with immediate problems on key routes and at congestion blackspots on the network.

The Highways Agency has a programme of work, known as the *Making Better Use (MBU)* programme, intended to bring quick relief from congestion on England's motorway and trunk road network. It spends over £200 million a year on the programme, mostly installing technology on the network or carrying out small scale construction schemes known as Local Network Management Schemes, which include altering the layout of road junctions and motorway slip roads, introducing free flow lanes at roundabouts and improving junctions between trunk roads and motorways. Since 1998-99, the Agency has spent almost £1.1 billion on the *MBU* programme. Further, in June 2003 the government announced that the Agency would develop its remit beyond building and maintaining the road network and take on greater responsibility for managing traffic as the network operator.

13.6. SUMMARY

Travel by 'Palkis' was ancient feature, non-motorized vehicles, like use horses and bullock cart was routine thing in past. Roads were built, Highways were made during British period. It had check posts at regular intervals and shade providing trees were on both sides to provide relief to traveler from hot sun rays in summers. Large – scale movement of trading caravans consisting of thousands of bull or horses caused rise of dust in the air. Sir Thomas Roe, first ambassador of East India Company to the Mughal Court, records that he had to wait at a road crossing for three hours while a herd of bulls was passing from Surat to Agra via Ajmer. SPM in the air and presence of soot and smoke due to burning of wood and cow doing was major cause of concern. In India oil exploration started in 1867 while, First oil well was started in 1859 in America and current oil production is less than 100 million barrel which will soon become 111 million barrel per day (Khameshra, 2015).

REFERENCES

Anonymous (2012) F5 Networks, Inc. All rights reserved. F5, F5 Networks, the F5 logo, BIG-IP, Local Traffic Manager, VIPRION, TCP Express, and I Rules are trademarks of F5 Networks, Inc. in the U.S. and in certain other countries.

Bhatt B. (2005) Pollution due to vehicles. In: Urban Air Pollution (eds.) Arya A., Bedi S.J. Patel V.S. and Jasrai Y.T. Nidhi Book Distributors. 230pp

Khameshra N. S. (2015) Tel Aevam Gas utpadan Ka Bhartiya Paridrashya (in Hindi) Bhuvigyan. 5:13-18

NAO (2004) Tackling congestion by making better use of England's motor ways and trunk roads. Report by comptroller auditor General HC 15 session 2004-2005, 26 Nov. 2004.

■■■■

Chapter - 14

Be Aware of Pollen, Fungal and Food Allergens !

Vinay M. Raole

Department of Botany, Faculty of Science,
The Maharaja Sayajirao University of Baroda, Vadodara, 390002 India

ABSTRACT

Our ancient civilization has provided evidence suggesting that foods can be effectively used for nutrition and medicine to treat and prevent disease. The 5000 year old ancient Indian health science, Ayurveda, have mentioned benefits of food for the therapeutic purpose. Even Hippocrates, considered by some to be the father of Western medicine, said that people should consider, Let food be thy medicine. When this food and fungi act as allergens in the polluted atmosphere the people may suffer and need an immediate attention.

Chinese Buddhist monk Faxian wrote that throughout the whole country the people do not kill any living creature, nor drink intoxicating liquor, nor eat onions or garlic. Hsuan-tsang told that fruits like mango, tamarind, mahua, melon jujube, wood apple, myrobalan, fig, coconut, pomegranate, jackfruit etc. No one was consuming onion. Raw vegetables were not eaten as it was thought that they will cause pain. People used to discard left out food. Boiled rice flavoured with tamarind and sesame seeds was used to make Poha. Food allergies are caused by particular protein present in a food item.

Key words: Human health, Pollen, Fungal, Food allergens

14.1 INTRODUCTION: ALLERGENICITY

In this paper account has been given about the allergy and allergens from the phytoworld, specifically the food particles, pollens and spores. The dominance of the individual centred perspective on human health is currently coming under challenge from two quarters. First, epidemiologists are giving greater emphasis to studying disease causation within a broader social context, including consideration of population-level phenomena. Second one is from the individualistic approach wherein number of plants and animals consumed as food, inhaled and touch create allergic response to human body due to rejection of some biomolecules present therein. Over and above, stimulus from global environmental changes has major implications for human health. In particular, it is anticipated that changes in stratospheric ozone concentration, world climate, and biodiversity stocks, food producing systems on land and sea, and freshwater supplies will have mainly adverse affects on health (McMichael 1993, Epstein 1995).

Allergies occur when an immune system that is hyper sensitive reacts to a common or unusual substance (also called allergens). In some people, allergies are hereditary but one can develop allergies at any time in his or her life. A person might be allergic to such allergens as chemical toxins, food, dust, animal hair, pollen or parasites. According to Ayurveda, ama (undigested toxins) and low immunity are the main cause of allergies. Ama can settle at the digestive level or in various tissues with different symptoms at each level. When improperly digested ama is carried through channels into the various tissues in the body, the resulting toxin accumulation imbalances the tissues. Ama toxicity aggravates Pitta or Kapha, and in each case, the symptoms may vary. In fact, you may be a Kapha, Pitta or Vata person, but an allergy when it occurs in any constitution, can throw up various symptoms.

According to Ayurveda, ama (undigested toxins) and low immunity are the main cause of allergies. Ama can settle at the digestive level or in various tissues with different symptoms at each level. When improperly digested ama is carried through channels into the various tissues in the body, the resulting toxin accumulation imbalances the tissues. Ama toxicity aggravates Pitta or Kapha, and in each case, the symptoms may vary. In fact, you may be a Kapha, Pitta or Vata person, but an allergy when it occurs in any constitution, can throw up various symptoms.

A Pitta based allergy, for instance, manifests itself through sore red eyes and an inflamed itchy nose. There could be low fever and the person might feel extremely irritable. Exposure to sunlight will set off sneezing attacks. Citrus blossoms are said to be main the culprits behind Pitta allergies.

A Kapha allergy, on the other hand, may induce a dull heavy feeling in the head, stuffy nose and sinuses, lethargy and drowsiness, swollen eyes and lots of mucous production. Damp related allergens like moulds can set off this type of allergy and even cause asthma in susceptible victims.

A Vata allergy attack may be caused due to dry dust in which case the throat may run dry and become hoarse. There would be a feeling of excessive dryness and pain in the nose and sinuses too. Asthma characterized by spasms may result in extreme cases.

The term allergy was coined in 1906 by von Pirquet to describe an altered reactivity in living beings (*i.e.*, an IgE mediated hypersensitivity) caused by a foreign substance (von Pirquet, 1907). Allergens, therefore, are the subset of antigens that stimulate an IgE-mediated response. Genetic factors are known to influence the ability to mount an IgE-mediated reaction (Marsh *et al.*, 1981), Antibodies (called immunoglobulins; Ig) are produced by the body that defend it against infections. In general, antibodies are developed toward macromolecules, which must have a mass of M, at least 4000 – 5000 in order to stimulate antibody production. One important antibody, known as IgE, is generated as part of the normal immune reaction to parasitic infections, such as malaria. However, predisposed individuals can develop IGE responses to macromolecules present in environmental agents, such as pollen grains, dusts and food items. These macromolecules recognized by the IgE are known as allergens. IgE can bind to the surface of certain cells, such as basophils or mast cells, which contains granules of performed inflammatory mediators, including histamine. As soon as a multi-valent allergen acts with such cells, it can cross-link the cellular-bound IgE, an event which causes the cells to release the mediators. These mediators cause the physiological changes which manifest themselves as an allergic reaction. Allergic symptoms include respiratory, skin- like rashes, eczema and atopic dermatitis and gastrointestinal reactions.

IgE-mediated reactions generally occur quite rapidly, they are quite distinct from food intolerances, which may take much longer (even days) to manifest themselves and do not necessarily have an immunological basis. Thus, lactose intolerance is a food intolerance, which actually results from the loss of a digestive enzyme meaning that individuals cannot consume cows' milk and some derived products. In contrast, the gluten intolerance syndrome known as coeliac disease is thought to involve a cellular based immune reaction, but without the development of an IgE-mediated response, to wheat gluten proteins. Although the gluten protein which trigger coeliac disease are not IgE-binding proteins some workers refer to them as allergens, and they are treated as such in the FAQ-WHO Codex Alimentarius Commission Food Labelling (FAQ-WHO, 1999).

Two Categories of Food Allergies

1. **Immunoglobulin E (IgE) mediated**. Symptoms result from the body's immune system making antibodies called Immunoglobulin E (IgE) antibodies. These IgE antibodies react with a certain food.
2. **Non-IgE mediated.** Other parts of the body's immune system react to a certain food. This reaction causes symptoms, but does not involve an IgE antibody. Someone can have both IgE mediated and non-IgE mediated food allergies.

IgE Mediated Food Allergies

The IgE mediated food allergies most common in infants and children are eggs, milk, peanuts, tree nuts, soy and wheat. The allergic reaction can involve the skin, mouth, eyes, lungs, heart, gut and brain. IgE mediated food allergies fall into two different categories. The first is foods which are thought to individuals via the gastrointestinal tract and can often result in severe sensitize symptoms, including anaphylaxis. Second, allergy to fresh fruits and vegetables, which develop as a result of any individual first developing inhalant allergies to agents such as pollen and latex.

- Skin rash, itching, hives
- Swelling of the lips, tongue or throat
- Shortness of breath, trouble breathing, wheezing
- Stomach pain, vomiting, diarrhoea
- Feeling like something awful is about to happen

Sometimes allergy symptoms are mild. Other times they can be severe. Take all allergic symptoms seriously. Mild and severe symptoms can lead to a serious allergic reaction called anaphylaxis. This reaction usually involves more than one part of the body and can get worse fast. Anaphylaxis must receive treatment right away because it can cause death.

IgE mediated food allergies fall into two different categories. The first is foods which are thought to individuals via the gastrointestinal tract and can often result in severe sensitize symptoms, including anaphylaxis. Second, allergy to fresh fruits and vegetables, which develop as a result of any individual first developing inhalant allergies to agents such as pollen and latex. These have been called cross-reactive allergy syndromes because of the high levels of homology between the sensitizing allergy found in pollen and latex and allergens in food such as apples, peaches, pears and celery.

Table 1. Food allergy: Target order and disorders

Target organ	IgE mediated disorder	Non IgE mediated disorder
Skin	Urticaria and angiodema	Atopic dermitis
	Atopic dermitis	Dermitis heptiformis
Gastrointestinal track	Oral allergy syndrome	Proctocolitis
	Gastrointestinal anaphylaxis	Enterocolitis
	Allergic eosinofilic Gastroenteritis	Allergic eosinofilic Gastroenteritis
		Enteropathy syndrome
		Celiac diseases
Respiratory track	Asthma	Heiner syndrome
Multisystem	Food induced anaphylaxis	
	Food associated exercise induced anaphylaxis	

This similarity means that IgE developed to pollen can bind homologous allergens in several plant foods. Consequently, consumption of fresh fruit can often trigger allergic reactions in individuals sensitized to pollen. Severe reactions can occur but generally the symptoms triggered in cross-reactive allergy syndromes are much milder (often being confined to the oral cavity) than in foods believed to sensitize via the oral route. This has led to pollen – fruit/vegetable allergies being associated with oral allergy syndrome (OAS). Of the foods classically described as causing the majority of foods allergies, four originate from plants: peanuts, tree nuts, wheat and soy.

14.2 PLANTS AS FOOD ALLERGENS

Plant tissues that are consumed by humans contain thousands of different proteins.. An official list of allergens is compiled by the International Union of Immunological Societies designated by the Allergen Nomenclature subcommittee (Hoffmann *et al.*, 1994; Larson and Lowenstin, 1996). Allergens included in this listing must induce IgE –mediated allergy in humans with a prevalence of IgE reactivity above 5%. However, many other IgE- binding proteins have also been identified in the literature which do not comply with this rigorous definition but are commonly known as allergens. An allergen is called 'major' if it is identified by IgE from at least 50% of a group of allergic individual but this is not related to allergenic potency. Other allergens are called 'minor'. Allergens from the same species with greater than 67% sequence identity are defined as iso-allergens. The allergen designation itself is based on the Latin name of the species from which it originates, being composed of the first three letters of the genus, followed by the first letter of the species finishing with an Arabic number. Thus, for example, jug r 1 relates to an allergen from *Juglans regia* (walnut).

14.2.1. Classification of Plant Food Allergens

Most of the major allergens in the main allergenic foods have now been identified and sequenced, providing information which has allowed the structural and evolutionary relationships of plant food allergens to be considered (Breinteneder and Ebner, 2000; Breinteneder and Redauer, 2004; Mills *et al.*, 2003, 2004a). It is now becoming evident that the vast majority of plant food allergens that sensitize via the gastro-intestinal tract belong to only two major plant protein superfamilies', the prolamin and cupin super families with other such as the cysteine protease family being less prevalent. In contrast, inhalant allergens that can trigger food allergies are more diverse in nature and belong to a larger number of protein families, although they are dominated by one family, which is known as the Bet v 1 family based on the major inhalant allergen in birch pollen.

Many pant food allergens can also be defined as 'pathogenesis- related' (PR) proteins, being expressed as part of a defence response following infection by microbes and fungi or in response to abiotic stress (van Loon and van Strien, 1999). Currently 17 different classes of PR proteins have been defined, with known food allergens belonging to seven and PR family membership is intriguing, and while no molecular basis for this has been defined, many PR allergens are small proteins which are resistant to denaturation at low pH and proteolysis in vitro (Hoffman-Sommergruber, 2002).

Most plant food allergens belong to a few protein families and superfamilies. Many allergens belong to the cupin superfamily (7S and 11S seed storage proteins) or the prolamin superfamily (2S albumins, nonspecific lipid transfer proteins [nsLTPs], a-amylase/trypsin inhibitors, and prolamin storage proteins of cereals. A summary of the three main allergen families: the prolamin and cupin superfamilies' that sensitize via the gastrointestinal tract and the Bet v1 family, together with a broad overview of many other plant food allergens identified till date.

14.2.2. The Prolamin Superfamily

Members of the prolamin superfamily, originally defined by Kreis *et al.* (1985), share a characteristics framework of cysteine residues, containing Cys and Cys 5, Cys mofits, where X represents any other residue. Subfamilies within this superfamily include the alcohol-soluble (prolamin) storage proteins of cereals, the 25 storage albumins, non specific lipid transfer proteins (nsLTP) and cereal inhibitors of cx –amylase and/or trypsin all but the nsLTPs being restricted to seeds. The conservation of structure extends to the protein fold, at least for the 25 albumins, nsLTP and cx amylase/trypsin inhibitors, which all share a related fold consisting of bundles of four cx-helices stabilized by disulphide bonds.

The cereal prolamins, named glutenins and gliadins in wheat, secalins in rye, and hordeins in barley, are the major storage proteins found in the endosperm of cereal grains. The sulphur-rich prolamins are composed of an N terminal domain that contains proline-and glutamine-rich repeats and a C-terminal non repetitive domain with even numbers of cysteine residues that form intrachain disulfide bonds.

Wheat can cause atopic dermatitis (Varjonen *et al.*, 1995, 2000) and a condition known as exercise-induced anaphylaxis (EIA) (Varjonen *et al.*, 1997). The latter is a severe allergic reaction that certain patients experience only when taking exercise following consumption of a problem food. Several allergens have been described as triggering such reactions, including a y- and cx-gliadin (Palosuo *et al.*, 1999, 2001a) and an co5 gliadin (Palosuo *et al.*, 2001b). Other prolamin storage proteins have been identified as major cereal allergens, including both the polymeric HMW and LMW subunits of glutenin, as well as the monomeric y- and cx- gliadins (Maruyama *et al.,* 1998; Simonato *et al.,* 2001b). Cooking appears to affect the type of symptoms observed (Guenard-Bilbault, Kanny and Moneret-Vautrin, 1999) and recent studies have indicated that baking may be essential for all allergenicity of cereal prolamins (Simonato *et al.*, 2001a).

2S storage albumins

These proteins are abundant in the seeds of many plant species and have been identified as major allergens in a number of plant foods including peanut (Arah 2, 6, 7: Kleber *et al.*, 1999), oriental and yellow mustards (Bra j 1, Sin a1: Menedez- Arias *et al.,* 1988: , oilseed rape (the napin, Bran 1), castor bean (Ric c 1, 3), walnut (Jug r 1), sesame (Ses i 1: Pastorello *et al.,* 2001b and Ses i 2: Beyer *et al.*, 2002a), almond (Poltronieri *et al.,* 2002) and sunflower (Kelly, *et al.* 2000). It has also been reported that the 25 albumins of soy (Shibasaki *et al.*, 1980 and chickpea (Vioque *et al.*, 1999) are allergens.

The nsLTPs are distributed even more widely in plants than 25 albumins, with over 100 sequence being available from a variety of plant organs including seeds, fruit and vegetative tissues. Major allergens have been identified in many Rosacea fruits including peach (Pru p 3, initially named Pru p 1: Pastorello *et al.*, 1999a) apple (Mal d 3. Pastorello, 1999c: Sanchez- Monge *et al.,* 1999). Apricot (Pruari: Pastorello *et al.*, 2000a) and cherry (Pru av 3: Scheurer *et al.*, 2003), hazelnut (Pastorello *et al.,* 2002), maize (zea m 14: Pastorello *et al.*, 2000b) and asparagus (Diaz-Perales *et al.*, 2002). This distribution has led to the family being termed as 'pan allergens' (Sanchez-Monge *et al.,* 1999). Like the 25 albumins, nsLTPs can sensitize by both inhalation and/or the oral route, since they are present in pollen and dusts as well as in foods. An individual has been reported with an occupational allergy to pelt (a type of wheat) who experienced symptoms on consuming a variety of fruits and where the major allergen was a nsLTP, indicating some true IgE cross-reactivity (Pastorello *et al.*, 2001a). nsLTPs have also been identified as pollen allergens. But as yet there is no clear evidence of a link between sensitization to pollen nsLTP and subsequent development of allergy to fruit.

Cereal cx-amylase inhibitors

These proteins can also sensitize individuals via the lungs by inhalation of dusts and are responsible for occupational allergies such as Bakers' asthma (wheat, barley and rye) as well as via the gastrointestinal tract (wheat, barley and rice). A single M, -15000 subunit of wheat has been identified as a food allergen (James *et al.*, 1997) while a second wheat inhibitor, termed CM3, has been identified as an allergen triggering atopic dermatitis (Kusaba – Nakayama *et al.*, 2000). An M, 16000 beer allergen has also been described, which originates from barley and appears to belong to the cx-amylase inhibitor family (Curioni *et al.*, 1999) together with an M, 16000 allergen in maize (Pastorello *et al.*, 2000b). Other well characterized allergenic inhibitors occur in rice with M, s of about 14000 – 16000. In particular an M, - 16000 subunit termed RA 17, having been identified (Nakase *et al.,* 1996).

14.2.3. IgE Reactivity and Sensitization potential of proteins

The methods for assessing cross-reactivity using human allergic sera are straightforward, although there are ethical issues regarding the development of serum banks on which such studies rely, together with the unequivocal diagnosis of food allergy in those individuals contributing sera. Cross–reactivity studies using suitable antibody preparations from experimental anima sources can also be used to substantiate the potential for cross-reactivity. The WHO/FAQ decision tree recommends that combinations of allergic sera from both food and inhalant (*e.g.* pollen) are used, which in the case of GMOs for food use encompass allergies to both monocotyledonous and dicotyledonous plant species.

14.3 FUNGI AS ALLERGENS

Fungi can cause a number of infectious diseases. These range from superficial skin lesions primarily of cosmetic concern to potentially sever systemic mycoses. Fungal infections are covered in numerous medical microbiology descriptions, although *Trichophyton* infections have been linked to asthma in some cases (Deuell *et al.,* 1991; Platts-Mills *et al.,* 1987). Many fungi produce potent toxins. More than 200 toxins are Table 1. Taxonomic distribution of selected genera of allergenic fungi. Most of which have been identified from growth on plant materials, including those used for food. Probably the best known are the aflatoxins (potent carcinogens) produced by *Aspergillus flavus* and *Aspergillus parasiticus* (Kendrick, 1985). Fungal toxins are best known from ingestion exposure. These toxic reactions to fungi must be differentiated from allergic responses that may occur following ingestion of mushrooms or of food contaminated with fungi (Lammpe and McCann, 1987). Allergic reactions to ingested fungi have been described infrequently, although one convincing case report of anaphylaxis due to mushroom consumption exists (Stricker *et al.,* 1986). Allergic reactions may occur within 30 min of exposure, whereas toxic reactions may not appear until 8 to 12 h after ingestion. Toxic reactions are nonimmunologic; thus, toxic reactions can occur with the first encounter. Allergic reactions, on the other hand, require a previous, sensitizing reaction. Fortunately, most significant mycotoxins are of low molecular mass (1 kDa) (Kendrick, 1985) and can be removed from fungal allergen extracts by dialysis. Hypersensitivity has been characterized immunologically into four types of reactions on the basis of clinical presentation and the underlying mechanisms (Cohen, 1988).

Allergic inhalant diseases such as asthma or rhinitis afflict 20% of the population of which 10% have significant or severe allergic disease (Gregen *et al.,* 1987; Heggy and Sentipen, 1969; Ogino *et al.,* 1990; Weeke, 1992). Skin tests suggest that at least 3 to 10% of adults and children worldwide are affected by fungal allergy (Beaumont *et al*, 1985) The exact prevalence has not been confirmed since reports of skin test reactivity to these fungal organisms range from 3 to 91%, depending upon the population studied, extracts used, and species tested (Sprenger *et al.,* 1988). Prevalence of reactivity is dependent upon the source and batch of allergen as well as the selection criteria for the test subjects. For example, reactions to commercial *Cladosporium herbarum* extracts varied from 12 to 65% in the same population (Ans, 1980). Likewise, an experimental *E. nigrum* mycelium extract detected sensitivity in 70% of another population, whereas only 6% of the same group reacted to a commercial extract (Portnoy, 1987).

Type I (immediate) hypersensitivity (allergy) is triggered by the interaction of immunoglobulin E (IgE)-specific antigens, which are called allergens, with IgE on the surface of mast cells or basophils.

Type II (cytotoxic) hypersensitivity occurs when antigens bound to cells are recognized by antigen-specific IgG or IgM. The surface-associated immune complexes are recognized by complement, and cell lysis occurs.

Type III reactions are due to circulating immune complexes that attach to vessel walls and activate complement. Neutrophils phagocytize the complexes and induce a diffuse inflammation called vasculitis.

Type IV (cell-mediated) reactions are mediated by antigen-specific T cells that release lymphokines in response to their encounter with antigen. A mononuclear cell infiltrate, which peaks at 24 to 72 h following exposure to antigen, occurs.

As with other hypersensitivity reactions, this phenomenon requires previous exposure to the antigen. Many hypersensitivity conditions are combinations of these four types of reactions.

A recent authoritative treatment of indoor allergens by the Institute of Medicine defines allergy as the state of immune hypersensitivity that results from exposure to an allergen and is distinguished by overproduction of immune system components (Pope *et al.,* 1993). Hypersensitivity pneumonitis and allergic bronchopulmonary aspergillosis (ABPA) are two respiratory disorders with type I hypersensitivity components, but neither syndrome is strictly an allergic phenomenon by our definition. Hypersensitivity pneumonitis (or extrinsic allergic alveolitis) is an immunologic lung disorder caused by sensitization to various inhalant antigens such as fungi, actinomycetes, organic dusts, and certain highly reactive inorganic chemicals. The pathogenesis of hypersensitivity pneumonitis includes a mixture of immune complex (type III) and cell-mediated (delayed, type IV) hypersensitivity responses.

ABPA is an inflammatory disease in which fungi (usually *Aspergillus fumigatus*) grow in mucous secretions in the lung. ABPA is thought to involve type I and type III hypersensitivities associated with IgE- and IgG-mediated immune response to *Aspergillus* species. Several allergens of *Aspergillus fumigatus* have been characterized (Moser *et al,* 1992). As discussed later, the major allergen, Asp f 1, is homologous with mitogillin (the protein sequences are approximately 95% identical), a potent cytotoxin. Although allergens are involved in ABPA, there are other hypersensitivity components and nonimmunologic components of ABPA. ABPA appears to be more prevalent than was previously thought, and other genera of fungi are recognized to cause similar or identical diseases.

14.3.1. IgE-Mediated Hypersensitivity to Fungi

The individuals with sustained elevated IgE levels are referred to as atopic. Type I allergic disease to fungal allergens is typically manifest either as rhinitis (hay fever) or as asthma. Allergic reactions, including respiratory allergy, may occur in two phases. The early-phase reaction occurs within minutes as a result of the release of preformed mediators. Late-phase responses occur 3 to 4 h after allergen exposure as a result of cellular infiltrates responding to early-phase mediators. A dual reaction involves both early- and late-phase reactions. Emerging evidence indicates that a significant, persistent inflammatory component in addition to IgE-triggered effects underlies the etiology of asthma. Fungal allergy is not as well defined as other seasonal allergies and probably causes problems indoors for fewer people than do major allergens from cats, mites, or cockroaches. A significant portion of the atopic population does have underlying sensitivities to fungal spore allergens, although subjects I Anaphylactic, allergic IgE antibodies Mediator release from IgE-sensitive mast cells or basophils.

In one study involving 6,000 patients with allergy, less than 1% were allergic to *Alternaria alternata* only, one of the species to which reactivity is most common. Thus, the common pattern is for subjects to react to many species. Furthermore, fungal allergies are very difficult to detect and treat than other allergies since fungi are far more numerous and antigenically variable than other allergens and are actually difficult to avoid. Thus, management of fungal allergy can be a real challenge to medical science.

14.3.2 Fungi as Aeroallergens

Fungal spores are universal atmospheric components indoors and outdoors and are now generally recognized as important causes of respiratory allergies. Allergic reactions associated with fungi involve the lower respiratory tract more frequently than do pollen allergies (Lehrer *et al.,* 1986). More than 80 genera of fungi have been associated with symptoms of respiratory tract allergy (Latge and Paris, 1991). Aerobiologic surveys show that fungal spores are present in the atmosphere worldwide. Multiple species may be observed at any time of the year, but in temperate climates, spore numbers peak during summer and fall, decrease with cooler temperatures, and are absent, at least outdoors, where snow cover occurs . Despite the clinical importance and abundant release of fungal spores, relatively few investigations have focused on the relationship between airborne spores and allergic disease. Allergic reactions normally occur at the site of allergen deposition. Most inhaled particles of .10 mm (most pollens and some larger spores) are deposited in the nasopharynx and are associated with nasal and/or ocular symptoms generally referred to as hay fever. Conversely, particles of10 mm, especially those of 5 mm, can penetrate the lower airways, where allergic reactions tend to manifest as asthma. Fungal spores differ in size and are associated with both upper and lower respiratory symptoms. Additionally, there is now evidence that secondary dispersal of allergens, i.e., on other, smaller particles, possibly spore fragments, may serve as a vehicle for allergens. This would permit the deposition of allergens even from large spores such as *E. nigrum* into the lower airways.

14.3.3. Correlations between Airborne Prevalence and Clinical Relevance

Ascomycetes, basidiomycetes, and zygomycetes are the major fungal groups that contain genera known to produce allergens and elicit allergic reactions. Collectively, these groups contribute most of the spores found in air. Early sampling methods for airborne fungal particles had several biases. These methods were nonvolumetric, which tends to favor large spores or spores that disperse in clumps. Recovered fungi were usually detected by their growth on culture medium, which selects for fungi able to grow on the medium used. Analysis of the sample usually consisted of identifying the cultures on the basis of sporulation characters, which excludes species that do not reproduce in culture. Many anamorphic fungi are easily recovered by these methods, however, and most early research on fungal allergens centered on allergenic species of these. As a result of more recent studies, in which nonviable, volumetric air sampling procedures have been used with direct microscopic analyses, it has been determined that conidia represent only 30 to 60% of airborne spores. The bulk of the remaining spores are ascospores and basidiospores. Recent evidence also indicates that the prevalence of hypersensitivity (skin test reactivity) to basidiospore and that to conidial allergens are also comparable. Thus, the intensity of exposure may indeed determine clinical relevance. The largest gap in our knowledge of aeroallergens pertains to ascospores, the sexual spores of ascomycetes. They constitute much of the balance of the airspora that is neither conidia nor basidiospores. Very little is known about ascospore allergens, although a certain degree of similarity to the allergens of conidial fungi might be inferred since the majority of the conidial fungi are anamorphs of ascomycete fungi.

14.4 ALLERGEN AND ALLERGEN-SPECIFIC IgE DETECTION

Skin testing is the simplest method to detect IgE directed against a specific allergen. Two methods commonly used to test allergies to inhalants, such as dust, mites, pollens, or animal dander's, are prick and intradermal skin testing. Skin prick testing correlates better with clinical history, radio allergosorbent test (RAST), and provocative inhalation challenge than the more sensitive intradermal test does. Intradermal tests also give false-positive results more frequently than prick testing does.

The RAST is an *in vitro* test designed to measure the circulating allergen-specific IgE antibodies of a patient. The RAST is generally considered less sensitive than skin testing, although a RAST conducted with partially purified fungal allergens can be comparable to skin tests in both sensitivity and specificity. However, the use of exquisitely specific skin test reagents may indicate a lower prevalence of reactivity since less-specific reagents will cross-react with more species of fungi. Thus, cross-reactive antigens could simultaneously detect allergies to several different species with only a single skin test and provide a more robust screening reagent for fungal allergy.

Crossed immune-electrophoresis (CIE) was used extensively for antigen analysis in the 1970s and 1980s, especially for complex mixtures such as allergen extracts. Crossed-lin immune-electrophoresis (CLIE) can identify common antigenic components in different sources and was used for early cross-. Allergen identification is possible with crossed radio immune electrophoresis (CRIE), a method in which a CIE agarose gel is overlaid with patient serum and any bound IgE is detected with radiolabeled anti-IgE. Although still useful to demonstrate antigenic purity, these methods have been replaced largely by immunoprinting (Western blotting [immunoblotting]) procedures. The basophil histamine release test is another valuable in vitro assay, but it has limited clinical usefulness since it requires fresh leukocytes. False-negative results can occur, so the clinical symptoms of the patient must be considered when interpreting test results. Other methods of *in vitro* allergy testing, based on enzyme or chemiluminescent detection, are becoming commercially available and are generating much interest. None is as yet widely accepted.

14.5 DIAGNOSIS OF FUNGAL ALLERGY

Diagnosis of any allergic disease is based on the clinical symptoms of the patient, determination of exposure to the allergen, results of skin and in vitro tests (particularly the RAST), and sometimes, if necessary, provocative inhalation challenge testing. Allergic symptoms may occur in definite patterns that can provide clues to the diagnostician. Fungal allergic symptoms are not usually considered seasonal, however, as are pollen related ailments. Moreover, many fungal spores derived from different species are similar morphologically; this can limit the completeness of spore identification in air sampling surveys.

Allergenic fungi is far greater than can feasibly be included in any panel of skin test reagents. This emphasizes the value of environmental sampling, since the range of fungi to be considered can be narrowed effectively if an aerobiological identification is available.

On the basis of aerobiologic surveys conducted in different locations of the world, skin test studies, and fungal allergen characterization, the consensus is that skin test panels should include, at the minimum, *Alternaria alternata*, *Aspergillus fumigatus*, *Cladosporium herbarum*, *E. nigrum*, *Fusarium roseum*, and *Penicillium chrysogenum*. An optimal screening panel of fungal allergens is not currently available. It is also important to note that this list does not include basidiomycetes, however, so experienced investigators must perform and interpret these tests, which include nasal, conjunctival. or bronchial provocation. Inhalation challenge can be an important tool in cases where conflicting or questionable data have been gathered, and it can confirm the clinical usefulness of allergen preparations in research settings as well. Conjunctival testing is less commonly used because of its poor reproducibility. Despite the introduction of provocative challenge and the development of sophisticated in vitro tests, making the diagnosis of mold allergy remains difficult. Standardized extracts are not available, the average person is constantly exposed to a wide variety of propagules, and many subjects are also allergic to nonfungal inhalant allergens. Additionally, the symptoms may be due to a fungus either whose spores are not restricted to a given season or whose particular spore type is not tallied independently by a local pollen or spore station, if one exists. These factors would obscure any correlation between spore count and symptoms that might indicate the offending species. Development of standardized fungal extracts and more extensive and detailed reporting of pollen or spore counts would contribute significantly to the reliability of diagnoses of fungal allergies.

14.6 CHARACTERIZATION OF FUNGAL EXTRACTS

Progress in understanding fungal allergy has been impeded by the great variety of potentially allergenic fungi, the instability and variability of their allergens, lack of agreement on current fungal nomenclature, and the selection of adequate source material for the preparation of extracts. For these and other reasons, characterization of fungal allergens has lagged behind characterization of other aeroallergens. The exceptions are *Aspergillus fumigatus*, *Cladosporium herbarum* and *Alternaria alternata.*

Fungal spores are structurally very different from pollen, but commercial fungal extracts are usually prepared like pollen extracts. Fungal extracts generally contain proteins, carbohydrates, and proteolytic and glycosidic enzymes in various amounts. Carbohydrate material was long considered to be essentially non allergenic, although reports have suggested the presence of carbohydrate allergens in *Alternaria alternata* (Portony *et al.,* 1987). Recent, convincing evidence has been presented for IgE epitomes on fungal mannan preparations from the yeast *Malessezia furfur* (*Pityrosporum ovale*). However, most of the Ig Ebinding activity probably resides in the protein components of the fungal extracts.

Fungal allergen extracts are usually made from cultures. Fungal growth in culture is affected by temperature, moisture, light, and culture media. Conditions for optimal growth and sporulation may differ within and among species. Many fungi have inducible enzyme systems, and thus the composition of fungal media must be clearly defined for the consistent production of the desired enzymes or allergens (Burge, 1985). The inducible enzyme systems of fungi are especially significant since many of the allergens may, in fact, be enzymes such as enolase and alcohol dehydrogenase from yeasts (Shen *et al.,* 1992). Strain variability within a given species presents an additional complicating factor. Even when a single strain is cultured under standard conditions, marked batch-to-batch differences may occur in protein composition and allergenic potency of extracts. Fungi also mutate readily in culture, contributing additional variables during *in vitro* cultivation. While the task is formidable, standardization of all phases of fungal extract production is an absolute requirement for significant progress in understanding fungal allergens. Biochemical variability within a given species has been documented for several important allergenic fungi. *Candida albicans*, for example, is known to modulate cell wall composition during the growth cycle. Extracts from different strains of *Aspergillus fumigatus* (Wallenbeck *et al.*, 1984), *Cladosporium herbarum, Alternaria alternata* (Vijay *et al.*, 1985), *E. nigrum, Candida albicans* (Savolainen *et al.*, 1989), and *Saccharomyces cerevisiae* (Baldo and Baker, 1988) vary greatly in their compositions. Some researchers and commercial suppliers seek to overcome the problem of variability by culturing several isolates or growing different batches of the specific fungus and then pooling extracted material from each culture.

14.7 VARIABILITY OF COMMERCIAL EXTRACTS

In spite of the difficulties in standardization, crude extracts are used routinely to diagnose and treat fungal allergies in clinical practice. These extracts contain many undefined components, are extracted by a variety of procedures, and, consequently, may differ qualitatively and/or quantitatively. In fact, it is not unusual for a particular allergen extract to contain specific allergens that are entirely absent in other extracts of the same fungus produced by another manufacturer. The variability of the fungal strains contributes to this problem (Vijay *et al.,* 1985).

14.7.2. Extractions

Ideally, extraction procedures for inhalant allergens should simulate the manner in which the allergens are released under natural conditions. However, the fungal cell wall of some species can reduce and slow the discharge of allergens into solution (Horner *et al.* 1991). Release of significant amounts of allergen from spores with certain types of cell walls may require physical disruption or extended elution, neither of which is natural. Other spore walls apparently present minimal resistance to the release of allergenic proteins. Regardless of spore wall effects, potent extracts have been prepared from spores, culture medium, and fresh and dried mycelial mats. This extracted material is centrifuged or filtered, and the supernatant or filtrate frequently undergoes further processing, such as dialysis, to remove low-molecular-weight nonallergenic material, including some toxins. Several reagents, such as phosphate-buffered saline (PBS), sodium bicarbonate ($NaHCO_3$), distilled water, or Coca's solution (0.5% NaCl, 0.275% $NaHCO_3$), can successfully remove allergens from undisrupted conidia of some fungi, and shaking or stirring the suspension may increase release (Lamp *et al* 1987). Other spores, *e.g.*, basidiospores, may require more rigorous extraction procedures.

The release of allergens from *Cladosporium cladosporioides* is affected by the culture medium, extraction time, and spore (conidium) concentration. The greatest yields were obtained when the culture had a low spore concentration (0.1 mg [dry weight] per ml) or was subjected to a brief ultra-sonication, but the yield varied little with the type of extraction solution used, including 0.1 N NaOH, Coca's solution, 50 mM Tris, or 0.9% NaCl. Mycelia of *Wallemia sebi* and two *Aspergillus* spp. were most active in the RAST within 20 min after homogenization in PBS. The yield of *Alternaria alternata* allergen from mycelium, as identified by CRIE, decreased when extraction continued beyond 6 h. Allergen yield from *Alternaria alternata* spores and mycelium was also greater when protease inhibitors such as phenylmethylsulfonyl fluoride or EDTA were included in the extraction buffer. Polyvinylpyrrolidone also aided extraction, presumably by binding polyphenols such as melanin and preventing co-precipitation of proteins. Polyvinylpyrrolidone would thus be most useful for fungi which are heavily pigmented with melanin, such as members of the Dematiaceae. Many different procedures and buffers have been used to extract allergens from fungi, and no standard procedure appears appropriate for all species. Factors known to be important and which should be optimized include the extraction buffer, length of extraction time, whether to physically or sonically disrupt the cells, and whether to add preservatives, protease inhibitors, or polyvinylpyrrolidone. These must be determined empirically for each species.

14.7.2. Stability

Extracts must remain stable to be useful clinically or for research. As with all proteinaceous mixtures, the stability of allergenic extracts depends on the type and quantity of allergen, the storage temperature, and the presence of preservative and other (nonallergenic) material in the mixture. At present, lyophilization is the best method to maintain the allergenic potency of fungal extracts. However, lyophilization is not an entirely innocuous procedure, and certain allergens may be permanently altered and/or rendered insoluble. Fractionation of lyophilized extracts may remove autolytic enzymes and further extend stability. Exposure to moisture and warmth will increase the degradation process; therefore, extracts must be desiccated and maintained at a cool temperature. Lyophilized *Alternaria alternata* extracts, for example, stored at 270, 220, and 48C were all stable for at least 21 months as measured by RAST inhibition. Similarly, the allergenicity of native Cla h 2 (antigen-54 [Ag-54]) from *Cladosporium herbarum* was unaltered for at least 5 years when it was stored at 2208C, although deglycosylated Cla h 2 lost most allergenic activity in 2 years.

14.8 APPLICATION OF GM TECHNOLOGY TO PLANT ALLERGENS

Although there is some concern that GM technology will result in the introduction of new allergens into the food chain, the risk assessment procedures discussed above mean that this is unlikely. However, it is worth noting in this respect that many biologically active proteins which have potential applications in protection against pests and pathogens are also either allergens or belong to protein families which contain allergenic proteins such as nsLTPs. Thus, the knowledge of protein families outlined above will prove to be valuable in selecting appropriate proteins for expression in transgenic plants.

The greatest impact of GM on allergenicity should be a positive one, by providing tools to reduce or eliminate allergenic proteins or epitopes. However, in practice this may be difficult to achieve, as many foods which cause severe reactions contain multiple allergenic proteins that may belong to separate families. Good examples of this are provided by peanut and soybean, which are related leguminous species. The major storage proteins in these species are 7S and 11S globulins, and in both cases these include major allergens (Ara h 1 in peanut is 7S, and Ara h 3 and 4 are 11S). In addition, peanut t contains allergenic 2S albumin storage proteins (Ara h 2, 6 and 7) and a profiling related protein (Ara h 5) while the soybean allergens include a cysteine proteinase (P34). Furthermore, most of these proteins have also been demonstrated to possess multiple IgE binding sites: at least 23 on Ara h 1, 16 on P34 and 11 on the soybean 11S globulin (Mills *et al.*, 2004b) for full details and references.

14.9 CONCLUSION

Most plant food allergens belong to a few protein families and super-families. Many allergens belong to the cupin superfamily (7S and 11S seed storage proteins) or the prolamin superfamily (2S albumins, nonspecific lipid transfer proteins [nsLTPs], α-amylase/trypsin inhibitors, and prolamin storage proteins of cereals. The identification of potential allergens, particularly those which have no structural or biological relationships with known allergens, is an uncertain process at present. It is clear that not all foods are equally allergenic, yet the precise molecular basis for these differences remains unclear. There is little evidence that particular protein sequences or structures promote the development of IgE responses although food allergens appear to be good antigens (*i.e* able to stimulate an antibody response). There is some evidence from animal studies that, based on a limited number of allergens tested so far, allergens are intrinsically able to elicit an IgE- response preferentially in experimental animals. Protein stability and resistance to digestion may also play a role in determining the allergenic potential of proteins when sensitizing via the GI tract. However, only when we have a greater understanding of the molecular pathology of food allergy which embraces both the aspects of food (allergen structure, interactions with the food matrix and the potential adjuvant effects of other food components) coupled with the genetic factors in man which predispose some individuals to becoming atopic, can a completely reliable prediction of novel protein allergenicity be made. Until this information is available, allergen identification will continue to be an uncertain process, requiring a great deal of information on many different aspects of protein structure and properties and relying on comparisons with known food allergens.

REFERENCES

Aas, K., J. Leegaard, L. Aukrust, and O. Grimmer. (1980). Immediate type hypersensitivity to common moulds. Allergy 35:443–451.

Baldo, B.A. and Baker R.S. (1988) inhalant allergies to fungi: reaction to baker's yeast and identification of baker's yeast enolase as an important allergen. Int. Arch. Allergy Appl. Immunol. 86; 201-208

Beaumont, F., H. F. Kauffman, H. J. Sluiter, and K. De Vries. 1985. Sequential sampling of fungal air spores inside and outside the homes of mould-sensitive, asthmatic patients: a search for a relationship to obstructive reactions. Ann. Allergy 55:740–746.

Beyer, K., Bardina, L., Grishina, G, and Sampson, H.A. (2002a) Identification of sesame seed allergens by 2 dimensional proteomics and Edman sequencing: seed storage proteins as common food allergens. Journal of Allergy and Clinical Immunology 110, 154 -159.

Breinteneder, H, and Ebner, C. (2000) Molecular and biochemical classification of plant derived food allergens, Journal of Allergy and Clinically Immunology 106, 27-36 .

Breinteneder, H, and Redauer, C, (2004) A classification of plant food allergens. Journal of Allergy and Clinical Immunology 113, 821-830.

Burge H.A. (1985) Fungus allergens. Clin Rev Allergy. 1985; 3(3):319–329.

Cohen, J. J. 1988. The immune system: an overview, p. 3–11. In E. Middleton, C. C. Reed, E. F. Ellis, N. D. Adkinson, and J. W. Yunginger (ed.), Allergy: principles and practice. The C.V. Mosby Co., St. Louis.

Curioni, A., Santucci, B., Cristado, A., Canistraci, C., Peitravalle, M., Simonato, B and Giannattasio, M. (1999) Urticaria from beer: an immediate hypersensitivity reaction due to a 10 kDa protein derived from beer. Clinical and Experimental Allergy 29, 407-413.

Deuell, B., L. K. Arruda, M. L. Hayden, M. D. Chapman, and T. A. E. Platts-Mills. 1991. Trichophyton tonsurans allergen. I. Characterization of a protein that causes immediate but not delayed hypersensitivity. J. Immunol. 147:96–101.

Diaz- Perales, A., Sanchez-Monge, R., Garcia, B.E., Gomez, B., Barber, D, and Salcedo, G, (2002a) What is the role of the hevein- like domain of fruit class I chitinases in their allergenic capacity? Clinically and Experimental Allergy 32, 448-454.

Epstein PR (1995) Emerging diseases and ecosystem instability: New threats to public health *American Journal of Public Health* 85 168 172

FAO-WHO (1999) Codex General Standard for the Labelling of Prepackaged Foods CODEX STAN 1-1985 (Rev. 1-1991), http://www.fao.org/DOCREP /05/Y2770E/ y2270e02.htm#bm02.

Gergen, P. J., P. C. Turkeltaub, and M. G. Kovar. 1987. The prevalence of allergic skin test reactivity to eight common aeroallergens in the U.S. Population: results from the second National Health and Nutrition Examination Survey. J. Allergy Clin. Immunol. 80:669–679.

Hagy, G. W., and G. A. Settipane. 1969. Bronchial asthma, allergic rhinitis, and allergy skin tests among college students. J. Allergy 44:323–332.

Hoffman, D., Lowenstein, H., Marsh, D. G., Platts- Mills, T.A.E. and Thomas, W. (1994) Bulletin of the World Health Organization 72, 796-806.

Hoffmann-Sommergruber, K (2002) Pathogenesis-related (PR) Proteins identified as allergens. Biochemical Society Transactions 30, 930-935.

James, J.M., Sixbey, J. P., Helm, R,M, Bannon, G.A and Burks, A, W, (1997) Wheat cx amylase inhibitor: a second route of allergic sensitisation. Journal of Allergy and Clinical Immunology 99, 239-244.

Kelly, J,D., Hlywka, J. J. and Hefle, S.L. (2000) Identification of sunflower seed IgE-binding proteins.

Kendrick, B. 1985. The fifth kingdom. Mycologue Publications, Waterloo, Canada

Kleber, J. T., Crameri, R., Appenzeller, U., Schlaak, M, and Becker, W.M (1999) Selective cloning of peanut allergens, including profiling and 2S albumins, by phage display technology International Archives of Allergy and Immunology 119, 265-274.

Kreis, M., Forde, B. G., Rahman, S., Miffin, B. J, and Shewry, PR.(1985) Molecular evolution of the seed storage proteins of barley, rye and wheat. Journal of Molecular Biology 183, 499-502.

Kusaba-Nakayama, M., KI, M., Iwamoto, M., Shibata, R., Sato, M, and Imaizumi K. (2000) CM3, one of the wheat alpha amylase inhibitor subunits, and binding of IgE in sera from Japanese with atopic dermatitis related to wheat. Food Chemistry and Toxicology 38, 179-185.

Larson, J.N. and Lowenstein, H. (1996) Allergen nomenclature, Journal of Allergy and Clinical Immunology 97, 577-578.

Lampe, K. F., and M. A. McCann. 1987. Differential diagnosis of poisoning by North American mushrooms, with particular emphasis on Amanita Phalloides- like intoxication. Ann. Emerg. Med. 16:956–962.

Latge, J. P., M. Moutaouakil, J.-P. Debeaupuis, J.P. Bouchara, K. Haynes, and M.C. Prevost. (1991). The 18-kilodalton antigen secreted by Aspergillus fumigatus. Infect. Immun. 59:2586–2594.

Lehrer, S. B., M. Lopez, B. T. Butcher, J. Olson, M. Reed, and J. E. Salvaggio. (1986). Basidiomycete mycelia and spore-allergen extracts: skin test reactivity in adults with symptoms of respiratory allergy. J. Allergy Clin. Immunol. 78:478–485.

Maruyama , N., Ichise, K., Katsube, T. Kishimoto, T., Kawase, SI., Matsumura, Y., Takeuchi, Y. Sawada, T, and Utsumi, S. (1998) Identification of major wheat allergens by means of the Escherichia coli expression system. European Journal of Biochemistry 255, 739-745.

Marsh, D. G., D. A. Meyers, and W. B. Bias. (1981). The epidemiology and genetics of atopic allergy. N. Engl. J. Med. 305:1221–1229.

McMichael AJ (1993) Planetary Overload: Global Environmental Change and the Health of the Human Species Cambridge (UK) Cambridge University Press

Mills, E.N.C., Madsen, C., Shewry, P.R, and Wichers, H.J. (2003) Food allergens of plant origin the relationship between allergenic potential and biological activity (PROTALL)., Trends in Food Science and Technology 14, 145-156.

Mills, E.N.C., Jenkins, J., Alcocer, M.J.C,.and Shewry, P.R. (2004a) Structural biological and evolutionary relationship of plant food allergens sensitising via the gastrointestinal tract. Critical Reviews in Food Science and Nutrition 44, 379 – 407.

Moser, M., R. Crameri, G. Menz, T. Schneider, T. Dudler, C. Virchow, M. Gmachl, K. Blaser, and M. Suter.(1992). Cloning and expression of recombinant Aspergillus fumigatus allergen I/a (rAsp f I/a) with IgE binding and type I skin test activity. J. Immunol. 149:454–460.

Nakase, M., Adachi, T., Urisu, A., Miyashita, T., Alvarez, A. M., Nagasaka, S., Aoki, N., Nakamura, r, and Matsuda, T. (1996) Rice (*Oryza sativa* L.) cx amylase inhibitors of 4 – 16 kDa are potential allergens and products of a multigene family. Journal of Agricultural and Food Chemistry 44, 2624 -2628.

Ogino, S., M. Irifune, T. Harada, T. Matsunaga, and M. Ishida. (1990). Nasal allergy in medical students. Rhinology 28:163–168.

Palosuo, K., Alenius, H., Varjonen, E., Kolvuluhta, M., Mikkola, J., Keskinen, H., Kalkkinen, N, and Reunala, T. (1999) A novel wheat gliadin as a cause of exercise induced anaphylaxis Journal of Allergy and Clinical Immunology 103, 912-917.

Palosuo, K., Varjonen, E., Kekki, O.M., Klemola, T., Kalkkinen, N., Alenius, H., and Reunala, T. (2001b) Wheat co 5 gliadin is a major allergen in children with immediate allergy to ingested wheat. Journal of Allergy and Clinical Immunology 108, 634-638.

Pastorello, E.A., Pravettoni, V., Farioli, L., Ispano, M., Fortunato, D., Monza, M., Giuffrida, M, G., Rivolta, F., Scibola, E., Ansaloni, R., Incorvala, C., Conti, A. (1999a) The major allergen of peach (*Prunus persica*) is a lipid transfer protein. Journal of Allergy and Clinical Immunology 103, 520-526.

Pastorello, E, A., Farioli, L., Pravettoni, V., Ispano, M., Scibola, E., Trambaioli, C., Giuffrida, M, G., Ansaloni, R., Godovac-Zimmermann, J., Conti, A., Fortunato, D. and Ortolani C. (2000b) The maize major allergen, which is responsible for food-induced allergic reactions, is a lipid transfer protein. Journal of Allergy and Clinical Immunology 106, 744-751.

Pastorello, E.A., Farioli, L., Robino, A.M., Trambaioli, C., Conti, A. and Pravettoni, V. (20001a) A lipid transfer protein involved in occupational sensitization to spelt. Journal of Allergy and Clinical Immunology 108, 145-146.

Pastorella, E, A., Vieths, S., Pravettoni, V., Farioli, L., Trambaioli, C., Fortunato, D., Luttkopf, D., Calamari M., Ansaloni, R., Scibilia, J., Ballmer-Weber, B. K., Poulsen, L. K., Wutrich, B., Hansen, K.S., Robino, A.M., Ortolani, C. and Conti, A. (2002) Identification of hazelnut major allergens in sensitive patients with positive double-blind, placebo-controlled food challenge results. Journal of Allergy and Clinical Immunology 109, 563-570.

Platts-Mills, T. A. E., G. P. Fiocco, M. L. Hayden, J. L. Guerrant, S. M. Pollart, and S. R. Wilkins. 1987. Serum IgE antibodies to Trichophyton in patients with urticaria, angioedema, asthma, and rhinitis: development of a radioallergosorbent test. J. Allergy Clin. Immunol. 79:40–45.

Poltronieri, P., Cappello,. M S., Dohmae, N., Conti, A., Fortunato, D., Pastorello, E.A., Ortolani, C and Zacheo, G. (2002) Identification and characterisation of the IgE-binding proteins 2S albumin and conglutin gamma in almond (*Prunus dulcis*) seeds. International Archives of Allergy and Immunology 128, 97-104.

Pope, A. M., R. Patterson, and H. Burge (ed.).(1993). Indoor allergens: assessing and controlling adverse health effects, p. 245. National Academy Press, Washington, D.C.

Portnoy, J., J. Chapman, H. Burge, M. Muilenberg, and W. Solomon. 1987. Epicoccum allergy: skin reaction patterns and spore/mycelium disparities recognized by IgG and IgE ELISA inhibition. Ann. Allergy 59:39–43.

Sanchez- Monge, R., Lombardero Garcia-, M., Selles, F.J., Barber, D. and Salcedo, G. (1999) Lipid-transfer proteins are relevant proteins in fruit allergy. Journal of Allergy and Clinical Immunology 103, 514-19.

Scheurer, S., Pastorello, E. A., Wangrsch, A., Kastner, M., Haustein, D. and Vieths, S. (2001) Recombinant allergens Pru av 1 and Pru av 4 and a newly identified lipid transfer protein in the *in vitro* diagnosis of cherry allergy. Journal of Allergy and Clinical Immunology 107, 724-731.

Shen H.D., Choo K.B. Chen J.H. Lin W.L. Chang Z.N. and Han S.H. (1992) Characterization of monoclonal antibody against the 68kD major allergen of *Penicillium notatum* Clin and exp. Allergy 22 ; 485-490.

Simonato, B., Pasini, G., Giannattasio, M., Peruffo, A. D., De Lazzari, F. and Curioni, A (2001a) Food allergy to wheat products: the effect of bread baking and in vitro digestion on wheat allergenic proteins. A study with bread dough, crumb, and crust. Journal of Agricultural and Food Chemistry 49, 5668-5673.

Sprenger, J. D., L. C. Altman, C. E. O'Neil, G. H. Ayars, B. T. Butcher, and S. B. Lehrer. (1988). Prevalence of basidiospore allergy in the Pacific Northwest. J. Allergy Clin. Immunol. 82:1076–1080.

Stricker, W. E., E. Anorve-Lopez, and C. E. Reed. (1986). Food skin testing in patients with idiopathic anaphylaxis. J. Allergy Clin. Immunol. 77:516–519.

Szantho, A., P. Osvath, Z. Horvath, E. K. Novak, and E. Kujalek.(1992).immediate type hypersensitivity to common moulds. Allergy 35:443-451

van Loon, L., C. and van Strein, E. A. (1999) The families of pathogenesis related proteins, their activities and comparative analysis of PR-1 type proteins. Physiological and Molecular Plant Pathology 55, 85-97.

Varjonen, E., Vainio, E. and Kalimo, K. (1997) Life-threatening, recurrent anaphylaxis caused by allergy to gliadin and exercise. Clinical and Experimental Allergy 27, 162-166.

Varjonen, E., Vainio, E. and Kalimo, K. (2000) Antigliadin IgE – indicator of wheat allergy in atopic dermatitis. Allergy 55, 386-391.

Varjonen, E., Vainio, E, Kalimo, K., Juntunen- Backman, K, and Savolainen, J. (1995) Skin prick test and RAST responses to cereals in children with stopic dermatitis.

Characterization of IgE binding components in wheat and oats by an immunoblotting method. Clinical and Experimental Allergy 25, 1100-1107

Vijay H.M. Young N.M. Jackson, GED White G.P. and Bernstein I.L. (1985) Studies on *Alternaria* allergens. Int. Arch. Allergy 55, 386-391.

Vioque, J., Sanchez-Vioque, R., Clemente, A., Pedroche, J., Bautista, J., and Millan, F. (1999) Purification and partial characterization of chickpea 2S albumin. Journal of Agricultural and Food Chemistry 47, 1405-1409.

von Pirquet, C. (1907). Die Allergieprobe. Zur Diagnose der Tuberculose im Kindesalter. Wien. Med. Wochenschr. 57:1369–1379.

Weeke, E. R. (1992). Epidemiology of allergic rhinitis, p. 167–172. In B. Wuethrich (ed.), Highlights in allergy and clinical immunology. Hogrefe & Huber Publishers, Bern, Switzerland.

Chapter - 15

Allelopathic effects of *Sesamum indicum* L. on *Cyperus rotundus* L.

N.B. Singh and Sunaina

Plant Physiology Laboratory, Department of Botany, University of Allahabad, Allahabad-211002.
E-mail: nbsingh2001@gmail.com

ABSTRACT

Cyperus rotundus is one of the most invasive and troublesome weeds. It affects the growth and development of crop plants by competition and allelopathic interactions. The effect of *Sesamum* leaf leachate of 25, 50, 75 and 100% concentrations on *C. rotundus* was studied. 21 days old *Cyperus* seedlings were used to study the biophysical and biochemical parameters. Photosynthetic pigment, protein and sugar content and activities of nitrate reductase and antioxidant enzymes were assayed. The leaf leachate of the donor plant significantly decreased photosynthetic pigments, protein, sugar content and nitrate reductase activity in as compared with control with maximum reduction in 100% concentration. Activities of antioxidant enzymes viz. superoxide dismutase, catalase and peroxidase increased in dose dependent manner. Plant height, root length, fresh and dry weight of the seedlings decreased under all concentrations. *Sesamum* exhibited inhibitory effects on growth and metabolism of *Cyperus*. Thus *Sesamum* has ability to control *Cyperus* by inhibiting the growth and development. The aim of the present study was to investigate the allelopathic effects of leaf leachate of *Sesamum* on growth and metabolism of *Cyperus* to aid control programmes of the obnoxious weed.

Keywords: Allelopathy, Antioxidant, *Cyperus rotundus, Sesamum indicum*

15.1 INTRODUCTION

Allelopathy is a biotic interactions mediated by the secondary bioactive metabolites released from plants and microorganisms into the environment. These bioactive metabolites produce favorable or unfavorable effects on growth and development of neighboring plants or microorganisms (Rice, 1984; Lodhi *et al.,* 1987). Allelopathy becomes more evident in crop rotation and in mixed or intercropping systems. In the environment allelochemicals are released as root exudates, leaf leachates, plant residues decomposition and microbial metabolic activities and cause allelopathic stress (Crutchfield *et al.,* 1985; Hyder *et al.,* 2002). Allelochemicals influence the growth and development of recipient plants in natural and agro ecosystems (Inderjit and Duck, 2003).

Weeds usually compete with crops for light, water and nutrients etc. and have allelopathic effect on the yield and quality of crops (Narwal *et al.,* 1992). *Cyperus rotundus* L. is one of the most invasive, aggressive and noxious weed. It is a perennial herb which is widely distributed throughout the world both in tropics and temperate regions (Holm *et al.*, 1979). The phytotoxic nature of *Cyperus* was studied against the growth and development of several fields and vegetable crops (Singh, 1968, Meissner *et al.,* 1979). Its existence in field significantly reduces crop yield because of tough competition for ground resources and release of allelopathic substances from the root, tubers and leaf which are harmful to neighbouring plants. It is difficult to control the weed because of its extensive system of rhizomes and tubers which are deep rooted and its resistance to most herbicides. Allelochemicals present in *Cyperus* trigger the production of reactive oxygen species (ROS) causing oxidative damage which leads to death of plant tissue. ROS and oxidative damage are the results of action of phytotoxins (Weir *et al.*, 2004). It is difficult to control *Cyperus* effectively through usual weed control methods because of its specific biological characteristics, i.e. having different means of propagation by rhizomes and tubers with varying degree of dormancy (Charles, 1997). Several crops including sorghum (Weston and Duke, 2003), rice (Seal *et al.,* 2004) and wheat (Huang *et al.,* 2003) have potential to inhibit certain weeds through allelopathic mechanisms. Iqbal and Cheema (2007) reported the effective control of *Cyperus* through the utilization of natural plant extracts from various allelopathic crops viz. sorghum, sunflower and brassica, etc. in the fields.

Sesamum indicum is an annual shrub with allelopathic activities which belongs to the family Pedaliaceae. It is grown mainly for the production of seeds which are rich in oil content. *Sesamum* is grown in tropical, subtropical and southern temperate regions of the world, particularly in India, China, South America and Africa. Sesamin, sesamol and sesamolin are the major phenolic constituents of *Sesamum* (Chakraborthy *et al.,* 2008). Seeds of *Sesamum* contain several other phenolic compounds *viz.* gallic, caffeic, cinnamic, benzoic, vanilic, ferulic, chlorogenic, p-coumaric, protocatechuic acids and catechol, catechin, caffeine, ellagic, chrysin and pyrogallol (Hassan, 2012).

Intercropping of field crops is an effective option for integrated weed management. Previously reported that intercropping of field crops cotton and intercropping of sorghum and sesame potentially control weeds (Iqbal and Cheema, 2007). Intercropping reduced herbicides input in agricultural land and also decrease the soil and ground water pollution. The present work was carried out to evaluate the allelopathic potential of *S. indicum* on growth and metabolism of *C. rotundus* to control the weed.

15.2 MATERIALS AND METHODS

A. Preparation of leachate

Plants of *S. indicum* L. were grown in field as donor crop in April, 2012. Leaves of 60 days old sesame plants were taken for leachate preparation. Leaves were cut into small pieces and soaked in double distilled water in the ratio 1:4w/v. After 3 days the leachate was filtered and centrifuged at 15000*g*. The supernatant was collected and raised to original volume and treated as 100% concentration and stored at 8°C temperature to avoid biodegradation. The mother leachate was diluted with distilled water to prepare leachate of 25, 50 and 75% concentrations.

The tubers of *C. rotundus* L. were collected from the Botanical Garden, Department of Botany, University of Allahabad, Allahabad (24°47′ and 50°47′ N latitude; 81°91′ and 82°21′ E longitude; 78 m above the sea level). The tubers of uniform size were screened. The tubers were washed under tap water to remove the soil. The roots of the tubers were removed with the help of scissors. The plastic pots (height 9 cm, width 17 cm, length 23 cm) were filled with garden soil mixed with sand (3:1). The tubers were sown in equal distance in plastic pots at the rate of 3 tubers per pot. The pots were irrigated with the leachates according to the treatments and kept in glass house with normal natural illumination. Pots irrigated with water were taken as control. The seedlings start coming out of soil within five days. The seedlings were irrigated with different concentrations of leachate at the interval of three days. The control seedlings were irrigated with water. The experiment was done in the replicates of three in the month of June, 2012. The twenty one days old seedlings were uprooted from the soil and sampled for biophysical parameters. First fully expanded leaves were sampled for biochemical analyses.

B. Measurement of pigment and protein content

Chlorophyll of experimental plant was extracted with 80% acetone. The amount of photosynthetic pigments was determined as per Lichtenthaler (1987). Ten mg fresh leaves were homogenized in 10 mL of 80% acetone and centrifuged. Supernatant was taken and optical density was measured at 663, 645 and 470 nm. Protein content was determined following Lowry *et al.* (1951). The amount of protein was calculated with reference to standard curve obtained from bovine serum albumin.

C. Measurement of sugar content

Sugar content was estimated following Hedge and Hofreiter (1962). About 0.25 g of the sample was homogenized in 2.5 mL of 95% ethanol. After centrifugation, the sugar content was determined in the supernatant. Supernatant (1mL) was mixed with 4 mL of anthrone reagent and heated on boiling water bath for 8 min. Absorbance was taken at 620 nm after rapid cooling. Standard curve was prepared from glucose.

D. Nitrate reductase activity

Nitrate reductase (EC 1.6.6.1) activity was assayed by modified procedure of Jaworski (1971) based on incubation of fresh tissue (0.25g) in 4.5 mL medium containing 100 mM sodium phosphate buffer (pH 7.5), 3% KNO_3 and 5% propanol. About 0.4 mL aliquot was treated with 0.3 mL 3% sulphanilamide in 3 N HCL and 0.3 mL 0.02% N-(1-Naphthyl) ethylene diamine dihydrochloride (NEDD). The absorbance was measured at 540 nm. NR activity was calculated with a standard curve prepared from $NaNO_2$ and expressed as μ mol NO_2 g^{-1} FW h^{-1}.

E. Extraction and assay of antioxidant enzymes

Enzyme extract was prepared by homogenizing 500 mg leaves in 10 mL of 0.1 M sodium phosphate buffer (pH 7.0). The homogenate was filtered and centrifuged at 15000*g* at 4° C for 30 min. The supernatant was collected and used for analyses of SOD (EC 1.15.11), CAT (EC 1.11.1.6) and POX (EC 1.11.1.7).

SOD activity was determined by the nitroblue tetrazolium (NBT) photochemical assay method following Beauchamp and Fridovich (1971). The reaction mixture (4mL) contained 63 μM NBT, 13 mM methionine, 0.1 mM ethylene diamine tetra acetic acid (EDTA), 13 μM riboflavin, 0.5 M sodium carbonate and 0.5 mL clear supernatant. Test tubes were placed under fluorescent lamps for 30 min and absorbance was recorded at 560 nm. One unit of enzyme was defined as the amount of enzyme which caused 50% inhibition of NBT reduction.

Catalase activity was assayed as per the method Cakmak and Marschner (1992). The reaction mixture (2mL) contained 25 mM sodium phosphate buffer (pH 7.0), 10 mM H_2O_2 and 0.2 mL enzyme extract. The activity was determined by measuring the rate of disappearance of H_2O_2 for 1min at 240 nm and calculated using extinction coefficient of 39.4 mM^{-1} cm^{-1} and expressed as enzyme unit g^{-1} fresh weight. One unit of CAT was defined as the amount of enzyme required to oxidize 1 μM H_2O_2 min^{-1}.

Peroxidase activities were assayed following Mc Cune and Galston (1959). Reaction mixture contained 2 mL enzyme extract, 2 mL sodium phosphate buffer, 1 mL 0.1 N pyrogallol and 0.2 mL 0.02% H_2O_2 and determined spectrophotometrically at 430nm. One unit of enzyme activity was defined as the amount which produced an increase of 0.1 OD per minute.

F. Statistical analysis

Standard errors of means were calculated in triplicates. In addition, analysis of variance was carried out for all the data generated from this experiment, employing one way ANOVA test using GPIS software 3.0 (GRAPHPAD California USA).

15.3 RESULTS

Allelopathic effects of *S. indicum* on biophysical and biochemical parameters were studied in control and stressed *Cyperus* plants. Remarkable changes on growth and biochemical parameters were observed in *Cyperus* plants under all treatments as compared with control. *Cyperus* growth significantly decreased under the influence of leaf leachate of *Sesamum*. The inhibition was intensified with increased concentrations of the leaf leachate. Our results exhibited the phytotoxic nature of *Sesamum* against *Cyperus*. The seedlings growth and fresh and dry weight of seedlings decreased with the increase in concentration of the leaf leachate. All the studied parameters of growth of seedlings viz. root and shoot length, fresh and dry weight, photosynthetic pigment, protein, sugar content and nitrate reductase activity successively decreased by the leachate of higher concentrations.

Root and shoot length of the seedlings were adversely affected by the leachate with maximum 43.51 and 40% inhibition in root and shoot length respectively under highest concentration as compared with control. Successive decrease in root and shoot length resulted in decline in fresh and dry weight of seedlings with maximum 80.34 and 34.51% inhibition respectively **(Table 1)**.

Table 1: Effects of *Sesamum* leaf leachate on root, shoot length, fresh and dry weight of *Cyperus*

	Seedling height			
Treatments	**Root length (cm)**	**Shoot length (cm)**	**Fresh weight (g/plant)**	**Dry weight (g/plant)**
C	16.2±0.981	28.5±0.461	5.29±0.037	0.565±0.001
T_1	12.20±0.230[a]	25.23±0.063[a]	4.31±0.014[a]	0.527±0.006[a]
T_2	10.9±0.230[a]	21.7±0.115[a]	2.88±0.066[a]	0.493±0.008[a]
T_3	9.8±0.057[a]	19.85±0.028[a]	1.41±0.096[a]	0.402±0.004[a]
T_4	9.15±0.028[a]	17.1±0.346[a]	1.04±0.013[a]	0.370±0.010[a]

Data are meanof three replicates ± SEM. [a] $p<0.001$versus C. C= control, T_1= 25%, T_2= 50%, T_3= 75% and T_4= 100% concentrations of *Sesamum* leaf leachate.

The pigment content decreased in dose dependent manner. Allelopathic stress drastically decreased the chlorophyll *a*, chlorophyll *b*, total chlorophyll and carotenoids with maximum inhibition of 46.15, 35.94, 41.88 and 45.25%, respectively. However, maximum amount of pigment content was recorded in the seedlings under control **(Table 2)**.

Table 2. Effects of *Sesamum* leaf leachate on the pigment content of *Cyperus*.

Treatments	Chlorophyll a (mg/g FW)	Chlorophyll b (mg/g FW)	Total chlorophyll (mg/g FW)	Carotenoids (mg/g FW)
C	0.377±0.003	0.281±0.005	0.659±0.002	0.137±0.012
T_1	0.356±0.016	0.258±0.005	0.614±0.011[c]	0.108±0.001[b]
T_2	0.301±0.020[a]	0.255±0.011[c]	0.556±0.009[a]	0.105±0.001[b]
T_3	0.284±0.028[a]	0.203±0.003[a]	0.488±0.024[a]	0.086±0.006[a]
T_4	0.203±0.007[a]	0.180±0.011[a]	0.383±0.003[a]	0.075±0.001[a]

Data are mean of three replicates ± SEM. [a] $p<0.001$, [b]$p<0.01$, [c]$p<0.05$ versus C.
C= control, T_1= 25%, T_2= 50%, T_3= 75% and T_4= 100% concentrations of *Sesamum* leaf leachate.

The amount of protein content gradually decreased in all treatments in dose dependent manner with maximum 50.27% inhibition in the highest concentration of the leachate. Sugar content gradually reduced under allelopathic stress. Sugar content significantly decreased with the increased concentration of the leachate as compared with control. Maximum 46.78% inhibition in sugar content was recorded under the highest concentration. Maximum NR activity was recorded in control.NR activity significantly decreased under the influence of the leachate. Maximum 2.07 folds inhibition in NR activity was recorded in the seedlings treated with the highest concentration of the leachate (**Table 3**).

Table 3. Effects of *Sesamum* leaf leachate on sugar and protein content and nitrate reductase activity of *Cyperus*.

Treatment	Protein (mg/g FW)	Sugar (mg/g FW)	NR (µmol NO_2 g^{-1} FW h^{-1})
C	19.79±0.360	14.92±0.056	10.28±0.826
T_1	17.10±0.375[b]	11.93±0.167[a]	9.47±0.005
T_2	15.20±0.311[a]	10.05±0.093[a]	6.13±0 003[a]
T_3	14.67±0.269[a]	8.34±0.018[a]	6.04±0.005[a]
T_4	9.84±0.291[a]	7.94±0.032[a]	5.43±0.003[a]

Data are mean of three replicates ± SEM. [a] $p<0.001$, [b] $p<0.01$ versus C. C= control, T_1= 25%, T_2= 50%, T_3= 75% and T_4= 100% concentrations of *Sesamum* leaf leachate.

Antioxidant enzyme activities were also analyzed. Antioxidant enzymes viz. SOD, CAT and POX activities were enhanced by the leachate in dose dependent manner. *Cyperus* seedlings treated with 100% concentration exhibited the highest activities of antioxidant enzymes while lowest activity was observed under control group. The activities of SOD, CAT and POX increased 1.58, 1.43 and 2.48 folds, respectively in100% concentration **(Table 4)**.

Table 4: Effects of *Sesamum* leaf leachate on antioxidant enzyme activities of *Cyperus.*

Treatment	SOD (EU g^{-1} FW)	CAT (EU g^{-1} FW)	POX (EU g^{-1} FW)
C	64.81±0.658	43.38±0.630	19.88±0.457
T_1	69.73±0.300 [a]	43.77±0.840	22.67±0.127 [a]
T_2	75.82±0.255[a]	47.88±0.006[a]	26.14±0.067[a]
T_3	91.61±0.320[a]	55.16±0.139[a]	39.19±0.053[a]
T_4	102.47±0.685[a]	62.11±0.670[a]	49.31±0.793[a]

Data are mean of three replicates ± SEM. [a] $p<0.001$versus C. C= control, T_1= 25%, T_2= 50%, T_3= 75% and T_4= 100% concentrations of *Sesamum* leaf leachate.

15.4 DISCUSSION

Our results clearly reveal the allelopathic nature of *S. indicum* which decreased the growth of *C. rotundus*. The aqueous leaf leachate of *Sesamum* variously affected the plant growth of the test plant. Allelochemical stress caused by aqueous leachate inhibited the root growth in crop plants (Romero-Romero *et al.*, 2005). The aqueous leachate of other phytotoxic plants also caused adverse effects on growth and protein expression of crop plants (Romero-Romero *et al.*, 2002). Seedling growth significantly decreased with increased concentrations of the leachate. The effect of allelochemical was concentration dependent (Singh and Singh, 2003). The decrease in growth was the manifestation of impaired metabolic activities due to several allelochemical present in the leaf leachate of the donor plant. The growth reduction of the seedlings under the influence of allelochemicals was well demonstrated in various plants (Batish *et al.*, 2006; Singh *et al.*, 2008; 2009). The decrease in FW and DW of seedlings may be due to decreased level of photosynthetic pigments.

In this experiment we observed that reduction in the photosynthetic pigments under allelopathins present in the leachate of the donor plant. Similar results were also reported by Venkateshwarlu *et al.* (2001) and Bagavathy and Xavier (2007). Allelochemicals reduce the biosynthesis of chlorophyll (Hejl *et al.*, 1993; Stiles *et al.,* 1994). The protein degradation or inhibition of gross protein synthesis under the influence of allelochemicals was reported by Thaper and Singh (2006). The decreased in protein content might be manifestation of cell damaged caused by allelochemical stress (Curz-Ortega, 2002). The impairment of metabolic activities leads to the inhibition of biosynthesis of protein. The decrease in sugar content is proportional to the level of photosynthetic pigments under the influence of allelochemicals present in the leaf leachate. NR activity decreased in allelopathic stressed seedlings. NR activity was found to be dependent on energy, e⁻ donor and carbon skeleton which were provided by the process of photosynthesis (Kaiser *et al.,* 1993). Nitrate reductase enzyme is induced by substrate. Inhibition of absorption of nitrate due to allelochemicals may be another possible reason for reduced activities of NR (Chen and Sung, 1983). In previous studies, inhibited NR activity was also reported in sorghum plants treated with eucalyptus leaf extract (Bagavathy and Xavier, 2007) and in *Vigna radiata* treated with *Dalbergea sissoo* (Tripathi *et al.*, 2000).

Under allelopathic stress several antioxidative enzymes of the defense system increased tremendously in order to avoid the oxidative damage caused by ROS (Foyer and Noctor, 2003). During stress condition these antioxidative enzymes act as stress markers. Allelochemicals stimulated the activities of SOD and CAT (Doblinski *et al.*, 2003). Present data show that the activities of antioxidative enzymes increased with increasing concentration of the leaf leachate. SOD, CAT and POX activities increased in response of free radicals production induced by the allelochemicals. Increased activity of CAT was also observed in tomato roots under the influence of leachate (Curz-Ortega, 2002). Leachate was inhibitor in its higher concentration as reported by Singh *et al.* (2008). SOD plays the leading role in detoxification of ROS (Gomez *et al.,* 2004).

Decreased seedling growth, fresh and dry weight, photosynthetic pigment, protein, sugar content and NR activity of *Cyperus* clearly indicated the allelopathic effects of *S. indicum.* It appears that activities of antioxidant enzymes cannot mitigate the allelopathic effect of *Sesamum.* This study showed that continuous cropping of *Sesamum* may be an effective system for integrated management of *Cyperus. Sesamum* has allelopathic potential to control the obnoxious weed *Cyperus.*

Acknowledgement

The authors are thankful to University Grants Commission, New Delhi, India for the award of UGC-RGNF to Sunaina.

REFERENCES

Bagavathy S and Xavier GSA (2007) Effects of aqueous extract of *Eucalyptus globules* on germination and seedling growth of sorghum. *Allelopathy J.* **20(2)** 395-402.

Batish DR, Singh HP, Rana N and Kohli RK (2006) Assessment of allelopathic interference of *Chenopodium album* through its leachates, debris extracts, rhizosphere and amended soil. *Archives of Agronomy and Soil Science.* **52** 705-715.

Beauchamp C and Fridovich I (1971) Superoxide dismutase: Improved assays and an assay applicable to acrylamide gels. *Anal. Biochem.* **44** 276-287.

Cakmak I and Marschner H (1992) Magnesium deficiency and highlight intensity enhance activities of superoxide dismutase ascorbate peroxidase, and glutathione reductase in bean leaves. *Plant Physiol.* **98** 1222–1227.

Chakraborty GS, Sharma G and Kaushik KN (2008) *Sesamum indicum*: A Review. *Herbal medicine and toxicology* J. **2(2)** 15-19.

Charles G (1997) Controlling nutsedge in cotton. CRS-Newsletter for the Research, Extension, Education Program 3.

Chen CL and Sung JM (1983) Effect of water stress on the reduction of nitrate and nitrite by soybean nodules. *Plant Physiol.* **73** 1065-1066.

Cruz-Ortega R, Ayala-Cordero G and Anaya AL (2002) Allelochemical stress produced by the aqueous leachates of *Calicarpa acuminate*: effects on roots of bean, maize and tomato. *Physiol Plant.* **116** 20–27

Doblinski PMF, Ferrarese MLL, Huber DA, Scapim CA, Braccini AL and Ferrarese-Filho O (2003) Peroxidase and lipid peroxidation of soybean roots in response to p-Coumaric and p-Hydroxybenzoic acids. *Braz Arch Biol Tech.* **46(2)** 193-198.

Foyer CH and Noctor G (2005) Oxidant and antioxidant signalling in plants: Is evaluation of the concept of oxidative stress in a physiological context. *Plant Cell Environ.* **28** 1056–1071.

Gomez JM, Jimenez A, Olmos E and Seveilla F (2004) Location and effects of long term NaCl stress on superoxide dismutase and ascorbate peroxidase isoenzymes of pea (*Pisum sativum* cv. Puget) chloroplast. *J Exp Bot.* **55** 119-130.

Heath RL and Packer L (1968) Photoperoxidation in isolated chloroplasts. I. Kinetics and stoichiometry of fatty acid peroxidation. *Arch Biochem Biophys.* **125** 189–198.

Hedge JE and Hofreiter BT (1962) Estimation of carbohydrate. In:Whistler, R.L. and Be Miller, J.N. (ed.), Methods in carbohydrate chemistry. *Academic Press, New York*, pp. 17-22.

Holm L, Pancho JV, Herberger JP and Plucknutt DL (1979) A geographical atlas of world weeds. *Wiley, New York*

Huang Z, Haig T, WuH, An M, Pratley J (2003) Correlation between phytotoxicity on annual ryegrass (*Lolium rigidum*) and production dynamics of allelochemicals within root exudates of an allelopathic wheat. J Chem Ecol. 29:2263–2279

Inderjit and Duke SO (2003) Ecophysiological aspects of allelopathy. *Planta* **217** 529-39.

Iqbal J, Cheema ZA (2007) Effect of allelopathic crops water extracts on glyphosate dose for weed control in cotton (*Gossypium hirsutum*). Allelopathy J. 19:403–410.

Jaworski E (1971) Nitrate reductase assay in intact plant tissue. *Biochem Biophys Res Commun.* **430** 1274–1279.

Kaiser WM, Spill D and Glaab J (1993) Rapid modulation of Nitrate reductase in leaf and roots: in direct evidence for the involvement of protein phosphorylation dephosphorylation. *Physiol Plant.* **89** 557–562

Lichtenthaler HK (1987) Chlorophyll and carotenoids: pigments of photosynthetic bio-membranes. In: Packer L, Douce R, editors. Methods Enzymology. *Academic Press Sandiego* pp. 350-382.

Lowry OH, Rosenbrough RJ, Farr AL and Randall RJ (1951) Protein measurement with Folin phenol reagent. *J Biol Chem.* **193** 265–275.

Manal AM, Hassan (2012) Studies on Egyptian sesame seeds (*Sesamum indicum* L.) and its products 1-physicochemical analysis and phenolic acids of roasted Egyptian sesame seeds (*Sesamum indicum* L.). *World J of Dairy and Food Science.* **7(2)** 195-201.

Romero-Romero T, Anaya AL and Cruz-Ortega R (2002) Screening for effects of phytochemical variability on cytoplasmic protein synthesis pattern of crop plants. *J Chem Ecol.* **28** 617-629.

Romero-Romero T, Sa´nchez-Nieto S, Sanjuan-Badillo A, Anaya AL and Cruz-Ortega R (2005) Comparative effects of allelochemical and water stress in roots of *Lycopersicon esculentum* Mill Plant (Solanaceae). *Plant Sci.* **168**, 1059-1066.

Seal AN, Pratley JE, Haig T, An M (2004) Identification and quantitation of compounds in a series of allelopathic and nonallelopathic rice root exudates. J Chem Ecol. 30:1629–1644.

Singh A, Singh D and Singh NB (2009) Allelochemical stress produced by aqueous leachate of *Nicotiana plumbaginifolia* Viv. *Plant Growth Regulation.* **58** 163-171.

Singh NB and Singh R (2003) Effects of leaf leachate of *Eucalyptus* on germination, growth and metabolism of green gram, black gram and peanut. *Allelopathy J.* **11** 43-52.

Singh NB, Singh A and Singh D (2008) Autotoxic effects of *Lycopersicon esculentum. Allelopathy J.* **22(2)** 429–442.

Thaper R and Singh NB (2006) Phytotoxic effect of Cassia tora on growth and metabolism of *Parthenium hysterophorus* L. *Allelopathy J.* **17(2)** 235–246.

Weir TL, Park S and Vivanco JM (2004) Biochemical and physiological mechanisms mediated by allelochemicals. *Curr Opin Plant Biol.* **7** 472-9.

Weston LA, Duke SO (2003) Weed and crop allelopathy. *Crit Rev Plant Sci.* 22:367.

■■■■

Chapter - 16

Clean up Strategy for the River Vishwamitri

V.S. Patel

Consulting Engineer and Visiting Professor,
The Maharaja Sayjirao University of Baroda, Vadodara 390002
E mail: vpatel_usa@hotmail.com

ABSTRACT

Some rivers are evolutionarily ephemeral or seasonal, while some rivers do not meet the sea, but disappear in the desert, like the Khari, Banas and Saraswati. These cannot be compared with rivers meeting the sea like Narmada, Tapi, Mahi, Sabarmati. The Vishwamitri river, a seasonal river originates in the hills of Pavagadh, flows west through city of Vadodara, joins the Dhadhar and Kanpur rivers (two small tributaries near the village of Kothawada), and empties into the Gulf of khambhat, near Khanpur village.

Historically, this system as a whole provided ecosystem (environmental) services such as natural flood water control, ground water recharge, habitats for different flora and fauna, promoting biodiversity, and ameliorating adverse climatic conditions. They also served various communities as alternative sources for irrigation, water supply, food, recreation, and livelihood. Vishwamitri River and its associated water bodies are also unique because they are home to threatened or endangered species of flora and fauna (Ravan taad, crocodile, etc.). Having a crocodile in the river habitat in the middle of Vadodara city is also extremely unique situation in itself. Instead of taking pride in it and protecting its habitat, future plans seems to be bent on destroying it.

The survey was conducted along the short section of Vishwamitri river that flows through the city of Vadodara. This 25 km river stretch was divided into five parts: (1) Sama to Fatehgunj; (2) Fatehgunj to Kalaghoda (3) Kala ghoda to Munj-mahuda; (4) Munj-mahuda to Kalali ; and (5) Kalali to Vadsar. Various reports and studies are carried out in the past 20 years for clean up river. We need workable innovative solution. The river is natural drainage for Vadodara city flows at the centre of the city from north east to south. Geographically river is shorted distance to carry the rain water from the city. Since last 30 years the important river is totally ignored and misused. It becomes dumping place. The Vishwamitri river become beyond repair. Various measures are suggested.

Keywords: Clean up, River Vishwamitri, Strategy, Vishwamitri Riverfront Development Project

16.1 INTRODUCTION

The rivers of India play an important role in the lives of the Indians. They provide potable water, cheap transportation, electricity, and the livelihood for a large number of people all over the country. This easily explains why nearly all the major cities of India are located by the banks of rivers. The rivers also have an important role in Hindu Dharma and are considered holy by all Hindus in the country.

Seven major rivers along with their numerous tributaries make up the river system of India. The largest basin system of the rivers pour their waters into the Bay of Bengal; however, some of the rivers whose courses take them through the western part of the country and towards the east of the state of Himachal Pradesh empty into the Arabian Sea. Parts of Ladakh, northern parts of the Aravalli range and the arid parts of the Thar Desert have inland drainage. All major rivers of India originate from one of the three main watersheds:

1. The Himalaya and the Karakoram ranges
2. Vindhya and Satpura ranges and Chotanagpur plateau in central India
3. Sahyadri or Western Ghats in western India

Himalayan glaciers in the Indian subcontinent are broadly divided into the three river basins, namely the Indus, Ganga and Brahmaputra. The Indus basin has the largest number of glaciers (3500), whereas the Ganga and Brahmaputra basins contain about 1000 and 660 glaciers, respectively. In agriculture development: several million acres of tropical forests have been destroyed all over the world, leaving many countries without sufficient amount of forest to support the nations requires annual rainfall.

16.2 RIVER ECOSYSTEM

The ecosystem of a river is the river viewed as a system operating in its natural environment, and includes biotic interactions amongst vegetation, animals and micro-organisms, as well nonliving physical and chemical interactions. River ecosystems are prime examples of lotic ecosystems. *Lotic* refers to flowing water, from the Latin *lotus*, washed. Lotic waters range from springs only a few centimeters wide to major rivers kilometers in width Much of this article applies to lotic ecosystems in general, including related lotic systems such as streams. The following unifying characteristics make the ecology of running waters unique from that of other aquatic habitats

- Flow is unidirectional.
- There is a state of continuous physical change.

- There is a high degree of spatial and temporal heterogeneity at all scales.
- Variability between lotic systems is quite high.
- The biota is specialized to live with flow conditions.

Water flow is the key factor in lotic systems influencing their ecology. The strength of water flow can vary between systems, ranging from torrential rapids to slow backwaters that almost seem like lentic systems. The speed of the water flow can also vary within a system and is subject to chaotic turbulence. This turbulence results in divergences of flow from the mean downslope flow vector as typified by eddy currents. The mean flow rate vector is based on variability of friction with the bottom or sides of the channel, sinuosity, obstructions, and the incline gradient. In addition, the amount of water input into the system from direct precipitation, snowmelt, and/or groundwater can affect flow rate. Flowing waters can alter the shape of the streambed through erosion and deposition, creating a variety of habitats, including riffles, glides, and pools.

Light is important to lotic systems, because it provides the energy for photosynthesis, and can also provide refuge for prey species in shadows it casts. The amount of light that a system receives can be related to a combination of internal and external stream variables. The area surrounding a small stream, for example, might be shaded by surrounding forests or by valley walls. Larger river systems tend to be wide so the influence of external variables is minimized, and the sun reaches the surface. These rivers also tend to be more turbulent, however, and particles in the water increasingly attenuate light as depth increases.[5] Seasonal and diurnal factors might also play a role in light availability because the angle of incidence, the angle at which light strikes water can lead to light lost from reflection. Known as Beer's Law, the shallower the angle, the more light is reflected and the amount of solar radiation received declines logarithmically with depth.[4] Additional influences on light availability include cloud cover, altitude, and geographic position (Brown 1987).

Most lotic species are **poikilotherms** whose internal temperature varies with their environment, thus temperature is a key abiotic factor for them. Water can be heated or cooled through radiation at the surface and conduction to or from the air and surrounding substrate. Shallow streams are typically well mixed and maintain a relatively uniform temperature within an area. In deeper, slower moving water systems, however, a strong difference between the bottom and surface temperatures may develop. Spring fed systems have little variation as springs are typically from groundwater sources, which are often very close to ambient temperature. Many systems show strong diurnal fluctuations and seasonal variations are most extreme in arctic, desert and temperate systems. The amount of shading, climate and elevation can also influence the temperature of lotic systems.

An entity which performs all the evolutionary (geomorphologic), ecological, social, cultural, economic functions of a river, giving life not only to a selected class of society, but the living world at large can be called as a **Living River**. The comparison can be only with the evolutionary character of the same river. For example, some rivers are evolutionarily ephemeral or seasonal, while some rivers do not meet the sea, but disappear in the desert. Australian definition (www.water.vic.gov.au/environment/rivers).

River has flow regimes, water quality and channel characteristics such that: natural ecosystem processes are maintained, major natural habitat features are represented and are maintained over time, in the river and along its banks, the majority of plant and animal species are native and the presence of exotic species is not a significant threat, native river bank vegetation is sustainable along the majority of the river's length, native fish and other animals can move and migrate up and down the river, linkages between river and floodplain and associated wetlands maintain ecological processes, natural linkages with the sea or terminal lakes are maintained, associated estuaries and terminal lake systems are productive ecosystems.

16.3 WATER CHEMISTRY

Water chemistry between systems varies tremendously. The chemistry is foremost determined by inputs from the geology of its water bodies, watershed, or catchment area, but can also be influenced by precipitation and the addition of pollutants from human sources. Large differences in chemistry do not usually exist within small lotic systems due to a high rate of mixing. In larger river systems, however, the concentrations of most nutrients, dissolved salts, and pH decrease as distance increases from the river's source.

Oxygen is the most important chemical constituent of a river or lotic systems, as all aerobic organisms require it for their survival. It enters the water mostly via diffusion at the water-air interface. Oxygen's solubility in water decreases as water pH and temperature increases. Fast, turbulent streams expose more of the water's surface area to the air and tend to have low temperatures and thus more oxygen than slow, backwaters. Oxygen is a byproduct of photosynthesis, so systems with a high abundance of aquatic algae and plants may also have high concentrations of oxygen during the day. These levels can decrease significantly during the night when primary producers switch to respiration. Oxygen can be limiting if circulation between the surface and deeper layers is poor, if the activity of lotic animals is very high, or if there is a large amount of organic decay occurring.

The inorganic substrate of lotic systems is composed of the geologic material present in the catchment that is eroded, transported, sorted, and deposited by the current. Inorganic substrates are classified by size on the Wentworth scale, which ranges from boulders, to pebbles, to gravel, to sand, and to silt. Typically, particle size decreases downstream with larger boulders and stones in more mountainous areas and sandy bottoms in lowland rivers. This is because the higher gradients of mountain streams facilitate a faster flow, moving smaller substrate materials further downstream for deposition-Substrate can also be organic and may include fine particles, autumn shed leaves, submerged wood, moss, and more evolved plants. Wastes from towns and villages can pollute river water. If industries discharge their untreated waste many harmful can be added to river water. Substrate deposition is not necessarily a permanent event, as it can be subject to large modifications during flooding events

16.4 BACTERIA AND LIVING PLANTS

Bacteria are present in large numbers in lotic waters. Free-living forms are associated with decomposing organic material. Algae, consisting of phytoplankton, are the most significant sources of primary production in most streams and rivers. Phytoplankton float freely in the water column and thus are unable to maintain populations in fast flowing streams. They can, however, develop sizable populations in slow moving rivers and backwaters. Periphyton are typically filamentous and tufted algae that can attach themselves to objects to avoid being washed away by fast currents. In places where flow rates are negligible or absent, periphyton may form a gelatinous, unanchored floating mat.

Plants exhibit limited adaptations to fast flow and are most successful in reduced currents. More primitive plants, such as mosses and liverworts attach themselves to solid objects. This typically occurs in colder headwaters where the mostly rocky substrate offers attachment sites. Some plants are free floating at the water's surface in dense mats like duckweed or water hyacinth. Others are rooted and may be classified as submerged or emergent. Rooted plants usually occur in areas of slackened current where fine-grained soils are found (Brown 1987). These rooted plants are flexible, with elongated leaves that offer minimal resistance to current.

Living in flowing water can be beneficial to plants and algae because the current is usually well aerated and it provides a continuous supply of nutrients. These organisms are limited by flow, light, water chemistry, substrate, and grazing pressure. Algae and plants are important to lotic systems as sources of energy, for forming microhabitats that shelter other fauna from predators and the current, and as a food resource (Brown 1987).

16.5 VANISHING BRANCHING PALMS : A RESULT OF HABITAT DESTRUCTION

Habitat destruction, which is also known as habitat loss by scientists is a natural or non-natural phenomenon where animals or plants lose their home- natural habitat- or the functionality of habitat comes to an end in such a way that it can no longer support the organisms depended on it. Scientists consider this to be a very serious issue which needs immediate attention: according to most scientists from different countries in the world, habitat destruction is the primary threat of 21st century, which can permanently harm the biodiversity of our planet. *Hyphaene dichotoma* (White) Furt. (syn *H. indica* Becc., *H. taprobanica*

Furtado) is a species of palm tree in Arecaeae family called as Doum Palm. It was reported growing in Bombay and its vicinity in 1839. It is also mentioned as *Borassus dichotomous* and is reported from India and Shri Lanka. It is occurring in banks of river Vishwamitri and also in Diu sea coast. Not many new plants can be located now as most of the seeds produced degenerate before germination. Habitat loss in the city area is also a main reason.

Earth may not be the largest planet in the universe but it surely is one of the most diverse planets and humans are lucky to be the inhabitants of this planet. The most significant result of habitat destruction is the elimination of many plants and animals, which have lower adaptation skills to the changing circumstances. However, some animals and plants may survive habitat destruction by changing their basic genetic structures, which is known as "the survival of the fittest" explained by Charles Darwin. Human interference and habitat destruction started in the Western world during the early 18th century when urban areas were heavily altered to meet the requirements of ever increasing urban population. The immediate result was the sudden vanishing of greenery, which was not a major concern earlier since the population was significantly low and the greenery which was still prevalent on the suburbs controlled the temperature and supplied sufficient fresh air to urban developments.

By 20th century, most countries in the Western World where economically developed and the high illiteracy rates made people aware about the importance of protecting the environment. But the fate of earth was not changing as it should be because many developing countries started exploiting the nature to the core. The entire African Continent and countries like Brazil, China and India elements several Asian countries explored the forbidden parts of the nature, eventually leading to mass habitat destruction. The issue fueled up by the ever rising population in the Developing World.

There are several species of palm that display braching habits--we all know the primitive customs of *Hyphaene*; the genus is unique in that there is mainly just one species that *doesn't* bifurcate (petersiana). Many other genus have just one or two represented species that display such characteristics (Vonitras: crinita, fibrosa, utilis) regularly. *Ptychosperma elegans* quite often grows with as many as four heads from one single stem! *Copernicia, Dypsis, Hedyscepe, Hyophorbe, Phoenix, Roystonea, Sabal,* and *Syagrus* have also been known to display such unique habits. Many people have told me this is due to an injury to the growing point/meristem. Other times the explanation is just a genetic anomaly. Are there any other reasons for this to occur?

Lastly, has anyone ever tried to produce bifurcating of a palm, not known to do so, naturally? *Ptychosperma elegans* is cheap, and I have often contemplated taking one to saw halfway down into the crown in an attempt to coax a split. Maybe if I had enough test subjects, could it be possible to exact my efforts?

Due to the high importance of the *Hyphaene compressa* poaching of its leaves in protected National Forests has become an issue. There has been evidence since the 1990s of overexploitation of the trees resources. Improper sap collection practices kill Palm trees and lack of knowledge on how to perform extraction is limiting the Palms longevity. *Hyphaene compressa* is in the process of being domesticated in some regions in Africa. Conservation laws to protect the tree from being overexploited have also been locally established as to ensure there is a lasting abundance of the East African Doum Palm (Amvatta, 2004).

16.6 INSECTS AND OTHER ANIMALS

Up to 90% of invertebrates in some lotic systems are insects. These species exhibit tremendous diversity and can be found occupying almost every available habitat, including the surfaces of stones, deep below the substratum, adrift in the current, and in the surface film. Insects have developed several strategies for living in the diverse flows of lotic systems. Some avoid high current areas, inhabiting the substratum or the sheltered side of rocks. Additional invertebrate taxa common to flowing waters include mollusks such as snails, limpets, clams, mussels, as well as crustaceans like crayfish and crabs. Like most of the primary consumers, lotic invertebrates often rely heavily on the current to bring them food and oxygen (Brown 1987). Invertebrates, especially insects, are important as both consumers and prey items in lotic systems.

Fish are probably the best-known inhabitants of lotic systems. The ability of a fish species to live in flowing waters depends upon the speed at which it can swim and the duration that its speed can be maintained. This ability can vary greatly between species and is tied to the habitat in which it can survive. Continuous swimming expends a tremendous amount of energy and, therefore, fishes spend only short periods in full current. Instead, individuals remain close to the bottom or the banks, behind obstacles, and sheltered from the current, swimming in the current only to feed or change locations.[1] Some species have adapted to living only on the system bottom, never venturing into the open water flow. These fishes are dorso-ventrally flattened to reduce flow resistance and often have eyes on top of their heads to observe what is happening above them. Some also have sensory barrels positioned under the head to assist in the testing of substratum (Brown, 1987).

Other vertebrate taxa that inhabit lotic systems include amphibians, such assalamanders, reptiles (e.g. snakes, turtles, crocodiles and alligators) various bird species, and mammals (e.g., otters, beavers, hippos, and river dolphins). With the exception of a few species, these vertebrates are not tied to water as fishes are, and spend part of their time in terrestrial habitats. Many fish species are important as consumers and as prey species to the larger vertebrates mentioned above.

Energy sources can be autochthonous or allochthonous.

Autochthonous energy sources are those derived from within the lotic system. During photosynthesis, for example, primary producers form organic carbon compounds out of carbon dioxide and inorganic matter. The energy they produce is important for the community because it may be transferred to higher trophic levels via consumption. Additionally, high rates of primary production can introduce dissolved organic matter (DOM) to the waters. Another form of autochthonous energy comes from the decomposition of dead organisms and feces that originate within the lotic system. In this case, bacteria decompose the detritus or coarse particulate organic material (CPOM; >1 mm pieces) into fine particulate organic matter (FPOM; <1 mm pieces) and then further into inorganic compounds that are required for photosynthesis. This process is discussed in more detail below.

Allochthonous energy sources are those derived from outside the lotic system, that is, from the terrestrial environment. Leaves, twigs, fruits, etc. are typical forms of terrestrial CPOM that have entered the water by direct litter fall or lateral leaf blow. In addition, terrestrial animal-derived materials, such as feces or carcasses that have been added to the system are examples of allochthonous CPOM. The CPOM undergoes a specific process of degradation. Allan gives the example of a leaf fallen into a stream. First, the soluble chemicals are dissolved and leached from the leaf upon its saturation with water. This adds to the DOM load in the system. Next, microbes such as bacteria and fungi colonize the leaf, softening it as the mycelium of the fungus grows into it. The composition of the microbial community is influenced by the species of tree from which the leaves are shed (Rubbo and Kiesecker 2004). This combination of bacteria, fungi, and leaf are a food source for shredding invertebrates, which leave only FPOM after consumption. These fine particles may be colonized by microbes again or serve as a food source for animals that consume FPOM. Organic matter can also enter the lotic system already in the FPOM stage by wind, surface runoff, bank erosion, or groundwater. Similarly, DOM can be introduced through canopy drip from rain or from surface flows

Invertebrates can be organized into many feeding guilds in lotic systems. Some species are shredders, which use large and powerful mouth parts to feed on non-woody CPOM and their associated microorganisms. Others are suspension feeders, which use their setae, filtering aparati, nets, or even secretions to collect FPOM and microbes from the water. These species may be passive collectors, utilizing the natural flow of the system, or they may generate their own current to draw water, and also, FPOM in Allan. Members of the gatherer-collector guild actively search for FPOM under rocks and in other places where the stream flow has slackened enough to allow deposition. Grazing invertebrates utilize scraping, rasping, and browsing adaptations to feed on periphyton and detritus. Finally, several families are predatory, capturing and consuming animal prey. Both the number of species and the abundance of individuals within each guild is largely dependent upon food availability. Thus, these values may vary across both seasons and systems.

Fish can also be placed into feeding guilds. Planktivores pick plankton out of the water column. Herbivore-detritivores are bottom-feeding species that ingest both periphyton and detritus indiscriminately. Surface and water column feeders capture surface prey (mainly terrestrial and emerging insects) and drift (benthic invertebrates floating downstream). Benthic invertebrate feeders prey primarily on immature insects, but will also consume other benthic invertebrates. Top predators consume fishes and/or large invertebrates. Omnivores ingest a wide range of prey. These can be floral, faunal, and/or detrital in nature. Finally, parasites live off of host species, typically other fishes. Fish are flexible in their feeding roles, capturing different prey with regard to seasonal availability and their own developmental stage. Thus, they may occupy multiple feeding guilds in their lifetime. The number of species in each guild can vary greatly between systems, with temperate warm water streams having the most benthic invertebrate feeders, and tropical systems having large numbers of detritus feeders due to high rates of allochthonous input.

16.7 THREAT TO INDIAN CROCODILES

We must understand the fact that the first step in any effort for the revival of the Vishwamitri River, we must have a acknowledge that a river is more than a channel carrying water; it is also a transporter of sediment; it consists of catchments, floodplains, river-bed, adjoining ravines, which along with the vegetation on both sides, is the river's natural mechanism to retain the additional water, prevent floods and provide habitat for wildlife. A river harbours and interacts with innumerable organisms like plants, animals and microbes. It is a natural, living, organic whole; it is a hydrological part of a larger ecological system. A river is also a network of tributaries and distributaries spread over its basin and the estuary.

The Vishwamitri River originating from the Pavagadh Hills has a unique ecosystem bearing a plethora of beautiful ravines right from its beginning till its end. Of paramount importance in its ecology is the presence of the highly protected species, the Indian crocodile. The crocodiles have been inhabiting and breeding in stretches of the Vishwamitri River for thousands of years. The importance of this species is illustrated by the multiple legal and policy efforts which have been developed by the Government of India to protect the crocodile population.

"Indian Crocodile Conservation Project" was launched as early as the late 1960's. Subsequently the crocodile has been included in Appendix-I of Convention of International Trade in Endangered Species (CITES) and brought under Schedule-I of 'The Wild Life (Protection) Act 1972', meaning that any activity which is against the survival of the highly protected species without having been approved by the State Wildlife Board / National Wildlife Board and the Government of India is patently illegal. The construction activities. have the potential to distract the natural ecosystem and habitat (its breeding grounds and nests) of the crocodile and other species.Capturing, coursing, snaring, trapping, driving or baiting any wild or captive animal and every attempt to do so;Injuring or destroying or taking any part of the body or any such animal or in the case of wild birds or reptiles, damaging the eggs or such birds or reptiles, or disturbing the eggs or nests of such birds or reptiles."

16.8 VISHWAMITRI : POLLUTED RIVER NEEDS ATTENTION

The Vishwamitri river, a seasonal river originates in the hills of Pavagadh, flows west through city of Vadodara, joins the Dhadhar and Kanpur rivers (two small tributaries near the village of Kothawada), and empties into the Gulf of Khambhat, near Khanpur village. This river system includes two reservoirs that hold water for drinking and irrigation: the Sayaji Sarovar, constructed on the Vishwamitri river near Ajawa, and the Dev Dam on the Dhadhar branch. The Vishwamitri river is subjected to the drainage of the city's sewage and effluents from nearby industries. It supports a small population of crocodiles.

The survey was conducted along the short section of Vishwamitri river that flows through the city of Vadodara. This 25 km river stretch was divided into five parts: (1) Sama to Fatehgunj; (2) Fatehgunj to Kalaghoda (3) Kala Ghoda to Munj-mahuda; (4) Munj-mahuda to Kalali and (5) Kalali to Vadsar

Table 1. Pollution status of Vishwamitri River, during October 2002.

Segment No	Vishwamitri river Segment name	Km surveyed	Remark
1	Sama to Fatehgunj	3.5	Highly polluted with rag. plastic bags and encroachment
2	Fatehgunj to Kalaghoda	2.5	Swamp and sewage stagnant water
3	Kala ghoda to Munj-mahuda	2.0	Stagnant water a breeding ground for malaria
4	Munj-mahuda to Kalali	3.0	Swamp and sewage stagnant water
5	Kalali to Vadsar	5.0	Flowing waste water with BOD 60

Sources in the Vadodara Municipal Corporation (VMC) reveal that drainage and sewage water from residential societies along the river are discharged directly into the river. And with last two successive failed monsoons, there is not much water in the river to flush out the drainage.

Various reports and studies are carried out in the past 20 years for clean up river. We need workable innovative solution . The river is natural drainage for Vadodara city flows at the centre of the city from north east to south. Geographically river is shorted distance to carry the rain water from the city. Since last 30 years the important river is totally ignored and misused. It has become a dumping place and looks beyond repair.

To cleanup river innovative dual response workable solution is needed. We have to use the river as natural drainage line and make pleasant surrounding. It is suggested that two no. 1 to 1.5 diameter meter RCC pipes should be laid at bottom of river. All existing sewage connection should be connected to new RCC drainage line to take care of dumping wastewater. The RCC drainage line should be connected to existing Atladra VMC sewage treatment plant by constructing intake well for collection waste water for further treatment. The proposed drainage line at bottom of river take care of all dumping wastewater and also help to take addition load of sewage from new developed Nizampura and Sama area. This is the shortest distance to transport waste water to Atladra. It will reduce the VMC sewage pumping load and breeding ground for mosquitoes. The river bottom should be graded with selected soil. Three to four new RCC check dam about 2 meters height should be constructed at suitable location.

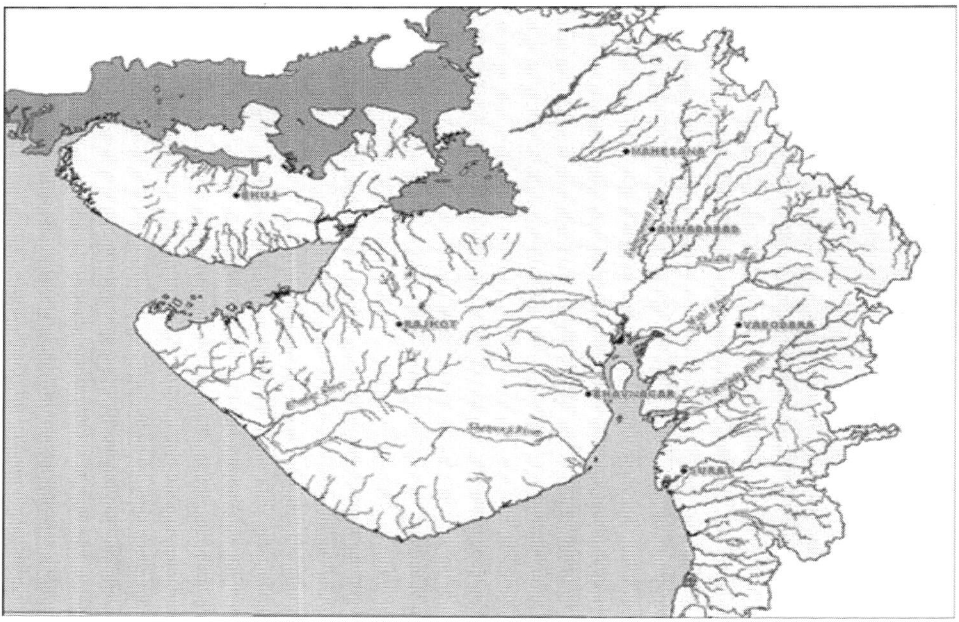

Fig. 1. Rivers in Gujarat state

The river should be filled new fresh water from Narmada channel project. The flow of water should be maintained at least 2 meter depth. Water carrying capacity of River : At present the river is not capable to take care of rainwater and overflow Ajawa Dam. The widening of river should be carried out to take care of rainwater. The existing obstacles in side of river should be removed. Private citizens, NGO, Institutions, VMC and Gujarat Government will contribute the project cost.

Gujarat Chief Minister Narendra Modi (Times of India, JUNE 10, 2002) said river development projects would be taken up in Vadodara (Vishwamitri river), Surat (Tapi river), Rajkot (Aaji river) and the project would be expanded later to cover cities on the banks of Mahi, Purna, Ambika and other rivers in Saurashtra as the part of Rs 17 crore Sant Sarovar Scheme.

Fig. 2 and 3. Crocodiles on river bank passing through the city area

Fig. 4. River Vishwamitri passing through the city area

16.9 SUMMARY

The river is the lifeline of the vast Indian plains. The highly meandering and slowly flowing Vishwamitri river system consists of the main river corridor and its associated tributaries, ravines, nalas / kaans (small streams/creeks), wetlands, ox-bows, and human-made ponds. The river, lake and water bodies systems offer a tremendous opportunity to reduce pollution, increase green cover and, in turn, our ground water table. However, in the absence of holistic planning, application of legal mechanisms and scientific thought this opportunity will be thrown to the wind and there will be neither river nor riverfront left to speak of. Environmental risk posed by the impugned Vishwamitri Project is further amplified by the fact that the same is being constructed on the wetlands of River Vishwamitri. Allowing the Project to continue without environment impact assessment and social impact assessment will result in irreversible damage to the ecology of the said wetlands which are sought to be protected by the Wetlands (Conservation and Management) Rules, 2010. The plan is suggested to make river more clean and ecofriendly with lot of green area around both sides of the river within city stretch.

REFERENCES

Amwatta, C.J.M. 2004. "Diversity of Use of Doum Palm (Hyphaene compressa) Leaves in Kenya" . PALMS **48** (4): 184–190. *Retrieved 25 April 2012.*

Anonymous 2010 *"Riverfront Development for Vishwamitri". Vadodara Property Centre. 15 March 2010. Archived from* the original *on 20 April 2010.*

http://www.depi.vic.gov.au/water

www.water.vic.gov.au/environment/rivers).

http://www.vishwamitririverproject.org/

Johnson, D. 1998. *Hyphaene dichotoma.* 2006 IUCN Red List of Threatened Species. Downloaded on 20 July 2007.

■■■■

Chapter - 17

Removal of Copper (II) from aqueous solution by using CSCMQ

Prapti U. Shah, Nirav Raval and Hitesh Solanki

Department of Environmental Science, School of Sciences, Gujarat University, Ahmedabad-380009, Gujarat, INDIA

E-Mail : prapti.shah1126@gmail.com, nirav.raval90@gmail.com, husolanki@yahoo.com

ABSTRACT

Removal of copper (II) ions onto cassava starch 5-choloromethyl-8-hydroxyquinoline polymer (CSCMQ) was examined with the variation in the parameters of pH, contact time, lead (II) ions concentration, temperature and the adsorbent dose. The Langmuir and Freundlich models have been applied. CSCMQ was characterized by FT-IR and SEM. Results showed that the adsorption process was better described by the Langmuir model. Adsorption kinetics data obtained for the metal ions sorption were investigated using pseudo-first-order, pseudo-second-order and intraparticle diffusion model. The maximum adsorption capacities (q_m) were 128.205, 131.579 and 139.986 at 25, 35 and 45°C, respectively. The dynamical data fit well with the first-order kinetic model. The results indicate that CSCMQ could be employed as low-cost material for the adsorption of Cu(II) ions from aqueous medium.

Keywords: Copper, Cassava Starch; 5-choloromethyl-8–hydroxyquinoline; Adsorption Isotherms; Adsorption Kinetics; Thermodynamics.

17.1 INTRODUCTION

Environmental pollution is the result of rapid industrialization, technological advancement and unprecedented increase in population. The increased technological progression, urbanization, industrial revolution, the intensive use of raw material and agricultural technology have, no doubt, improved our life styles, which could not be imagined a decade ago, but these advancements have simultaneously polluted the natural environment. Pollution is reduction in the quality of the environment by introduction of impurities. Smoke and dust pollute air; junks pollute land and industrial wastes, municipal sewage and domestic waste cause water pollution. Most of the industries discharge their waste directly (without any treatment) into the stream, lakes and oceans as well as in the open land and that contaminate the ground water. There are number of pollutants like fertilizers, pesticides, heavy metals, etc. which seriously affect human life by entering into the system directly or indirectly through food material. Amidst all pollutants, heavy metals are most perilous, non-biodegradable and perpetual in the environment (Bharti, 2012).

Heavy metals are naturally occurring elements that are found throughout the earth's crust. Most environmental contamination and human exposure result from anthropogenic activities such as mining and smelting operations, industrial production and use, as well as domestic and agricultural use of metals and their alloys. (Murphy *et al.,* 1980; Shallari *et al.*, 1998; Herawati, S. Suzuki *et al.*, 2000; He *et al.*, 2005). Environmental contamination can also occur through metal corrosion, atmospheric deposition, soil erosion of metal ions and leaching of heavy metals, sediment re-suspension and metal evaporation from water resources to soil and ground water. Natural phenomena such as weathering and volcanic eruptions have also been reported to significantly contribute to heavy metal pollution (Nriagu, 1989). Industrial sources contributing to heavy metal pollution mainly include metal processing in refineries, coal burning in power plants, petroleum combustion, nuclear power stations and high tension lines, plastics, textiles, microelectronics, wood preservation and paper processing plants (Pacyna, 1996; Sträter *et al.,* 2010; Arruti *et al.,* 2010).

It has been reported that metals such as cobalt (Co), copper (Cu), chromium (Cr), iron (Fe), magnesium (Mg), manganese (Mn), molybdenum (Mo), nickel (Ni), selenium (Se) and zinc (Zn) are essential nutrients that are required for various biochemical and physiological functions. They are important constituents of several key enzymes and play important roles in various oxidation-reduction reactions but their deficiency or excess exposure could lead to number of disorders (WHO 1996). The heavy metals entering the ecosystem have led to geo-accumulation, bioaccumulation and bio magnification (Lokeshwari and Chandrappa, 2006). The bio-uptake of metals includes the following steps:

1. DIFFUSSION: Trace metals and their complexes first diffuses from the external medium to the surface of the organism (mass transport).
2. COMPLEXATION: Metals may dissociate and reassociate (complexation/dissociation) during diffusion to the biological surface.
3. INTERNALIZATION: Metals reacts with a sensitive site on the biological membrane (adsorption/desorption) followed by biological transport (internalization).

Most metals are insoluble in water, soluble in fats and resistant to biological and chemical degradation - properties that result in their accumulation in the tissues of organisms and their biological amplification in food chain and food webs (Golam, 2014). Therefore, a better understanding of heavy metal sources, their accumulation and the effect of their presence in environment seem to be particularly an important issue of present-day research on risk assessments.

Amongst all heavy metal ions toxicity of copper is in limelight and thus copper^{+2} was selected as pollutants for the present study.

Copper is a reddish metal with a face-centred cubic crystalline structure. Its atomic number is 29 and it belongs to 11th group of periodic table. Copper forms a rich variety of compounds, usually with oxidation states +1 and +2, which are often called cuprous and cupric, respectively. It is a soft, malleable and ductile metal with very high thermal and electrical conductivity. Copper is found as a pure metal in nature, and this was the first source of the metal to be used by humans in 8000 BC (Wiberg *et al.*, 2001).

Today, Copper is used in several industries like automotive, battery manufacturing, metal finishing electroplating, plastic, mining, steel and refining. It is also used in construction for roofing. Copper is highly used in electrical wiring. Effluent from all these industries, containing various concentration of copper, is released in water streams and thus causing detrimental impacts.

Copper toxicity has led various harmful effects on human health as well as marine ecosystem. In human, chronic exposure of copper leads to irritation of nose, mouth and eyes. It also causes stomach aches, dizziness, vomiting and diarrhoea. High uptakes of Cu may cause liver and kidney damage and even death (Tumin *et al.,* 2008). In mammals, Cu toxicity causes wide range of effects such as liver cirrhosis, necrosis in kidney and brain, gastrointestinal distress, lesions, low blood pressure and fatal mortality. Excess of copper in marine ecosystem damages gills, liver, kidney and nervous system of marine organisms (Ware, 1994; Kabata-Pendias, 2010). Therefore, it is necessary to treat copper contaminated wastewater prior to its discharge in the environment so as to meet the National Regulatory Standards as well as to protect public health (Barakat, 2011).

Multifarious methods have been used for the removal of heavy metal ions such as Chemical precipitation, Coagulation, Ion exchange, Solvent extraction, Flocculation, Flotation, Ultrafiltration, Adsorption. Although all wastewater treatment techniques can be employed to remove heavy metals, they have their inherent advantages and limitations. Amongst all these methods, adsorption process is considered better than other methods because of convenience, easy operation and simplicity of design. Further, this process can remove/ minimize different types of pollutants and thus have wider applicability in water pollution control (Varma *et al.,* 2013).

Abundant approaches have been studied for the development of cheaper and more effective adsorbents containing natural polymers. Many natural polysaccharides and their derivatives containing required functional groups may have some potential. Recently, starch derivations (Ameye *et al.,* 2001; Kweon *et al.,* 2001; Zhang and Chen, 2002; Xu *et al.,* 2004; Delval *et al.*, 2005) modified cellulose (Saito and Isogai, 2005) chitin derivations (Kang *et al.*, 1999; Yang and Shao, 2000; Varma *et al.,* 2004; Dhakal *et al.*, 2005) and low-cost agricultural by-products (Chamarthy *et al.,* 2001; Shukla and Pai, 2005; Srivastava *et al.*,

2006) have been studied with respect to their ability to remove toxic heavy metal ions from wastewater. Of all these natural polysaccharides, starch appears to be a very fascinating field for research with high possibilities for removing metal ions, thereby enlarging the range of its utilization.

Starch is one of important agricultural biopolymers used in the industry due to its low cost, renewability and biodegradability. There are different types of starch, derived from potato, maize, sago, cassava, etc., available but cassava starch (*Manihot esculenta*, also known as manioc or yucca) has been chosen for the present study as it has many remarkable characteristics like high paste viscosity, high paste clarity, and high freeze – thaw stability, which are advantageous.

Cassava is a woody shrub of the Euphorbiaceae (spurge family) native to South America. It is grown as an annual crop in tropical and subtropical regions for its edible starchy tuberous root. Cassava starch is also called tapioca. Starch is a combination of amylose and amylopectin.

Cassava starch contains 17% of amylose and 83% of amylopectin (2011). The molecular structure of Cassava starch is shown in Figure 1.

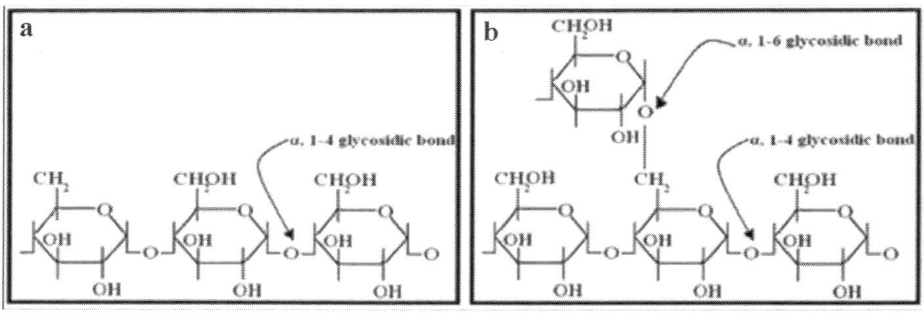

Figure 1. Amylose (a) and Amylopectin (b) of Cassava starch

17.2 MATERIALS AND METHODS

17.2.1 Reagents

Cassava starch (CAS No.: 9057-07-2 and Linear formula: $(C_6H_{10}O_5)_n$), utilized for the adsorption of metal ions, was procured from Sanstar biopolymer Ltd., Ahmedabad and Sigma Aldrich. Other chemicals like 8-hydroxyquinone, formaldehye, hydrochloric acid, tetrahydrofuran and triethyl amine were obtained from ACS Chemicals Pvt. Ltd., Ahmedabad. All reagents used were of analytical grade and without further purification.

17.2.2 Synthesis of Adsorbent

Adsorbent was prepared by blending of dried cassava starch (10 g) with an appropriate amount of Triethyl amine and Tetra hydro Furan. After the agitation of 30 minutes, 5-chloromethyl-8-hydroxyquinoline (CMQ), which was synthesized, purified and characterized as reported by Burckhalter *et al.* (1961). Then the mixture was allowed to heat for 6 hours at 60ºC. The product synthesized was filtered and sun dried. Synthetic pathway for preparation of adsorbent [Cassava Starch/5-chloromethyl-8-hydroxyquinoline] (CSCMQ) is shown in Figure 2.

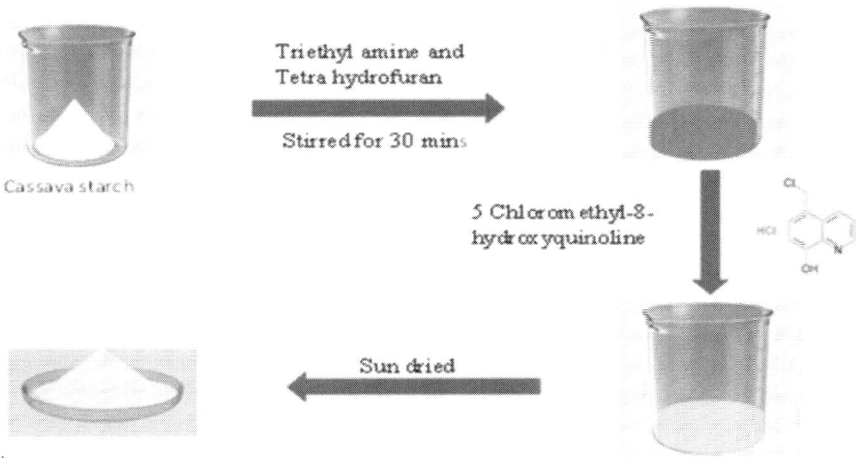

Figure 2. Synthetic pathway for the preparation of CSCMQ

17.3. CONTRIVANCE USED

Atomic absorption spectrophotometer (Model novAA® 400 P, Analytic Jena, Germany) was used for the determination of metal ion concentrations before and after adsorption. FT-IR, Nicolet IR-200, USA, potassium bromide was utilize to analyze the functional groups present on the adsorbent. Scanning electron micrographs was obtained by using SEM (SNE-15 00M, Korea and CarlZeiss, EVO MA 15, England).

17.4 BATCH ADSORPTION STUDY

Adsorption experiments were performed under room temperature. For each batch adsorption experiment, 0.1g of CSCMQ polymer and 100 ml copper solution of different initial concentration was agitated in 250 ml Erlenmeyer flask at 200 rpm at three different temperatures 25ºC, 35ºC and 45ºC. At the end of predetermined time interval the reaction mixture were filtered out. The concentration of respective metal ions were analyzed by using Atomic Absorption Spectrometer, AAS (Model novAA® 400 P, Analytic Jena, Germany). The amount of metal adsorbed by the solid (q) and the percent of metal removal were calculated using following equations:

$$\mathbf{R}\% = \frac{(C_i - C_e)}{C_i} \times 100\% \qquad (1)$$

$$q_e = \frac{(C_i - C_e)V}{W} \qquad (2)$$

where q_e is the adsorption capacity (mg/g), C_i is the initial concentration of metal ion in the solution (mg/L), C_e is the equilibrium concentration of metal ion (mg/L), V is the volume of metal ion solution (mL) and W is the weight of the adsorbent (mg).

To optimize this adsorption process following studies has been carried out.

17.5 EFFECT OF pH

Hydrogen ion concentration is a substantial parameter affecting the surface charge of the adsorbent. The pH is one of eminent environmental factors influencing not only site dissociation, but also the chemistry of solution containing heavy metals (Chowdhury and Saha, 2010). In order to determine the optimum pH for metal adsorption, the adsorption was studied at various pH. Adjustments to pH were made with NaOH and HCl.

17.6 EFFECT OF INITIAL METAL ION CONCENTRATION AND CONTACT TIME

Maximum uptake of metal ion was analyzed by using AAS technique. The effect of initial concentration of metal ions using CSCMQ was studied at different concentrations (10, 20, 30, 40, 50, and 100 mg/L) at variable contact time for adsorption of selected metal ions by CSCMQ. Test solutions were prepared using stock solution of respective metal ions. The flasks containing test solution were agitated on a shaker at 200 rpm for 1 hour. After the establishment of an equilibration the adsorbent was separated by filtration through Whatman paper no. 40.

17.7 EFFECT OF ADSORBENT DOSE

The significance of dosage in adsorption studies is to determine the capacity of adsorbent for a given initial concentration of metal ions solution. In addition, applying an optimum dosage of adsorbent for adsorbate removal in a batch adsorption system is crucial for its cost–effective application (Wang *et al.*, 2014).

17.8 EFFECT OF TEMPERATURE

The equilibrium capacity of the adsorbent for adsorbate will change with a change in temperature. It has been believed that temperature has two major effects on the adsorption process. Increasing the temperature will increase the rate of diffusion of the adsorbate molecules across the external boundary layer and in the internal pores of the adsorbent particle, owing to the decrease in the viscosity of the solution. On the other hand, with the increase in temperature, the mobility of the metal ions increases while the retarding forces acting on the diffusing ions decrease, thereby increasing the adsorption capacity of adsorbent (El-Sayed *et al.,* 2011).

17.9 ADSORPTION ISOTHERM STUDIES

Equilibrium adsorption isotherms described the possible mechanism occurring between adsorbate and adsorbent. Therefore, the correlation of equilibrium data by either theoretical or empirical equations is important for predicting the adsorption capacity of the adsorbent (Wong *et al.,* 2004). The two most common models used to study adsorption isotherms are: (i) Langmuir Isotherm and (ii) Freundlich Isotherm.

17.10 LANGMUIR ISOTHERM MODEL

Langmuir proposed a theory to describe the adsorption of gas molecules onto metal surfaces. The Langmuir adsorption isotherm has found successful applications in many other real adsorption processes of monolayer adsorption. The Langmuir model assumes that the each adsorbate molecule is located at specific homogenous sites within the adsorbent. Thus, it predicts the formation of a monolayer of the adsorbate on the homogenous adsorbent surface and does not consider surface heterogeneity of the adsorbent. The linear form of the Langmuir adsorption isotherm (Langmuir 1918) is represented as:

$$\frac{C_e}{q_e} = \frac{1}{K_L} + \frac{a_L}{K_L}.C_e \quad (3)$$

Where, C_e (mg/L) is the equilibrium concentration of the metal ion in the solution. q_e (mg/g) is the equilibrium adsorption capacity per gram dry weight of the adsorbent corresponding to complete coverage of the adsorptive sites. a_L (L/mg) and K_L (L/g) are the Langmuir isotherm constants. The values of a_L and K_L are calculated from the slope $\left(\frac{a_L}{K_L}\right)$ and intercept $\left(\frac{1}{K_L}\right)$ of the plot of $\frac{C_e}{q_e} \rightarrow C_e$. Ration of $\frac{K_L}{a_L}$ gives the value of monolayer sorption capacity of the adsorbent, q_m (mg/g).

The essential characteristic feature of Langmuir isotherm can be expressed in terms of "R_L", a dimensionless constant referred to as "separation factor" or "equilibrium parameter". The value of R_L is calculated using the following equation:

$$R_L = \frac{1}{(1 + a_L.C_o)} \quad (4)$$

The R_L value indicates whether the adsorption is unfavorable ($R_L>1$), linear ($R_L=1$), favorable ($0<R_L<1$) or irreversible ($R_L=0$) (Futalan *et al.,* 2011).

17.11 FREUNDLICH ISOTHERM MODEL

The Freundlich isotherm model is the earliest known relationship describing the non-ideal and reversible adsorption, not restricted to the formation of a monolayer. This empirical model can be applied to multilayer adsorption with non-uniform distribution of adsorption heat and affinities over the heterogeneous surface (Foo and Hameed, 2010). The linear form of Freundlich isotherm model (Freundlich, 1906) is as follows:

$$\log q_e = \log K_F + \frac{1}{n} \log C_e \tag{5}$$

Where, K_F (L/g) is the Freundlich adsorption isotherm constant, relating to the extent of adsorption. n (g/L) is the Freundlich exponent. The values of K_F and 1/n are calculated from the slope and intercept of the plot of $\log q_e \rightarrow \log C_e$.

17.12 ADSORPTION KINETIC STUDIES

The kinetic study is helpful to select the optimum operating conditions for the full-scale batch process as well as it also provides valuable insights into the reaction pathways and in turn controls the residence time of adsorbate uptake at the solid-solution interface (Ahmad and Kumar, 2010). To investigate the adsorption kinetics of selected heavy metal ions onto CSCMQ, the linear forms of pseudo-first-order, pseudo-second-order and intraparticle diffusion models were used.

17.13 PSEUDO-FIRST-ORDER RATE EQUATION

Lagergren (1898) presented a first-order rate equation to describe the kinetic process of liquid-solid phase adsorption of oxalic acid and malonic acid onto charcoal, which is believed to be the earliest model pertaining to the adsorption rate based on the adsorption capacity.

To distinguish kinetic equations based on adsorption capacity from solution concentration, Lagergren's first order rate equation has been called pseudo-first-order (Ho and McKay 1998).

$$\log (q_e - q_t) = \log q_e - \frac{K_1}{2.303} t \tag{6}$$

Where q_e and q_t (mg/g) are the sorption capacity at equilibrium and at time t, respectively, k1 (min^{-1}) is the equilibrium rate constant of pseudo-first-order chemical sorption.

17.14 PSEUDO-SECOND-ORDER RATE EQUATION

For the rate constant of the pseudo-second order chemical sorption process,

$$\frac{dqt}{dt} = k_2 (q_e - q_t) \tag{7}$$

For the boundary conditions t = 0 to t = t and $q_t = 0$ to $q_e = q_t$, the integrated form of equation (8) becomes:

$$\frac{1}{q_e - q_t} = \frac{1}{q_e} + k_2 t \tag{8}$$

where, q_e and q_t (mg/g) are the sorption capacity at equilibrium and at time t, respectively, k_2 (g/mg.min) is the equilibrium rate constant of pseudo-second-order chemical sorption.

Equation (8) is the integrated rate law for a pseudo-second-order reaction. which can be rearranged to obtain:

$$q_t = \frac{t}{\frac{1}{k_2 q_e^2} + \frac{t}{q_e}} \tag{9}$$

The has a linear form of equation (9) can be represented as:

$$\frac{t}{q_t} = \frac{1}{k_2 q_e{}^2} + \frac{t}{q_e} \tag{10}$$

Ho's second-order rate equation has been called pseudo-second-order rate equation to distinguish kinetic equations based on adsorption capacity from concentration of solution (Ho 2006a). The pseudo-second-order rate constants can be determined experimentally by plotting t/q_t against **t** (equation 10). Although there are many factors which influence the sorption capacity, including the initial sorbate concentration, the reaction temperature, the solution pH, the sorbent particle size and the nature of the solute whereas a kinetic model is concerned only with the effect of observable parameters on the overall rate (Ho 2006b).

17.15.A. Intraparticle diffusion model

The pseudo-first-order and pseudo-second-order kinetic models could not identify the diffusion mechanism hence the kinetic results were further analyzed by using the intraparticle diffusion model. Weber-Morris found that in many adsorption cases, solute uptake caries almost proportionally with $t^{1/2}$ rather than with contact time 't' (Alkan *et al.*, 2007).

The adsorbate species are most probably transported from the bulk of the solution into the solid phase through intraparticle diffusion/transport process, which is often the rate limiting step in many adsorption processes. The possibility of intraparticle diffusion was explored by using the intraparticle diffusion model, which is commonly expressed as:

$$q_t = K_{id}\, t^{\frac{1}{2}} + 1 \tag{11}$$

Where, q_t (mg/g) is the sorption capacity at time t. K_{id} is the intraparticle diffusion rate constant (mg/g.min$^{1/2}$) and **I** (mg/g) is the intercept.

Intraparticle diffusion can be described by the three consecutive steps:

1. The transport of adsorbate from the bulk solution to the outer surface of the adsorbent by molecular diffusion.
2. The internal diffusion, the transport of adsorbate from the particles surface into interior sites.
3. The adsorption of solute particles from the active sites into the interior surface of the pores.

The value of intercept (I), provides information about the thickness of the boundary layer. If the intraparticle diffusion is involved in the adsorption process, then a plot of the square root of time versus the uptake would result in a linear relationship and the particle diffusion would be the controlling step if this line passed through the origin. When the plots do not pass through the origin, this is indicative of some degree of boundary layer control which further show that the intraparticle diffusion is not the only rate controlling step, but also other processes may control the rate of adsorption, all of which may be operating simultaneously (Santhi *et al.,* 2010).

17.15.B. Error Analysis

For the evaluation of the best fitting isotherm equation to the experimental equilibrium data, an error function has to be performed. In this study, linear coefficient of determination (R^2) and a non-linear Chi-square test (χ^2) were performed for both the single component isotherms as well as for two kinetic rate models (pseudo-first-order and pseudo-second-order). The Chi-square statistic test refers to the sum of the squares of the differences between the experimental data and the data obtained by calculating from isotherm models, with each squared difference divided by the corresponding data obtained by calculating from models.

The mathematical equation can be represented as:

$$\chi^2 = \Sigma \frac{(q_{e(exp)} - q_{e(cal)})^2}{q_{e(cal)}} \tag{12}$$

where, $q_{e(exp)}$ (mg/g) is the experimental data of the equilibrium capacity; $q_{e(cal)}$ (mg/g) is the equilibrium capacity obtained by calculating from the model. If data from the model are similar to the experimental data, χ^2 will be a smaller number, and if they differ, χ^2 will be a bigger number. Therefore it is necessary to analyze the data using the non-linear Chi-square test to confirm the suitable isotherm and kinetic model for present adsorption system (Unuabonah *et al.*, 2007).

17.15.C. Adsorption Thermodynamic Study

The thermodynamic parameters such as change in Gibb's free energy (ΔG^o), change in enthalpy (ΔH^o) and change in entropy (ΔS^o) were also studied for better understanding of the effect of temperature on the adsorption process.

The Gibbs free energy of adsorption ΔG^o is calculated from the following equation:

$$\Delta G^o = -RT \ln K_C \tag{13}$$

where K_c is the adsorption equilibrium constant. The value of K_c is calculated from the following equation:

$$K_C = \frac{C_{ae}}{C_e} \tag{14}$$

where C_{ae} is the equilibrium adsorbate concentration on the adsorbent (mg/L) and C_e is the equilibrium adsorbate concentration in solution (mg/L).

Standard enthalpy and standard entropy of adsorption can be estimated from Van't Hoff equation:

$$\ln K_C = \frac{\Delta H^o}{RT} + \frac{\Delta S^o}{R} \tag{15}$$

The value of ΔH^o and ΔS^o were calculated from the slope and intercept of Van't Hoff plot of $\ln K_C$ vs. 1/T.

17.16 RESULTS AND DISCUSSION

3.1. FT-IR (FOURIER TRANSFORM INFRARED SPECTROSCOPY)

The characterization of cassava starch, CMQ and CSCMQ were done by FT-IR spectrophotometer. The FT-IR spectrum of cassava starch shown in Figure. 3(a) narrated that there is O-H stretching absorption in the region 3550-3200 cm^{-1}(broad) and single intermolecular polymeric band C-H stretching at 2930 cm^{-1} (m) and 1076 cm^{-1}. The bending mode of water is shown at 1800-1600 cm^{-1}. The wave number at 1658 cm^{-1} was assigned as a peak of the first overtone of OH bending. On the other hand, 1150 cm^{-1} was assigned as C-O-C stretching.

In Figure 3(b) (CMQ) a sharp peak is seen at 1491 cm^{-1} which confirms C=N aromatic nitro compound in CMQ. The peak found in the region 3550-3200 cm^{-1} (broad) showed the existence of the O-H stretching which was attributed to water molecule. Further, the peaks observed at 1274.99 cm^{-1} represented C-C stretch where as peak 1018.45 cm^{-1} and 866.07 cm^{-1} confirmed the presence of aromatic C-H. The band observed at 785.05 cm^{-1} is attributed to C-Cl.

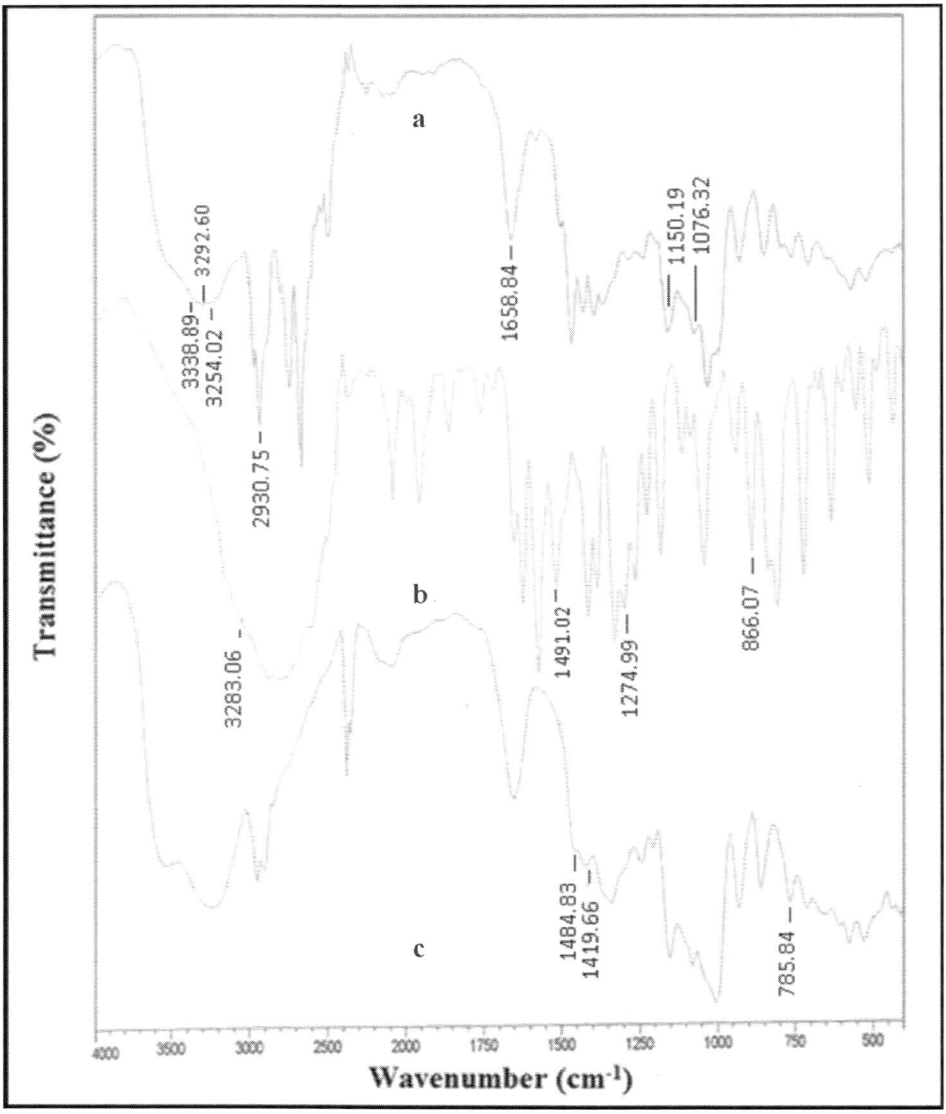

Figure: 3 Fourier Transform Infrared Spectroscopy (a) Cassava Starch (b) CMQ (c) CSCMQ

The spectra of CSCMQ showed the existence of sharp peak at 785.84 cm^{-1} as observed in Figure 3(c), which arises from the stretching vibration mode of C-Cl bond and a small peak is observed at 1484.83 cm^{-1} which confirms the presence C=N aromatic nitro compound. A weak symmetrical stretching band

1431-1396 cm^{-1}, characteristics of C=O group indicated the occurrence of grafting. Most other peaks are related to starch backbone.

17.18 SCANNING ELECTRON MICROSCOPY

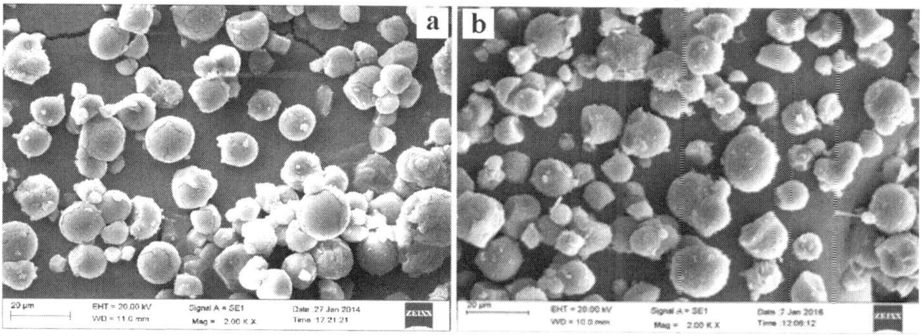

Figure: 4 SEM (a) before adsorption and (b) after adsorption

The SEM analysis was conducted to explicate the surface morphology of CSCMQ before and after adsorption of heavy metal ions. Results depicted that SEM images of CSCMQ bearing 2.0 KX magnification presented in Figure:4 (a) had intercellular voids, roughness and porous structure whereas Figure:4 (b) reveals the SEM images of CSCMQ, after adsorption of copper metal ions, was having smooth and even surface. After the adsorption of metal ion, porosity of the surface was filled. Thus, from the morphological study of each metal loaded CSCMQ it can be confirmed that the adsorption had taken place in the inner walls of CSCMQ surface.

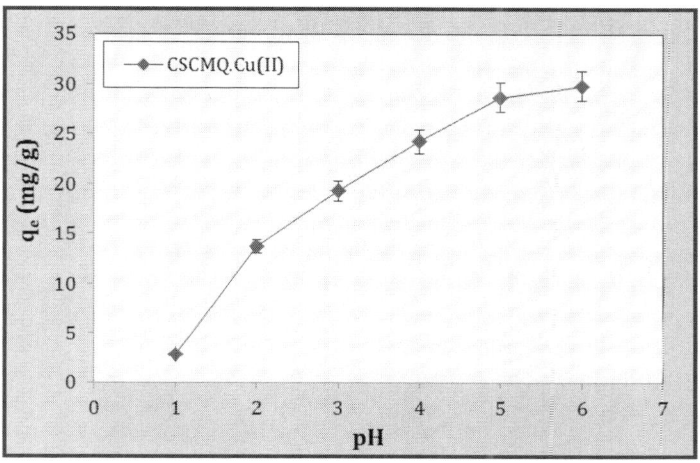

Figure 5: Effects of pH on the adsorption of Cu(II) ions by CSCMQ. [C_0 = 50 mg/L; contact time = 90 min; dosage of adsorbent = 100g; temperature = 318 K; agitation speed = 120 rpm]

17.19 EFFECT OF pH

The effect of pH for the removal of copper (II) by CSCMQ from an aqueous solution is reported in Figure 5. It was observed from the figure that in strong acidic medium the adsorption capacity decreased due to the partial protonation of active groups which caused repulsion of groups bearing positive charge. With an increase in pH from 3.0 to 6.0, adsorption of copper ions increased since more binding sites of adsorbent were exposed to metal ions. Maximum adsorption was found at pH 6 due to the fact that metal precipitation appeared at higher pH values and interfered with the accumulation or deterioration of adsorbent.

17.20 EFFECT OF INITIAL CU(II) IONS CONCENTRATION AND CONTACT TIME

The effect of initial Cu(II) ions concentration and contact time for adsorption of copper ions on CSCMQ is presented in Figure 6 (a-c). The adsorption of Cu(II) increased with an increase in contact time and attained equilibrium within 90 minutes. Beyond 90 minutes, there was no increase in adsorption capacity of adsorbate. Thus optimum time for establishment of equilibrium was considered to be 90. It was observed that with an increase in initial Cu (II) concentration from 10 to 100 mg/L, there was an increase in the amount of copper adsorbed from 5.439 to 44.908 mg/g at 298 K, 5.932 to 49.032 mg/g at 308 K and 6.391 to 53.121 mg/g at 318 K due to the driving force of concentration gradient. It was significantly observed that at an initial concentration of 50mg/L a steady equilibrium was obtained and hence for studies 50mg/L was considered to be optimum metal ion concentration.

17.21 EFFECT OF DOSAGE OF ADSORBENT

The effect of adsorbent dosage for removal of copper by CSCMQ is shown in Figure 21(a,b). It was observed from figure that the amount of copper adsorbed (mg/g) decreased from 43.47 to 29.47 mg/g with an increase in adsorbent dosage from 10 mg to 100 mg. Further increase in adsorbent dosage exhibited minor depreciation in adsorption capacity. Normally it was observed that the decrease in adsorption capacity (mg/g) with increase in adsorbent dosage is due to the high number of unsaturated adsorption sites. It was observed that the percentage removal of copper increased from 8.694% to 59.694% for adsorbent dosage of 10 to 100 mg due to the availability of more binding sites on adsorbent. Thus 100 mg dosage of CSCMQ was considered to be optimum for further studies.

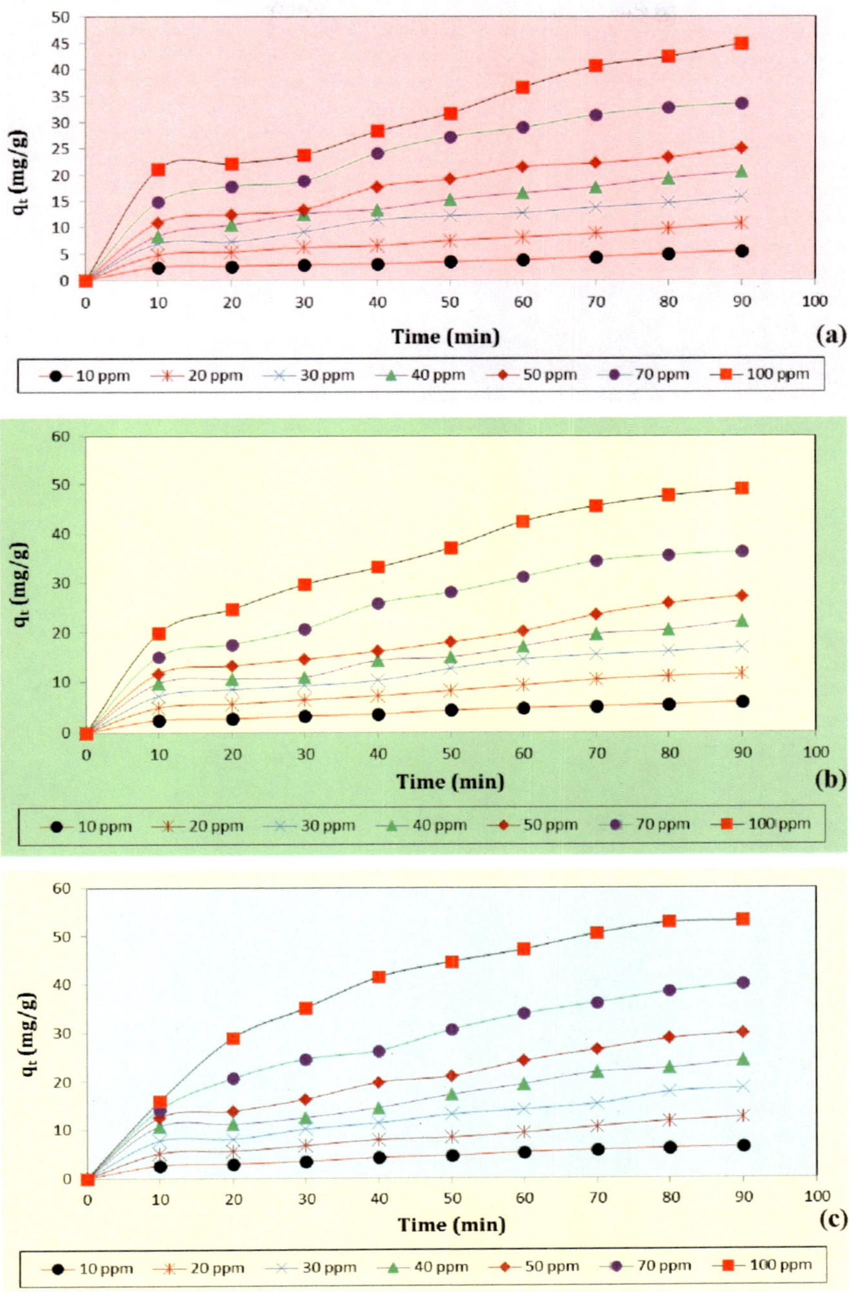

Figure 6: Effects of initial metal ions concentrations and contact time on the adsorption of Cu(II) ions by CSCMQ at 298 K (**a**), 308 K (**b**) and 318 K (**c**).[pH = 6.0; dosage of adsorbent = 100 mg; agitation speed = 120 rpm]

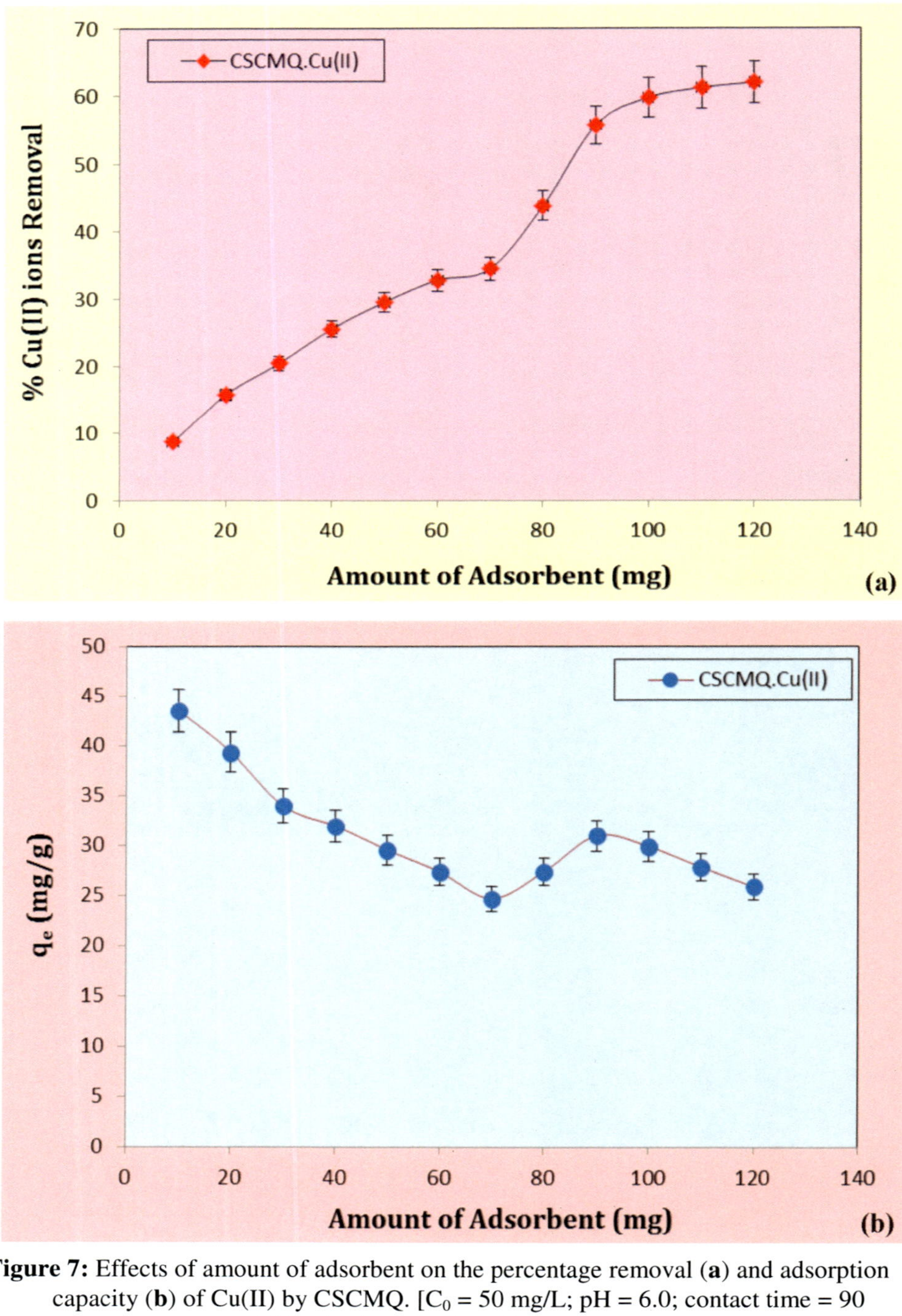

Figure 7: Effects of amount of adsorbent on the percentage removal (**a**) and adsorption capacity (**b**) of Cu(II) by CSCMQ. [C_0 = 50 mg/L; pH = 6.0; contact time = 90 min; temperature = 318 agitation speed = 120 rpm]

17.22 EFFECT OF TEMPERATURE

Temperature plays an important role on the process of adsorption. Effect of temperature on the adsorption of Cu (II) ions was studied at different temperatures (25, 35 and 45°C) and results are presented in Figure: 8. It was observed from the graph that the adsorption of copper ions increased with an increase in temperature due to the enlargement of pore size resulting in an increase in the number of active sites on the surface of adsorbent available for interaction with Cu(II) ions. The maximum adsorption of copper ions was found at 45°C.

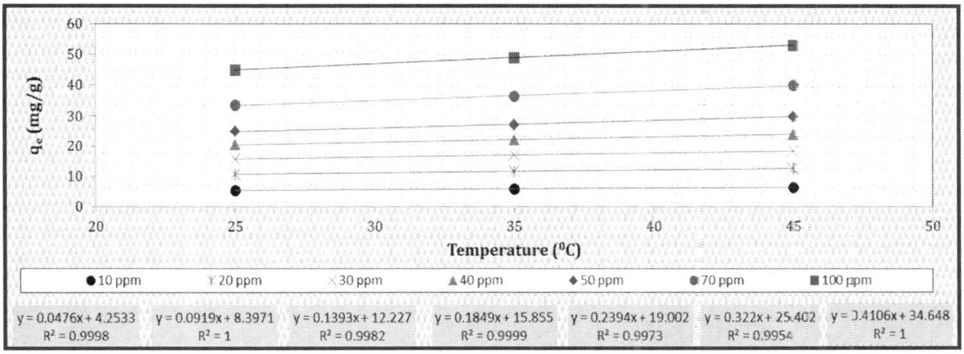

Figure 8: Effects of temperature on the adsorption of Cu(II) ions by CSCMQ. [pH = 6.0; contact time = 90 min; agitation speed = 120 rpm]

17.23 ADSORPTION ISOTHERM STUDY

The results of the isotherm study for adsorption of Cu(II) onto CSCMQ are shown in Table 7, and the plots modeled isotherms are shown in Figure 9 (a,b). The adsorption of Cu(II) ions followed both the isotherms well because all the plots indicated a straight line (Figure 9). The comparison between relatively high R^2 (~ 1) and low χ^2 (~ 0) asserted that the Langmuir isotherm suited well as compare to the Freundlich isotherm. The basic assumption of Langmuir adsorption isotherm is based on monolayer coverage of the adsorbate on the surface of adsorbent, which was an indication of the fact that the adsorption of Cu(II) ions onto CSCMQ generated monolayer formation. The values of $R_L < 1$ as reported in Table 1 confirmed that the adsorbent was favorable for the adsorption of Cu(II) ions.

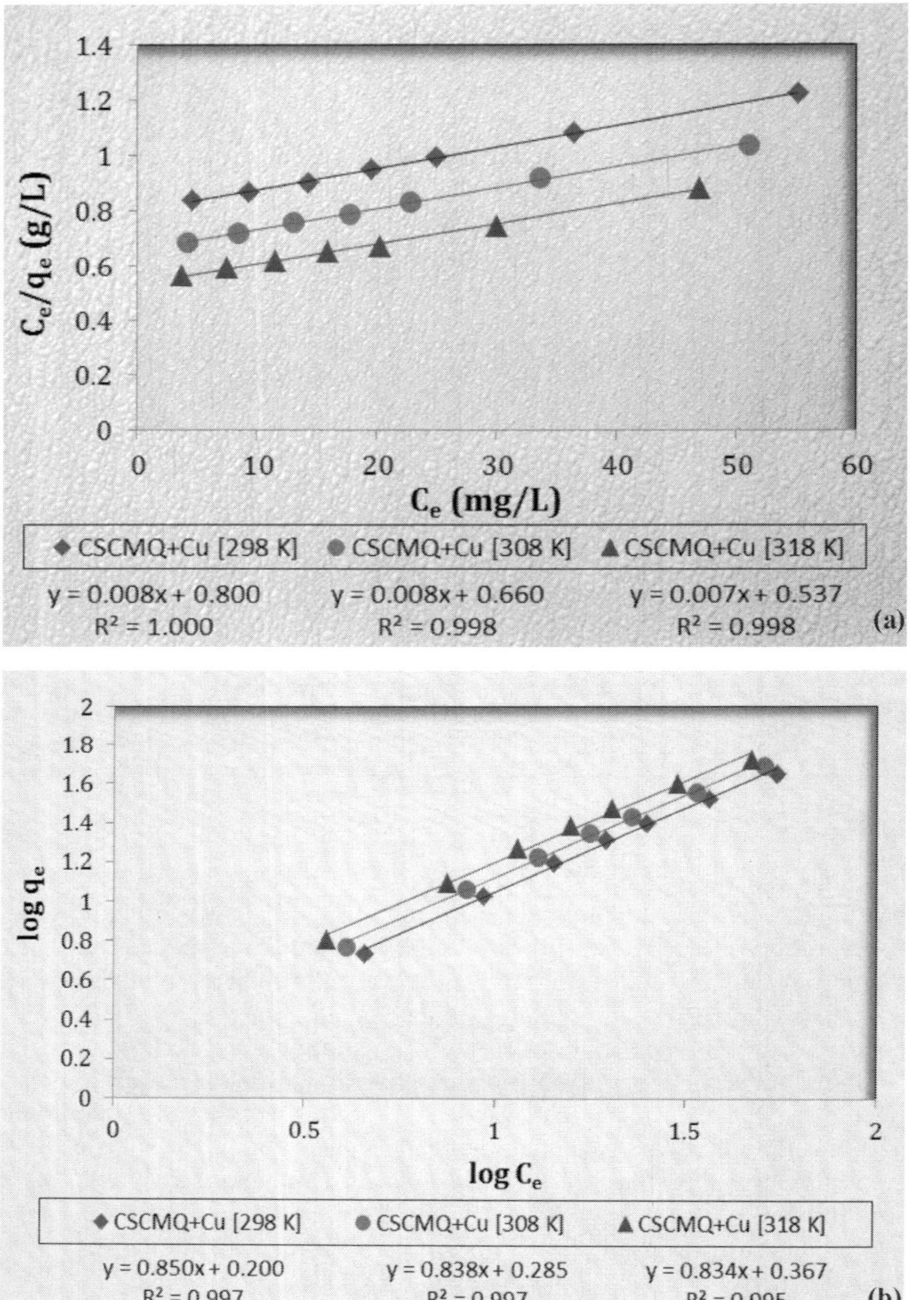

Figure 9: Langmuir (**a**) and Freundlich (**b**) isotherm plots for the adsorption of Cu(II) ions onto CSCMQ. [pH = 6.0; contact time = 90 min; dosage of adsorbent = 100 mg; and agitation speed = 120 rpm]

Table 1. Constants for equilibrium isotherm models with error analysis values for the adsorption of Cu(II) ions by CSCMQ.

Cu	Isotherm Parameters		CSCMQ		
			298 K	308 K	318 K
Langmuir Isotherm	q_m (mg/g)		128.205	131.579	139.986
	a_L (L/mg)		0.010	0.012	0.014
	R_L	10 (mg/L)	0.911	0.897	0.880
		20 (mg/L)	0.837	0.813	0.786
		30 (mg/L)	0.774	0.743	0.710
		40 (mg/L)	0.719	0.685	0.648
		50 (mg/L)	0.672	0.635	0.595
		70 (mg/L)	0.594	0.554	0.512
		100 (mg/L)	0.506	0.465	0.424
	R^2		1.000	0.998	0.998
	χ^2		0.001	0.006	0.01
Freundlich Isotherm	K_F (L/g)		1.581	1.928	2.323
	1/n		0.849	0.837	0.833
	n		1.178	1.195	1.200
	R^2		0.997	0.997	0.995
	χ^2		0.258	0.259	0.508

17.23 ADSORPTION KINETIC STUDY

The straight line plots for the pseudo-first-order and pseudo-second-order models are presented in Figure 10, 11 to obtain the rate parameters which are given in Table 2. The linear plot of log (q_e-q_t) versus time of pseudo-first-order provided low R^2 values, ranging from 0.822 to 0.943 for CSCMQ where as in pseudo-second-order, R^2 values were higher in the range of 0.886 to 969 which indicated that the pseudo-first-order model had limited applicability on the adsorption of divalent copper. On the contrary χ^2 value of pseudo-first-order model was lower compared to pseudo-second-order model and the difference between the experimental adsorption capacity ($q_{e(exp)}$) of Cu(II) and the calculated value (q_e) from pseudo-first-order model was also lower as compare to the difference between experimental adsorption capacity ($q_{e(exp)}$) of Cu(II) and the calculated value (q_e) from pseudo-second-order model which indicated that pseudo-first-order model fitted well for adsorption of copper ions.

The kinetic data were analyzed by the intraparticle diffusion model to elucidate the diffusion mechanism. Intraparticle diffusion is the sole rate-determining step if the plot is linear and passes through the origin. From Figure:12, it is indicated that the plots did not pass through the origin. This deviation from the origin is due to the difference in the rate of mass transfer in the initial and final stages of the adsorption process (Zulfikar *et al.*). The good regression of metal ions (≤1) showed the existence of intraparticle diffusion mechanism on the sorption of Cd(II) onto CSCMQ. Moreover, the presence of the boundary layer effect (I) showed the existence of the surface sorption indicating that intraparticle diffusion was not the only rate-limiting step.

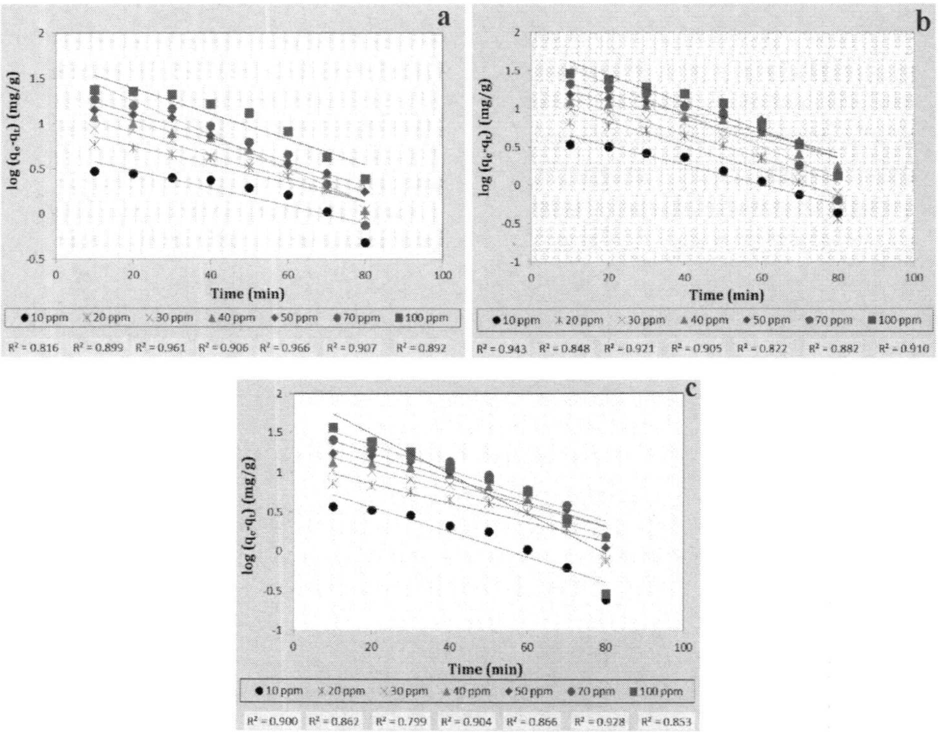

Figure 10: Pseudo-first-order plot for the adsorption of Cu(II) ions onto CSCMQ at 298 K (**a**), 308 K (**b**) and 318 K (**c**), respectively. [pH = 6.0; contact time = 90 min; dosage of adsorbent =100 mg; and agitation speed = 120 rpm]

17.24 ADSORPTION THERMODYNAMIC STUDY

The values of ΔH° and ΔS° were calculated from the slope and intercept of the plots of ln K_c vs. 1/T (Figure 13). The negative values of ΔG° observed from Table 4 indicated the spontaneous and favourable nature of the adsorption of Cu(II) onto CSCMQ. The positive values of ΔH° and ΔS° suggested endothermic nature of the adsorption and increase in randomness at solid/solution interface during adsorption process. The adsorbed water molecules, which are displaced by the adsorbate species gain more translational energy than lost by the adsorbate ions which allowed the prevalence of randomness in the system (Dali-youcef *et al.*, 2006).

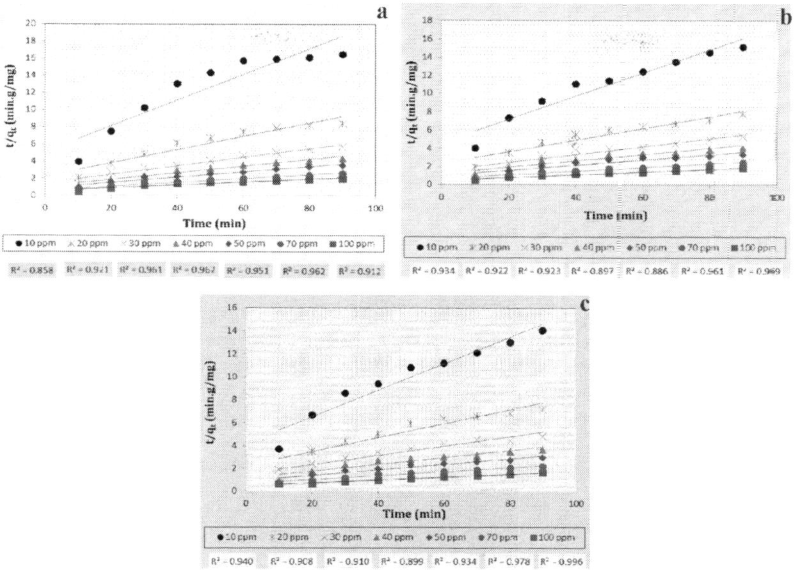

Figure 11: Pseudo-second-order plot for the adsorption of Cu(II) ions onto CSCMQ at 298 K (**a**), 308 K (**b**) and 318 K (**c**), respectively. [pH = 6.0; contact time = 90 min; dosage of adsorbent = 100 mg; and agitation speed = 120 rpm]

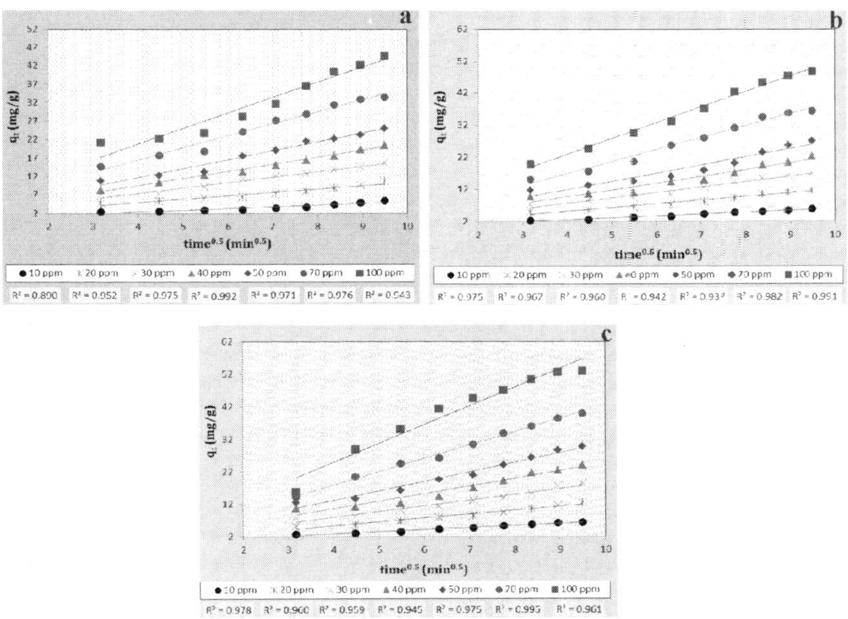

Figure 12: Intraparticle diffusion plot for the adsorption of Cu(II) ions onto CSCMQ at 298 K (**a**), 308 K (**b**) and 318 K (**c**), respectively. [pH = 6.0; contact time = 90 min; dosage of adsorbent = 100 mg; and agitation speed = 120 rpm]

Table 2. Parameters of kinetic rate models and intraparticle diffusion model for the adsorption of various concentrations of Cu(II) ions on CSCMQ.

Adsorption Kinetics	Kinetic Parameters (308 K)	Initial metal ion concentration (mg/L)						
		10 mg/L	20 mg/L	30 mg/L	40 mg/L	50 mg/L	70 mg/L	100 mg/L
	$q_{e\,(exp)}$ (mg/g)	5.928	11.619	17.034	22.299	27.238	36.419	49.032
Pseudo-first-order model	k_1 (L/min)	0.029	0.038	0.036	0.028	0.030	0.047	0.041
	q_e (mg/g)	5.959	14.595	19.688	21.903	29.067	54.840	62.589
	R^2	0.943	0.848	0.921	0.905	0.822	0.882	0.910
	χ^2				14.372			
Pseudo-second-order model	k_2 (g/mg.min)	0.0035	0.0018	0.0012	0.0009	0.0008	0.0006	0.0005
	q_e (mg/g)	7.868	15.601	22.936	29.240	35.336	49.505	65.360
	R^2	0.934	0.922	0.923	0.897	0.886	0.961	0.969
	χ^2				18.751			
Intra particle Diffusion Model	K_{id} (mg/g.min$^{1/2}$)	0.581	1.146	1.691	2.105	2.549	3.727	4.908
	I (mg/g)	0.264	0.590	0.943	1.363	1.695	2.010	3.326
	R^2	0.975	0.967	0.960	0.942	0.937	0.982	0.991

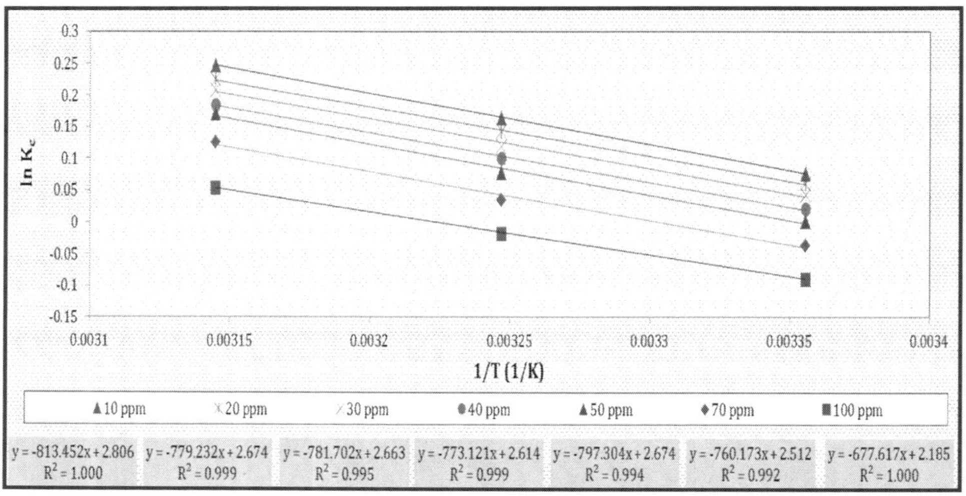

Figure 13: Van't Hoff plot of ln Kc vs. 1/T for the adsorption of Cu(II) ions onto CSCMQ.

Table 3. Adsorption thermodynamic parameters for Cu(II) ions adsorption by CSCMQ.

Thermodynamic Constants of CSCMQ + Cu								
Sr. No.	**C_o (mg/L)**	**K_c**	**ln K_c**	**ΔG° (kJ/mol)**	**1/T (K^{-1})**	**ΔS° (kJ/mol/K)**	**ΔH° (kJ/mol)**	**R^2**
		1.193	0.076	-0.436	0.00336			
1	10	1.456	0.163	-0.962	0.00325	0.054	15.574	1.000
		1.771	0.248	-1.511	0.00314			
		1.149	0.060	-0.345	0.00336			
2	20	1.386	0.142	-0.837	0.00325	0.051	14.919	0.999
		1.678	0.225	-1.369	0.00314			
		1.104	0.043	-0.246	0.00336			
3	30	1.314	0.119	-0.699	0.00325	0.051	14.967	0.995
		1.615	0.208	-1.268	0.00314			
		1.050	0.021	-0.121	0.00336			
4	40	1.260	0.100	-0.591	0.00325	0.050	14.803	0.999
		1.529	0.185	-1.124	0.00314			
		1.005	0.002	-0.012	0.00336			
5	50	1.197	0.078	-0.460	0.00325	0.051	15.266	0.994
		1.481	0.171	-1.039	0.00314			
		0.922	-0.035	0.201	0.00336			
6	70	1.085	0.035	-0.208	0.00325	0.048	14.554	0.992
		1.335	0.125	-0.764	0.00314			
		0.815	-0.089	0.506	0.00336			
7	100	0.962	-0.017	0.099	0.00325	0 042	12.974	1.000
		1.133	0.054	-0.331	0.00314			

17.3 CONCLUSION

The present study showed that the novel CSCMQ polymer can be employed as a potentially low cost sorbent for the removal of Cu(II) ions from industrial wastewaters. The adsorption of Cu(II) is found to be greatly reliant on the initial pH of the solution, its concentration in the solution and adsorbent mass. The maximum biosorption of Cu(II) was found at pH value 6.0. The kinetic experiments show that the adsorption was rapid and the adsorption equilibrium was achieved within 90 min of contact time. The equilibrium data are well described by the Langmuir isotherm model and the maximum adsorption capacity of Cu (II) on CSCMQ polymer was 139.98 mg/g. Further, thermodynamic parameters which were calculated from the temperature-dependent isotherms signify that the adsorption reaction is a spontaneous process. These cessation demonstrates that the adsorbent has a very low economical value can be used efficaciously for removal of Cu(II) ions from an aqueous systems for the protection of environment.

Acknowledgements

I am grateful to ***AIRF (JNU- New Delhi)*** and ***SICART (Anand)*** for extending the facility of sample analysis. Finally, I am highly thankful to **University Grant Commission-Maulana Azad National Fellowship *(UGC-MANF-Delhi)*** (Award letter number: F1-17.1/2014-15/MANF-2014-15-JAI-GUJ-35064) for providing the funds so as to undertake my research work.

REFERENCES

Ahmad R, and Kumar R (2010) Adsorptive removal of congo red dye from aqueous solution using bael shell carbon. Appl Surf Sci 257: 1628–1633. doi: 10.1016/j.apsusc.2010.08.111

Alkan M, Demirbaş Ö, Doğan M (2007) Adsorption kinetics and thermodynamics of an anionic dye onto sepiolite. Microporous Mesoporous Mater 101:388–396. doi: 10.1016/j.micromeso.2006.12.007

Ameye D, Voorspoels J, Foreman P, (2001) Trypsin inhibition, calcium and zinc ion binding of starch–g-poly(acrylic acid) copolymers and starch/poly(acrylic acid) mixtures for peroral peptide drug delivery. J Controlled Release 75:357–364. doi: 10.1016/S0168-3659(01)00408-4

Arruti A, Fernández-Olmo I, Irabien Á (2010) Evaluation of the contribution of local sources to trace metals levels in urban PM2.5 and PM10 in the Cantabria region (Northern Spain). J Environ Monit 12:1451–1458. doi: 10.1039/B926740A

Barakat MA (2011) New trends in removing heavy metals from industrial wastewater. Arab J Chem 4:361–377. doi: 10.1016/j.arabjc.2010.07.019

Bharti PK (2012) Heavy metals in Environment, First. Lambert Academic Publishing GmbH & Co. KG, Germany

Burckhalter JH, Leib RI (1961) Amino- and Chloromethylation of 8-Quinolinol. Mechanism of Preponderant ortho Substitution in Phenols under Mannich Conditions1a,b. J Org Chem 26:4078–4083. doi: 10.1021/jo01068a104

Chamarthy S, Seo CW, Marshall WE (2001) Adsorption of selected toxic metals by modified peanut shells. J Chem Technol Biotechnol 76:593–597. doi: 10.1002/jctb.418

Chowdhury S, Saha P (2010) Sea shell powder as a new adsorbent to remove Basic Green 4 (Malachite Green) from aqueous solutions: Equilibrium, kinetic and thermodynamic studies. Chem Eng J 164:168–177. doi: 10.1016/j.cej.2010.08.050

Dali-youcef N, Ouddane B, Derriche Z (2006) Adsorption of zinc on natural sediment of Tafna River (Algeria). J Hazard Mater 137:1263–1270. doi: 10.1016/j.jhazmat. 2006. 03.068

Delval F, Crini G, Bertini S, et al (2005) Preparation, characterization and sorption properties of crosslinked starch-based exchangers. Carbohydr Polym 60:67–75. doi: 10.1016/j.carbpol.2004.11.025

Dhakal RP, Ghimire KN, Inoue K, et al (2005) Acidic polysaccharide gels for selective adsorption of lead (II) ion. Sep Purif Technol 42:219–225. doi: 10.1016/j.seppur. 2004.07.016

El-Sayed GO, Mohammed TY, El-Sayed OE (2011) Removal of basic dyes from aqueous solutions by sugar can stalks. Adv Appl Sci Res 2:283–290.

Foo KY, Hameed BH (2010) Insights into the modeling of adsorption isotherm systems. Chem Eng J 156:2–10. doi: 10.1016/j.cej.2009.09.013

Freundlich HM. (1906) Over the adsorption in solution. J Phys Chem 57:385–470.

Futalan CM, Kan C-C, Dalida ML, et al (2011) Comparative and competitive adsorption of copper, lead, and nickel using chitosan immobilized on bentonite. Carbohydr Polym 83:528–536. doi: 10.1016/j.carbpol.2010.08.013

Golam K. (2014) Trace/heavy Metals and Its Impact on Environment, Biodiversity and Human Health- A Short Review. doi: 10.13140/RG.2.1.3102.2568

Herawati S. Suzuki N, Hayashi K, Rivai IF (2000) Cadmium, Copper, and Zinc Levels in Rice and Soil of Japan, Indonesia, and China by Soil Type. Bull Environ Contam Toxicol 64:33–39. doi: 10.1007/s001289910006

He ZL, Yang XE, Stoffella PJ (2005) Trace elements in agroecosystems and impacts on the environment. J Trace Elem Med Biol 19:125–140. doi: 10.1016/j.jtemb. 2005.02.010

Ho Y-S (2006a) Review of second-order models for adsorption systems. J Hazard Mater 136:681–689. doi: 10.1016/j.jhazmat.2005.12.043

Ho Y-S (2006b) Second-order kinetic model for the sorption of cadmium onto tree fern: A comparison of linear and non-linear methods. Water Res 40:119–125. doi: 10.1016/j.watres.2005.10.040

Ho YS, McKay G (1998) A comparison of chemisorption kinetic models applied to pollutant removal on various sorbents. Process Saf Environ Prot 76:332–340.

Kabata-Pendias A (2010) Trace elements in soils and plants. CRC press

Kang DW, Choi HR, Kweon DK (1999) Stability constants of amidoximated chitosan-g-poly(acrylonitrile) copolymer for heavy metal ions. J Appl Polym Sci 73:469–476. doi: 10.1002/(SICI)1097-4628(19990725)73:4<469::AID-APP2>3.0.CO;2-I

Kweon D-K, Choi J-K, Kim E-K, Lim S-T (2001) Adsorption of divalent metal ions by succinylated and oxidized corn starches. Carbohydr Polym 46:171–177. doi: 10.1016/S0144-8617(00)00300-3

Langmuir I (1918) THE ADSORPTION OF GASES ON PLANE SURFACES OF GLASS, MICA AND PLATINUM. J Am Chem Soc 40:1361–1403. doi: 10.1021/ja02242a004

Lokeshwari H, Chandrappa GT (2006) Impact of heavy metal contamination of Bellandur Lake on soil and cultivated vegetation. Curr Sci 91:622–627.

Murphy SD, Doull J, Klaasen CD, Amdur MO (1980) Toxicology: the basic science of poisons.

Nriagu JO (1989) A global assessment of natural sources of atmospheric trace metals. Nature 338:47–49. doi: 10.1038/338047a0

Pacyna JM (1996) Monitoring and assessment of metal contaminants in the air. Toxicol Met CRC Boca Raton FL 9–28.

Polat H, Erdogan D (2007) Heavy metal removal from waste waters by ion flotation. J Hazard Mater 148:267–273. doi: 10.1016/j.jhazmat.2007.02.013

Saito T, and Isogai A (2005) Ion-exchange behavior of carboxylate groups in fibrous cellulose oxidized by the TEMPO-mediated system. Carbohydr Polym 61:183–190. doi: 10.1016/j.carbpol.2005.04.009

Santhi T, Manonmani S, Smitha T (2010) Removal of malachite green from aqueous solution by activated carbon prepared from the epicarp of Ricinus communis by adsorption. J Hazard Mater 179:178–186. doi: 10.1016/j.jhazmat.2010.02.076

Shallari S, Schwartz C, Hasko A, Morel JL (1998) Heavy metals in soils and plants of serpentine and industrial sites of Albania. Sci Total Environ 209:133–142. doi: 10.1016/S0048-9697(98)80104-6

Shukla SR, Pai RS (2005) Adsorption of Cu(II), Ni(II) and Zn(II) on dye loaded groundnut shells and sawdust. Sep Purif Technol 43:1–8. doi: 10.1016/j.seppur.2004.09.003

Srivastava VC, Swamy MM, Mall ID (2006) Adsorptive removal of phenol by bagasse fly ash and activated carbon: Equilibrium, kinetics and thermodynamics. Colloids Surf Physicochem Eng Asp 272:89–104. doi: 10.1016/j.colsurfa.2005.07.016

Sträter E, Westbeld A, Klemm O (2010) Pollution in coastal fog at Alto Patache, Northern Chile. Environ Sci Pollut Res 17:1563–1573. doi: 10.1007/s11356-010-0343-x

Tumin ND, Chuah AL, Zawani Z, Rashid SA (2008) Adsorption of copper from aqueous solution by Elais Guineensis kernel activated carbon. J Eng Sci Technol 3:180–189.

Unuabonah EI, Olu-Owolabi BI, Adebowale KO, Ofomaja AE (2007) Adsorption of lead and cadmium ions from aqueous solutions by tripolyphosphate-impregnated Kaolinite clay. Colloids Surf Physicochem Eng Asp 292:202–211. doi: 10.1016/j.colsurfa.2006.06.024

Varma AJ, Deshpande SV, Kennedy JF (2004) Metal complexation by chitosan and its derivatives: a review. Carbohydr Polym 55:77–93. doi: 10.1016/j.carbpol.2003.08.005

Varma G, Singh R, Sahu V (2013) A comparative study on the removal of heavy metals by adsorption using fly ash and sludge: a review. Int J Appl Innov Eng Manag 2:45–56.

Wang H, Yuan X, Zeng G, et al (2014) Removal of malachite green dye from wastewater by different organic acid-modified natural adsorbent: kinetics, equilibriums, mechanisms, practical application, and disposal of dye-loaded adsorbent. Environ Sci Pollut Res 21:11552–11564. doi: 10.1007/s11356-014-3025-2

Ware GW (1994) The pesticide book. Fresno.

WHO (1996) Trace elements in human nutrition and health. Nutrition

Wiberg N, Holleman AF, Wiberg E (eds) (2001) Holleman-Wiberg's Inorganic Chemistry, 1 edition. Academic Press, San Diego: Berlin; New York

Wong YC, Szeto YS, Cheung WH, McKay G (2004) Adsorption of acid dyes on chitosan — equilibrium isotherm analyses. Process Biochem 39:695–704. doi: 10.1016/S0032-9592(03)00152-3

Xu S-M, Feng S, Yue F, Wang J-D (2004) Adsorption of Cu(II) ions from an aqueous solution by crosslinked amphoteric starch. J Appl Polym Sci 92:728–732. doi: 10.1002/app.13677

Yang Y, Shao J (2000) Synthesis of sulfhydryl chitin and its adsorption properties for heavy metal ions. J Appl Polym Sci 77:151–155. doi: 10.1002/(SICI)1097-4628(20000705)77:1<151::AID-APP20>3.0.CO;2-W

Zhang L-M, Chen D-Q (2002) An investigation of adsorption of lead(II) and copper(II) ions by water-insoluble starch graft copolymers. Colloids Surf Physicochem Eng Asp 205:231–236. doi: 10.1016/S0927-7757(02)00039-0

Zulfikar MA, Bahri A, Nasir M (2002) Adsorption of Humic Acid from Aqueous Solution onto PMMA Nanofiber: Kinetics Study.

ATSDR - Toxicological Profile: Copper. http://www.atsdr.cdc.gov/toxprofiles/ tp.asp?id=206&tid=37. Accessed 6 Nov 2016

Tapioca | Cassava Starch. In: IngredientsWizard Healthy Aging. http://ingredientswizard.com/home/tapioca-cassava-starch/. Accessed 1 Jul 2016.

■■■■

Chapter - 18

The role of Polyamines during environmental stresses in crop plants

Manisha Farsodia and Sunil Kumar Singh

Plant Cell and Molecular Biology Laboratory, Department of Botany, Faculty of Science, The M.S. University of Baroda, Vadodara-390 002, Gujarat, India. Email: sunilkrsingh@hotmail.com

ABSTRACT

Global warming predicted to increase evapotranspiration, increasing the frequency and intensity of environmental stresses which likely to severely affect growth and productivity of crop plants. To overcome these, crops with high yield potential and better stress tolerance needs to be developed. An effective modern approach is targeted identification of specific gene/s responsible for stress tolerance and their precise integration into crop plants. One such interesting candidate for stress tolerance is gene/s involved in the biosynthesis of polyamines (PAs) in plants, as polyamines are shown to be vital for cell viability and has protective role under various stress conditions. The protective role of polyamines cannot be generalized for all the polyamines. In many cases accumulation of free and conjugated polyamines as well as their oxidation has been observed when enhanced stress tolerance of plants is reported. This review focuses on summarizing the role of polyamines during environmental stress tolerance in plants.

Keywords: Polyamines, environmental stress, putrescine, spermidine, spermine, thermospermine

18.1 INTRODUCTION

Environmental stresses such as heat, drought, flood and salinity are some of the major stresses limiting crop production worldwide. These stresses are severely affecting growth, development and productivity of agricultural crops and global climate change is likely to exaggerate this situation (Gill and Tuteja, 2010). As plants are sessile, they constantly face and respond to change of their microenvironment. It may be the change in intensity of light, temperature or maybe flood or drought like conditions. Individually, these stresses have been the subject of intense research for many years (Bray *et al.*, 2000). Each plant species behaves differently, in a particular stress condition; some species may be stressful, and some may not be even though living in same conditions. In most cases, the physiology of plants play a crucial role to allow the plants to live in a wide range of environmental conditions (Minocha *et al.*, 2014a). The developmental stage of plant also plays a significant role in its response to changing environmental conditions. Therefore, maintaining crop yield under adverse environmental stresses is probably the most challenging aspect in today's agriculture.

Polyamines (PAs) ubiquitously distributed in all the living organisms, are considered as new class of plant growth substances also functions as the regulators of environmental stresses in plants (Gill and Tuteja, 2010). PAs are known for their anti-senescence and anti-stress effects due to their antioxidant and cationic nature. Also they have membrane stabilizing and cell wall stabilizing abilities (Zhao and Yang, 2008). PAs also play a crucial role in modulating defense responses, especially during oxidative stress (Rider *et al.*, 2007), drought (Yamaguchi *et al.*, 2007) and chilling conditions (Cuevas *et al.*, 2008; Groppa and Benavides, 2008; Chattopadhayay *et. al,* 2002; Ndayiragije and Lutts, 2006). Moreover, exogenous applications of PAs such as putrescine have successfully been used to increase plant tolerance against salinity (Ndayiragije and Lutts, 2006, Chattopadhayay *et al.*, 2002), cold (Nayyar and Chander, 2004, Nayyar, 2005), drought (Zeid and Shedeed, 2006), high temperature (Murkowski, 2001), osmotic stress (Liu *et al.*, 2004) and flooding (Yiu *et al.*, 2009). Furthermore, genetic transformation of plants with PA biosynthetic genes such as arginine decarboxylase (ADC), ornithine decarboxylase (ODC) s-adenosyl methionine decarboxylase (SAMDC), spermidine synthase (SPDS) enhanced their tolerance to varieties of environmental stresses (Liu *et al.*, 2007). Such multiple stress tolerance is of practical importance as plants face all these stresses concurrently during their life cycle. In this review, we have summarized the current knowledge about PAs metabolism and their role during environmental stress regulation in plants.

18.2 POLYAMINES IN PLANTS

PAS are small aliphatic amines present in all living cells. The most common PAs in plants are putrescine, spermidine and spermine as they are present relatively in larger quantity. Besides these, tetramine thermospermine is also reported in plants in addition to spermine (Fuell *et al.* 2010; Minguet *et al.*, 2008). At neutral pH, PAs exist as cations and due to their cationic nature, they affect important biological activities (Valero *et al.*, 2002). PAs occurs freely or in a conjugated form with other macromolecules such as nucleic acids and proteins, but their concentration depends on the need and type of cells and tissues (Duan *et al.*, 2008). Moreover, total and individual content of PAs varies with plant species, organs, developmental stage and are much higher than the phytohormones (Pal *et al.*, 2015). Thus, it is not surprising that fluctuation in their cellular content is often related to a varied response of plants to different forms of stress (Minocha *et al.*, 2014a).

18.3 POLYAMINE METABOLISM AND FUNCTIONS IN PLANTS

Precursor of all PAs is putrescine, which synthesized from ornithine by the enzyme ornithine decarboxylase (ODC) (EC 4.1.1.17) in all organisms, except in plants, where putrescine is alternatively formed via the activity of arginine decarboxylase (ADC) (EC4.1.1.19) through arginine. Further putrescine is converted to triamine spermidine through spermidine synthase (SPDS) (*EC* 2.5.1.16) and subsequently to tetraamines spermine or thermospermine via spermine synthase (SPMS) (*EC*2.5.1.22) and thermospermine synthase (TSPMS) (EC2.5.1.79) respectively. Enzyme S-adenosylmethionine decarboxylase (SAMDC) (*EC*4.1.1.50) transfers an aminopropyl group from s-adenosyl methionine (SAM) to both SPDS and SPMS sequentially to produce spermidine and spermine, respectively.

Catabolism of PAs is typically carried out by two enzymes; Diamine oxidases (DAO) and Polyamine oxidases (PAO). DAO breaks down putrescine, while PAO degrades spermidine and spermine to release ammonia and H_2O_2 (Bagni and Tassoni, 2001; Cohen, 1998; Smith and Best, 1978) (Fig. 1). Degradation of PAs gives rise to different compounds such as 4-Aminobutanal and 1,3-Diaminopropane which are biochemically active during different sets of reactions (Bouchereau *et al.*, 1999; Federico and Angelini, 1991; Šebela *et al.*, 2001; Walters, 2003) and can be further metabolized to β-alanine and GABA, respectively, which are associated with stress tolerance in plants (Bouche and Fromm, 2004; Koc *et al.*, 1998). Thus, these enzymes are important in preventing PA levels from getting very high in cells (Cervelli *et al.*, 2003; Minois *et al.*, 2011; Wang *et al.*, 2004). A controlled regulation of PAs levels is vital as PAs are involved in various plant developmental processes such as embryogenesis, reproductive organ development, root development, tuberization, fruit development, fruit ripening and leaf senescence as well as environmental stresses (Bais and Ravishankar, 2002; Evans, 1989; Galston *et al.*, 1997).

18.4 POLYAMINES AND ENVIRONMENTAL STRESSES

One of the most difficult problems relating the PAs roles during environment induced stress is lack of our understanding of the mechanisms underlying their functions (Hussain *et al.*, 2011; Minocha *et al.*, 2014).

PAs are known to have a direct protective role during stress. As they are cationic in nature, they can bind to various negatively charged macromolecules such as nucleic acids and proteins and stabilize their structure, especially under stress conditions. They are also able to bind with phospholipid head groups of membranes influencing their permeability (Lightfoot and Hall, 2014). Besides their direct protective role, PAS also regulate various cellular processes as signaling molecules (ref). Evidence suggests that plant PAs function in adaptive responses in various stresses, and this phenomenon is supported by extensively varying level of PAs during different stress conditions (Pal *et al.*, 2015). Varying levels of accumulation of PAS subjected to various environmental stresses have been reported in several plant species (Alcázar *et al.*, 2010a; Hussain *et al.*, 2011; Kusano *et al.*, 2008; Liu *et al.*, 2007; Shi and Chan, 2014). This accumulated size of cellular PAs pool can also be correlated with the stress tolerance capacity of the plants, further explaining the efficiency of PAs during stress conditions (Hatmi *et al.*, 2015). Moreover, accumulation of PAs under stress conditions is largely due to increased *de novo* synthesis of free PAs as the synthesis of PAs is mainly regulated at the transcriptional level (Liu *et al.*, 2015). This is further supported from the fact that most of PA biosynthetic genes including ADC or ODC, SPDS, SPMS and SAMDC have also shown to up regulated during stress conditions (Liu *et al.*, 2006, 8, 9 and 11). Stress tolerance of plants is mainly influenced by the role of PAs in signaling cascade rather than by their accumulation (Pal *et al.*, 2015). In addition to these, there are growing evidences that so called uncommon PAs such as norspermine or thermospermine also play an important role in stress tolerance (Cvikrová *et al.*, 2012).

Transgenic analysis have also supported the role of PAs during plant stress responses. Upregulation of PAs biosynthesis via overexpressing PAs gene generally increases plants tolerance to various stresses, whereas their knockout mutants were found less tolerant to abiotic stresses (Fariduddin *et al.*, 2013; Minocha *et al.*, 2014b; Pal *et al.*, 2015).

PAs acts as both reactive oxygen species ROS producer and ROS scavenger, regulating the redox homeostasis in plants (Saha *et al.*, 2015). Both the catabolism and backconversion of PAs by amine oxidases results into production of H_2O_2 in apoplast and peroxisome (Pottosin and Shabala, 2014). H_2O_2 have been known as a signal molecule since long which can activate stress specific signaling process through Mitogen Activated Protein Kinases (MAPKs) (Moschou *et al.*, 2008). PAs are related to the production of not only ROS but also to NO which functions as intercellular as well intracellular messenger induces various cellular responses, including stress responses in plants. Further it has been reported that spermidine and spermine treatment lead to rapid accumulation of NO in Arabidopsis (Tun *et al.*, 2006). Plants adaptation to stress is closely related to their ability to regulate and maintain ion channels. One of the mechanisms of PAs action is their ability to control ion channels, which they exert both by direct binding to ions or through signaling molecules such as ROS and NO (Pal *et al.*, 2015).

18.5 SALT STRESS

Various attempts have been made to provide salinity tolerance to plants by traditional breeding programs but it has not got much success (Duan *et al.*, 2008). In recent years, exogenous application of PAs in agriculture field and over expression of PAs biosynthetic genes have been implicated to provide salt tolerance to crop plants. Transgenic approaches have shown enhanced salinity tolerance to rice, tobacco, tomato and *Arabidopsis* by overexpressing various PAs biosynthtic genes (Kasukabe *et al.*, 2004; Neily *et al.*, 2011; Roy and Wu, 2001; Wi *et al.*, 2006).

Exogenously application of spermidine to *Cucumis sativus* affected the PAs metabolism and enhanced the short term salinity tolerance. Treated plants showed marked increase in ADC, ODC, SAMDC and DAO activities. Also an accumulation of free spermidine, spermine and putrescine conjugate was observed in roots during salt stress (Duan *et al.*, 2008). Role of putrescine in salt tolerance have also been investigated in arabidopsis (Urano *et al.,* 2004). In arabidopsis, there was induction of ADC2 upon salt stress causing accumulation of free putrescine. This conclude that ADC2 is the key enzyme for producing putrescine under salt stress. Analysis of effect of salt stress on plant growth and PAs level in *Lycopersicum esculentum, Brassica oleracea* and *Beta vulgaris* have revealed that PAs level changed with the salinity stress and mostly increased level of spermidine and spermine observed during salinity tolerance (Zapata *et al.*, 2004).

18.6 DROUGHT AND WATER STRESS

Drought is one of the most important environmental stresses affecting the productivity of crops. The level of drought tolerance of plants depends on their genetic makeup and how fast defense mechanisms activated against water deficits. Recent evidence suggested that metabolic pathways associated with PA metabolism also play a key role in drought stress tolerance (Hatmi *et al.*, 2015).

Arabidopsis ectopically overexpressing *Capsicum annuum* SAMDC significantly showed enhanced drought tolerance, also the levels of main PAs were significantly increased in overexpressing lines (Wi *et al.*, 2014). Similarly rice plants overexpressing *Datura* ADC produced much higher level of putrescine under drought stress which promoted more spermidine and spermine synthesis and ultimately protecting the plants from drought stress (Capell *et al.,* 2004). Furthermore, the transgenic *Arabidopsis* plants containing high level of putrescence and increased ADC2 activity showed drought tolerance by reducing water loss caused by transpiration (Alcázar *et al.*, 2010b). In contrast, *Arabidopsis* PA biosynthetic spms mutant was found to be hypersensitive to drought stress and this phenotype could only be rescue by spermine pretreatment, which shows that the drought hypersensitivity exhibited by the mutant is due to the spermine deficiency. Thus, spermine plays a protective role against the drought condition in *Arabidopsis* (Yamaguchi *et al.*, 2007).

Evidence suggests that exogenously applied PAs stabilize plasma membrane integrity and protected them under damage caused by flood stress as exogenously applied putrescine reduces the flood induced oxidative damage in *Allium fistulosum* by increasing the antioxidant capacity (Yiu *et al.*, 2009). Exogenous application of putrescine resulted in reduced ROS content, thus less oxidative stress in plant cells during flood.

18.7 HEAT AND CHILLING STRESS

Heat stress is known to have a detrimental effect on crops. High temperature leads to more energy loss through respiration, thus rapid respiration costs more to maintain the plants causing less productivity. Very little work has been done to understand the role of PAs during heat stress in plants. A recent study suggested that elevated PAs and proline content facilitate rapid response to heat stress and activates defense mechanism (ref). Heat stress triggers induced activity of both ADC and ODC and also an increase in the level of spermidine and spermine was observed due to their de novo synthesis in the leaves and root of tobacco. Similarly tomato plants, overexpressing SAMDC showed higher level of spermidine and spermine and was more heat tolerant than the control plants, through increasing the activities of antioxidant enzymes and by protecting lipid peroxidation (Cheng *et al.*, 2009). Additionally, Arabidopsis plants overexpressing SPMS were more thermo tolerant than spms knockout mutant by activating expression of heat shock (HS) related transcription factors and heat shock protein (HSP) related genes (Sagor *et al.*, 2013). Uncommon PAs and their biosynthesis were thoroughly studied in thermophilic bacteria and were shown to be involved in protecting them under extreme environment conditions (Oshima, 2007). These uncommon PAs were also reported to be increased in the appreciable amount in plants during heat stress (Roy and Ghosh, 1996).

PAs content also increases when plants are challenged with low temperature, however, the role of PAs in regulating cold stress tolerance is not so clear. It has been observed that during cold stress, PAs biosynthetic enzyme ADC is mainly induced in plants (Hummel *et al.*, 2004). Moreover, plants expressing ADC1 shown better resistance to cold stress than the control plants (Tiburcio *et al.*, 2015). Accumulation of putrescine is essential for arabidopsis to survive under freezing temperature as ADC1 and ADC2 mutants in Arabidopsis displayed reduced chilling tolerance (Cuevas *et al.*, 2008). Evidence suggests that PAs play important an role in protecting cucumber plants against chilling stress an by acting as antioxidative compound (Zhang *et al.*, 2009).

18.8 OXIDATIVE STRESS

Global anthropological activities have led to the release of halogenated hydrocarbons resulting in the depletion of the ozone which in turn led to an increase in the amount of UV-B radiation. Increased UV-B has the deleterious effect on plants (Krupa and Kickert, 1989; Teramura, 1983; Tevini and Teramura, 1989) as cell membrane is it potential targets (Murphy, 1983). As observed in other stresses, PAs titre also shown to increase in response to UV-B radiation in cucumber cultivars (Ref). This finding is consistent with the observation that exogenously applied PAs resulted in protection from injury during ozone (Bors *et al.*, 1989; Ormrod and Beckerson, 1986). Such tolerance ability of plants is may be due to the capacity of PAs to stabilize membrane structure and inhibit lipid peroxidation (Kitada *et al.,* 1979; Slocum *et al.*, 1984; Smith, 1985; Tadolini, 1988).

18.9 CONCLUSION

Rising temperature and drought specifically are negatively affecting the crop productivity worldwide. These stresses hinder the plant performance considerably by reducing their growth, development and ultimately the productivity. There is a need to develop crops with enhanced tolerance to these stresses as the extent of agriculture damage caused by a combination of these stresses is very high.

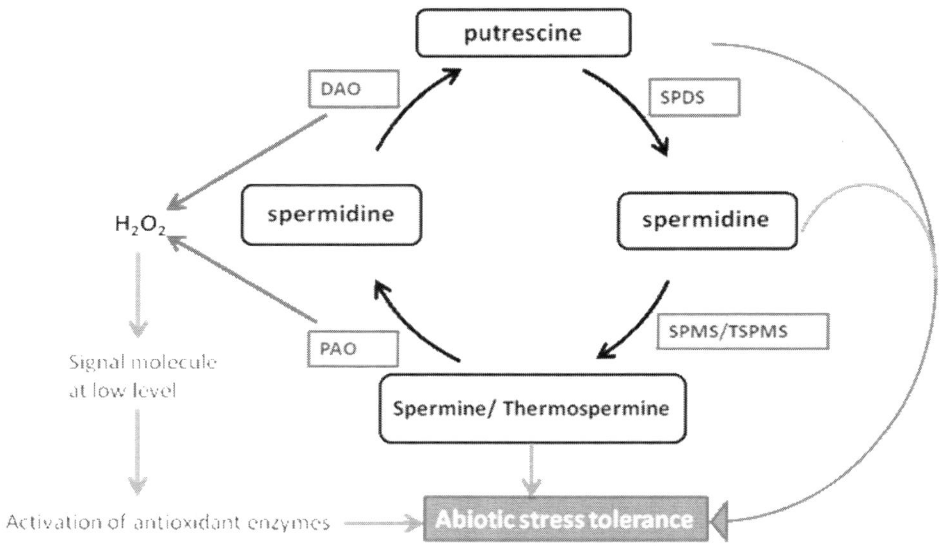

Figure 1. Polyamine cycle: Synthesis of higher polyamines through spermidine synthase (SPDS) and Spermine (SPMS) or Thermospermine Synthase (TSPMS) (shown in blue). Catobolism and back conversion of polyamines via Polyamine oxidase (PAO) and diamine oxidase (DAO) (shown in red) generates ROS such as H_2O_2. H_2O_2 produced from PAs play key role in abiotic stresses. Polyamines directly and indirectly providing stress tolerance.

Stress tolerance is complex and multigenic trait and most of its regulatory gene networks are largely unknown. It has become evident from the review that PAs are playing an important role in plant stress tolerance even though its underlying molecular regulation is still not very well known. Therefore, more research efforts are required to understand the molecular mechanism behind putrescine, spermidine and spermines protective roles under environmental stresses. Also it is clear from the review that by increasing or modulating in PAs endogenous level, more tolerance against heat, cold, drought, flood and oxidative stresses can be achieved crops. In various crops to meet the global food demands, this approach of increasing crop tolerance with respect to environmental stress is very important to further improve agricultural productivity and enhance food security.

Acknowledgements

MF thank the DST for INSPIRE Fellowship (No. DST/ INSPIRE Fellowship/ 2012/ 729). SKS thank the DBT (BT/PR13957/BRB/10/882/2010) and SERB (SERB/SR/SO/PS/53/2012) for grant-in-aid support.

REFERENCES

Alcázar, R., Altabella, T., Marco, F., Bortolotti, C., Reymond, M., Koncz, C., Carrasco, P. and Tiburcio, A.F. (2010a). Polyamines: molecules with regulatory functions in plant abiotic stress tolerance. Planta *231*, 1237-1249.

Alcázar, R., Planas, J., Saxena, T., Zarza, X., Bortolotti, C., Cuevas, J., Bitrián, M., Tiburcio, A.F. and Altabella, T. (2010b). Putrescine accumulation confers drought tolerance in transgenic Arabidopsis plants over-expressing the homologous Arginine decarboxylase 2 gene. Plant Physiology and Biochemistry *48*, 547-552.

Bagni, N. and Tassoni, A. (2001). Biosynthesis, oxidation and conjugation of aliphatic polyamines in higher plants. Amino acids *20*, 301-317.

Bais, H.P. and Ravishankar, G. (2002). Role of polyamines in the ontogeny of plants and their biotechnological applications. Plant Cell, Tissue and Organ Culture *69*, 1-34.

Bors, W., Langebartels, C., Michel, C. and Sandermann, H. (1989). Polyamines as radical scavengers and protectants against ozone damage. Phytochemistry *28*, 1589-1595.

Bouche, N. and Fromm, H. (2004). GABA in plants: just a metabolite? Trends in Plant Science *9*, 110-115.

Bouchereau, A., Aziz, A., Larher, F. and Martin-Tanguy, J. (1999). Polyamines and environmental challenges: recent development. Plant Science *140*, 103-125.

Bray, E.A., Bailey-Serres, J. and Weretilnyk, E. (2000). Responses to abiotic stresses. Biochemistry and molecular biology of plants *11581249*.

Capell, T., Bassie, L. and Christou, P. (2004). Modulation of the polyamine biosynthetic pathway in transgenic rice confers tolerance to drought stress. Proceedings of the National Academy of Sciences of the United States of America *101*, 9909-9914.

Cervelli, M., Polticelli, F., Federico, R. and Mariottini, P. (2003). Heterologous expression and characterization of mouse spermine oxidase. Journal of Biological Chemistry *278*, 5271-5276.

Chattopadhayay, M.K., Tiwari, B.S., Chattopadhyay, G., Bose, A., Sengupta, D.N. and Ghosh, B. (2002). Protective role of exogenous polyamines on salinity-stressed rice (Oryza sativa) plants. Physiologia Plantarum *116*, 192-199.

Cheng, L., Zou, Y., Ding, S., Zhang, J., Yu, X., Cao, J. and Lu, G. (2009). Polyamine accumulation in transgenic tomato enhances the tolerance to high temperature stress. Journal of integrative plant biology *51*, 489-499.

Cohen, S. (1998). A guide to the polyamines. Oxford University Press.

Cuevas, J.C., López-Cobollo, R., Alcázar, R., Zarza, X., Koncz, C., Altabella, T., Salinas, J., Tiburcio, A.F. and Ferrando, A. (2008). Putrescine is involved in Arabidopsis freezing tolerance and cold acclimation by regulating abscisic acid levels in response to low temperature. Plant physiology *148*, 1094-1105.

Cvikrová, M., Gemperlová, L., Dobrá, J., Martincová, O., Prásil, I.T., Gubis, J. and Vanková, R. (2012). Effect of heat stress on polyamine metabolism in proline-over-producing tobacco plants. Plant Science *182*, 49-58.

Duan, J., Li, J., Guo, S. and Kang, Y. (2008). Exogenous spermidine affects polyamine metabolism in salinity-stressed Cucumis sativus roots and enhances short-term salinity tolerance. Journal of plant physiology *165*, 1620-1635.

Evans (1989). Do polyamines have roles in plant development? Annual Review of Plant Biology *40*, 235-269.

Fariduddin, Q., Varshney, P., Yusuf, M. and Ahmad, A. (2013). Polyamines: potent modulators of plant responses to stress. Journal of Plant Interactions *8*, 1-16.

Federico, R. and Angelini, R. (1991). Polyamine catabolism in plants. Biochemistry and Physiology of Polyamines in Plants, In: Solum, R.D. and Flores, H.E. (Eds.). CRC Press, Boca Raton, FL, 41-56.

Fuell, C., Elliott, K.A., Hanfrey, C.C., Franceschetti, M. and Michael, A.J. (2010). Polyamine biosynthetic diversity in plants and algae. Plant Physiology and Biochemistry. 48, 513-520.

Galston, A. Kaur-Sawhney, R., Altabella, T. and Tiburcio, A. (1997). Plant polyamines in reproductive activity and response to abiotic stress. Botanica Acta *110*, 197-207.

Gill, S.S. and Tuteja, N. (2010). Polyamines and abiotic stress tolerance in plants. Plant signaling and behavior *5*, 26-33.

Groppa, M. and Benavides, M. (2008). Polyamines and abiotic stress: recent advances. Amino acids *34*, 35-45.

Hatmi, S., Gruau, C., Trotel-Aziz, P., Villaume, S., Rabenoelina, F., Baillieul, F., Eullaffroy, P., Clément, C., Ferchichi, A. and Aziz, A. (2015). Drought stress tolerance in grapevine involves activation of polyamine oxidation contributing to improved immune response and low susceptibility to Botrytis cinerea. Journal of experimental botany *66*, 775-787.

Hummel, I., Bourdais, G., Gouesbet, G., Couée, I., Malmberg, R.L. and El Amrani, A. (2004). Differential gene expression of arginine decarboxylase ADC1 and ADC2 in Arabidopsis thaliana: characterization of transcriptional regulation during seed germination and seedling development. New Phytologist *163*, 519-531.

Hussain, S.S., Ali, M., Ahmad, M. and Siddique, K.H. (2011). Polyamines: natural and engineered abiotic and biotic stress tolerance in plants. Biotechnology advances *29*, 300-311.

Kasukabe, Y., He, L., Nada, K., Misawa, S., Ihara, I. and Tachibana, S. (2004). Over-expression of spermidine synthase enhances tolerance to multiple environmental stresses and up-regulates the expression of various stress-regulated genes in transgenic Arabidopsis thaliana. Plant and Cell Physiology *45*, 712-722.

Kitada, M., Igarashi, K., Hirose, S. and Kitagawa, H. (1979). Inhibition by polyamines of lipid peroxide formation in rat liver microsomes. Biochemical and Biophysical Research Communications *87*, 388-394.

Koc, E., Bagga, S., Songstad, D., Betz, S., Kuehn, G. and Phillips, G. (1998). Occurrence of uncommon polyamines in cultured tissues of maize In Vitro Cellular & Developmental Biology-Plant *34*, 252-255.

Krupa, S.V. and Kickert, R. (1989). The greenhouse effect: impacts of ultraviolet-B (UV-B) radiation, carbon dioxide (CO2), and ozone (O3) on vegetation. Environmental Pollution *61*, 263-393.

Kusano, T., Berberich, T., Tateda, C. and Takahashi, Y. (2008). Polyamines: essential factors for growth and survival. Planta *228*, 367-381.

Lightfoot, H.L. and Hall, J. (2014). Endogenous polyamine function - the RNA perspective. Nucleic acids research, 42, 11275-11290.

Liu, H., Dong, B., Zhang, Y., Liu, Z. and Liu, Y. (2004). Relationship between osmotic stress and the levels of free, conjugated and bound polyamines in leaves of wheat seedlings. Plant Science *166*, 1261-1267.

Liu, J.-H., Inoue, H. and Moriguchi, T. (2008). Salt stress-mediated changes in free polyamine titers and expression of genes responsible for polyamine biosynthesis of apple in vitro shoots. Environmental and Experimental Botany *62*, 28-35.

Liu, J.-H., Kitashiba, H., Wang, J., Ban, Y. and Moriguchi, T. (2007). Polyamines and their ability to provide environmental stress tolerance to plants. Plant Biotechnology *24*, 117-126.

Liu, J.H., Nada, K., Honda, C., Kitashiba, H., Wen, X.P., Pang, X.M. and Moriguchi, T. (2006). Polyamine biosynthesis of apple callus under salt stress: importance of the arginine decarboxylase pathway in stress response. Journal of experimental botany *57*, 2589-2599.

Liu, J.-H., Wang, W., Wu, H., Gong, X. and Moriguchi, T. (2015). Polyamines function in stress tolerance: from synthesis to regulation. Frontiers in plant science *6*.

Liu, J., Nakajima, I. and Moriguchi, T. (2011). Effects of salt and osmotic stresses on free polyamine content and expression of polyamine biosynthetic genes in Vitis vinifera. Biologia plantarum *55*, 340-344.

Liu, J.H., Ban, Y., Wen, X.-P., Nakajima, I. and Moriguchi, T. (2009). Molecular cloning and expression analysis of an arginine decarboxylase gene from peach (Prunus persica). Gene *429*, 10-17.

Minguet, E.G., Vera-Sirera, F., Marina, A., Carbandl, J. and Blazquez, M.A. (2008). Evolutionary diversification in Polyamines biosynthesis. Molecular Biology and Evolution 25, 2119-2128.

Minocha, R., Majumdar, R. and Minocha, S.C. (2014). Polyamines and abiotic stress in plants: a complex relationship. Frontiers in Plant Science 5, 175.

Minois, N., Carmona-Gutierrez, D. and Madeo, F. (2011). Polyamines in aging and disease. Aging (Albany NY) *3*, 716.

Moschou, P.N., Paschalidis, K.A., Delis, I.D., Andriopoulou, A.H., Lagiotis, G.D., Yakoumakis, D.I. and Roubelakis-Angelakis, K.A. (2008). Spermidine exodus and oxidation in the apoplast induced by abiotic stress is responsible for H2O2 signatures that direct tolerance responses in tobacco. The Plant cell *20*, 1708-1724.

Murkowski, A. (2001). Heat stress and spermidine: effect on chlorophyll fluorescence in tomato plants. Biologia plantarum *44*, 53-57.

Murphy, T.M. (1983). Membranes as targets of ultraviolet radiation. Physiologia Plantarum *58*, 381-388.

Nayyar, H. (2005). Putrescine increases floral retention, pod set and seed yield in cold stressed chickpea. Journal of Agronomy and Crop Science *191*, 340-345.

Nayyar, H. and Chander, S. (2004). Protective effects of polyamines against oxidative stress induced by water and cold stress in chickpea. Journal of Agronomy and Crop Science *190*, 355-365.

Ndayiragije, A. and Lutts, S. (2006). Do exogenous polyamines have an impact on the response of a salt-sensitive rice cultivar to NaCl? Journal of plant physiology *163*, 506-516.

Neily, M.H., Baldet, P., Arfaoui, I., Saito, T., Li, Q.I., Asamizu, E., Matsukura, C., Moriguchi, T. and Ezura, H. (2011). Overexpression of apple spermidine synthase 1 (MdSPDS1) leads to significant salt tolerance in tomato plants. Plant Biotechnology, 28, 33-42.

Ormrod, D. and Beckerson, D. (1986). Polyamines as antiozonants for tomato. HortScience (USA).

Oshima T. 2007. Unique polyamines produced by an extreme thermophile, Thermus thermophilus. Amino Acids 33: 367-372.

Pal, M., Szalai, G. and Janda, T. (2015). Speculation: Polyamines are important in abiotic stress signaling. Plant Science *237*, 16-23.

Pottosin, I. and Shabala, S. (2014). Polyamines control of cation transport across plant membranes: implications for ion homeostasis and abiotic stress signaling. Frontiers in Plant Science, 5, 154.

Rider, J., Hacker, A., Mackintosh, C., Pegg, A., Woster, P. and Casero Jr, R. (2007). Spermine and spermidine mediate protection against oxidative damage caused by hydrogen peroxide. Amino acids *33*, 231-240.

Roy, M. and Ghosh, B. (1996). Polyamines, both common and uncommon, under heat stress in rice (*Oryza sativa*) callus. Physiologia Plantarum *98*, 196-200.

Roy, M. and Wu, R. (2001). Arginine decarboxylase transgene expression and analysis of environmental stress tolerance in transgenic rice. Plant Science *160*, 869-875.

Sagor, G., Berberich, T., Takahashi, Y., Niitsu, M. and Kusano, T. (2013). The polyamine spermine protects Arabidopsis from heat stress-induced damage by increasing expression of heat shock-related genes. Transgenic research *22*, 595-605.

Saha, J., Brauer, E.K., Sengupta, A., Popescu, S.C., Gupta, K. and Gupta, B. (2015). Polyamines as redox homeostasis regulators during salt stress in plants. Frontiers in Environmental Science *3*, 21.

Šebela, M., Radová, A., Angelini, R., Tavladoraki, P., Frébort, I. and Peč, P. (2001). FAD-containing polyamine oxidases: a timely challenge for researchers in biochemistry and physiology of plants. Plant Science *160*, 197-207.

Shi, H. and Chan, Z. (2014). Improvement of plant abiotic stress tolerance through modulation of the polyamine pathway. Journal of Integrative Plant Biology *56*, 114-121.

Slocum, R.D., Kaur-Sawhney, R. and Galston, A.W. (1984). The physiology and biochemistry of polyamines in plants. Archives of Biochemistry and Biophysics *235*, 283-303.

Smith, T.A. (1985). Polyamines. Annual Review of Plant Physiology *36*, 117-143.

Smith, T.A. and Best, G.R. (1978). Distribution of the hordatines in barley. Phytochemistry *17*, 1093-1098.

Tadolini, B. (1988). Polyamine inhibition of lipoperoxidation. The influence of polyamines on iron oxidation in the presence of compounds mimicking phospholipid polar heads. Biochemical Journal *249*, 33-36.

Teramura, A.H. (1983). Effects of ultraviolet B radiation on the growth and yield of crop plants. Physiologia Plantarum *58*, 415-427.

Tevini, M. and Teramura, A.H. (1989). UVB effects on terrestrial plants. Photochemistry and Photobiology *50*, 479-487.

Tiburcio, A.F., Artigas, T.A. and Monleón, A.F. (2015). Plant having resistance to low-temperature stress and method of production thereof. U.S. Plant No. 9, 139, 841.

Tun, N.N., Santa-Catarina, C., Begum, T., Silveira, V., Handro, W., Floh, E.I.S. and Scherer, G.F. (2006). Polyamines induce rapid biosynthesis of nitric oxide (NO) in *Arabidopsis thaliana* seedlings. Plant and Cell Physiology *47*, 346-354.

Urano, K., Yoshiba, Y., Nanjo, T., Ito, T., Yamaguchi-Shinozaki, K. and Shinozaki, K. (2004). Arabidopsis stress-inducible gene for arginine decarboxylase AtADC2 is required for accumulation of putrescine in salt tolerance. Biochemical and biophysical research communications *313*, 369-375.

Valero, D., Pérez-Vicente, A., Martínez-Romero, D., Castillo, S., Guillen, F. and Serrano, M. (2002). Plum storability improved after calcium and heat postharvest treatments: role of polyamines. Journal of food science *67*, 2571-2575.

Walters, D.R. (2003). Polyamines and plant disease. Phytochemistry *64*, 97-107.

Wang, X., Ikeguchi, Y., McCloskey, D.E., Nelson, P. and Pegg, A.E. (2004). Spermine synthesis is required for normal viability, growth, and fertility in the mouse. Journal of Biological Chemistry *279*, 51370-51375.

Wi, S.J., Kim, S.J., Kim, W.T. and Park, K.Y. (2014). Constitutive S-adenosylmethionine decarboxylase gene expression increases drought tolerance through inhibition of reactive oxygen species accumulation in Arabidopsis. Planta *239*, 979-988.

Wi, S.J., Kim, W.T. and Park, K.Y. (2006). Overexpression of carnation S-adenosylmethionine decarboxylase gene generates a broad-spectrum tolerance to abiotic stresses in transgenic tobacco plants. Plant cell reports *25*, 1111-1121.

Yamaguchi, K., Takahashi, Y., Berberich, T., Imai, A., Takahashi, T., Michael, A.J. and Kusano, T. (2007). A protective role for the polyamine spermine against drought stress in Arabidopsis. Biochemical and biophysical research communications *352*, 486-490.

Yiu, J.C., Juang, L.D., Fang, D.Y.T., Liu, C.W. and Wu, S.J. (2009). Exogenous putrescine reduces flooding-induced oxidative damage by increasing the antioxidant properties of Welsh onion. Scientia horticulturae *120*, 306-314.

Zapata, P.J., Serrano, M., Pretel, M.T., Amorós, A. and Botella, M.Á. (2004). Polyamines and ethylene changes during germination of different plant species under salinity. Plant Science *167*, 781-788.

Zeid, I. and Shedeed, Z. (2006). Response of alfalfa to putrescine treatment under drought stress. Biologia Plantarum *50*, 635-640.

Zhang, W., Jiang, B., Li, W., Song, H., Yu, Y. and Chen, J. (2009). Polyamines enhance chilling tolerance of cucumber (*Cucumis sativus* L.) through modulating antioxidative system. Scientia Horticulturae *122*, 200-208.

Zhao, H. and Yang, H. (2008). Exogenous polyamines alleviate the lipid peroxidation induced by cadmium chloride stress in *Malus hupehensis* Rehd. Scientia Horticulturae *116*, 442-447.

C: Environmental Problems and Solutions

प्रगति की ऊचाईयाँ

काश हम बनाते ऐसे नगर
जिनमें सड़कों पर गति होती
शोर और प्रदूषण नहीं,
भवनों एवं गगनचुम्बी अट्टालिकाओं में
प्यार और सहिष्णुता बसती
आपसी बैर नहीं,
काश हम बनाते ऐसे नगर
जिनमें कल कारखाने होते प्रदूषण रहित।
काश हम मानव के भविष्य को
नये अन्वेषणों, नवाचारों से
और उज्जवल बना पाते
और साथ ही
करते नये मानव मूल्यों का निर्माण
परस्पर सौहार्द और
आपसी द्वेष को भुला, पौधों की तरह
पवन में हर दम झूमना, लहराना
खुशी में कलियों का खिल खिलाना
आत्म विश्वास से भरे
गले लगाना सभी को
आपस में घुलमिल कर रहना,
सदा ऊचाईयों की और बढ़ते रहना।

Chapter - 19

Forests and Biodiversity Conservation : A Religious Perspective

Prof. S.D. Sabnis

Ex. Pro-Vice Chancellor, The Maharaja Sayajirao University of Baroda, Vadodara 390002

ABSTRACT

A world treaty, first of its kind, was signed by 150 states at the UN conference on Environment and development, held at Rio de Juniero, Brazil in June 1992. The strategies included human capacity for conserving and using biodiversity sustainably and efforts were made to strengthen it in developing countries. Conservation actions were taken through international cooperation and national planning. Habitat protection was ensured for conserving plant and wildlife. Number of National parks, wildlife sanctuaries and Biosphere reserves were formed to protect the biodiversity. Conservation efforts are needed, use of plants in religious ceremonies and protection of plants in sacred groves is a praiseworthy effort. Lord Buddha has said, "The forests makes no demands for its sustenance and extends protection to all beings offering shade even to the axe man who destroys it."

Keywords: Forests, Biodiversity, Conservation, Sacred Groves.

19.1 FORESTS: LIFE IN HARMONY

A few Ancient Indian texts equate forests and trees with God and call upon people to venerate them and worship them. Why this Godly status has been given to forests and trees when there are countless number of Gods and deities in the Hindu pantheon?

In a forest, every living being has its own place and function. From the mightiest animal to the lowliest bug or butterfly, each lives in balance and harmony with the others. None of its inhabitants take more than what it needs for its sustenance and well-being. Present day man being the only exception.

In spite of the fact that the forests support numerous and diverse species, these are pure and pristine and are provided with natural cleansing agents. What may be a waste for one is useful or essential to another. There is thus a natural rhythm which involves all the living species including the Man in the system.

The forests contribute to the welfare of the entire planet, giving out the life-supporting oxygen into the atmosphere. These are also rich sources of a variety of materials: and medicinal herbs and micro-organisms. The various ecological services provided by the forests are still little understood by the common man.

Without realizing the importance of forests or gaining total knowledge about their inhabitants, Man has, in his quest for more and more, unleashed an undeclared war on them. Forests have been and are being wantonly destroyed. The rain forests, very rich in biodiversity, at one time covered 14% of earth surface. Presently these occupy less than 6%. If this trend continues, the rain forests will disappear from the planet within the next few decades. Many efforts at forest and biodiversity conservation have not brought the desired results. Hence it is suggested that old Indian tradition of 'Sacred Groves' be subtly introduced to safeguard forests and their very valuable biodiversity.

During the process of evolution of human societies and their urge for permanent settlements, the idea of preserving portions of good forests and setting them apart for a variety of purposes took deep roots. Traditional literature is replete with reference indicating the practices of preservation of these natural woodlots on ethical and aesthetics grounds. The practice appears to be prevalent even during pre-Harappan and Harappan periods. The practice seems to have been recognized and widely accepted.

19.2 SACRED GROVE: RELIGIOUS CONCEPT

The religious concept of 'sacred grove' emerged during the last few centuries pockets of more or less climax vegetation preserved on religious grounds in remote areas". These are revered by local inhabitants and even outsiders as dwellings of Gods and deities to be left untouched and undisturbed.

These repositories of plant and animal diversity represent a traditional form of nature worship or may even be a clever ploy to protect proprietary right of the priests managing the temples within. Whatever may have been the compelling reason, the facts stands that the "sacred groves are performing their functions quite admirably in many parts of the country. It is very likely that the fear of deity's wrath rather than just reverence which has left the grooves fairly undisturbed. Even picking twigs from forests floor is considered a sacrilege and an offence, which invites deity's anger.

The sacred groves are widely scattered in many parts of the country. Their expanse ranges from 5 ha to a few hundred hectares. Nearly 350 such groves are listed along the Western Ghats in Maharashtra alone. 28 of these are fairly large and should be protected at all costs for their roles as preservers of biodiversity. The sacred groves or Devrai as commonly understood in parts of Maharashtra are conspicuous by their absence in Gujarat. Reasons for this do not come to mind immediately as both these regions have shared close historical and cultural links. It is possible that Gujarat being more or less a flat, country, agricultural lands have been more valuable than the forests lands. A king therefore would offer agricultural land as 'baksheesh' to his sardars and temple priests. It is also possible that forest lands were cleared by these recipients for agriculture. National parks, game reserves, sanctuaries are areas where good forests develop because of the protection provided over the years. Gujarat boasts of such protected areas which provide shelter to its dwindling wildlife. Will it not be appropriate to cleverly create and operate 'sacred grove' psychology and philosophy in these areas to ensure their permanence? It should allow the diverse life forms to live in peace and proliferate. If such unconventional measures are subtly initiated with active co-operation of leaders of various religions and their sects or cults, it will be possible to inculcate among the masses love and respect for nature as God's valuable creation. After all, we are interested in the 'end' rather than the 'means'. Everything should be pardonable if that leads to forest and biodiversity conservation in this forest-deficient state. The 'sacred grove' philosophy will have to be woven into the social fabric of the Indian Society.

19.3 SHALL WE BE ABLE TO DO THIS? A DEBATABLE QUESTION INDEED!

What is true for the forests is equally true for a number of trees venerated and/or worshipped by followers of various religious faiths.

The Hindus revere and even worship quite a few trees: Ashwatha (Peepal), Vat, Oudumbar, Bilva (*Aegle marmelos*), Coconut (*Cocos nucifera*), Mango (*Mangifera india*), Sita Ashok (*Saraca ashoka*), Palash (*Butea monosperma*), Neem (*Azadirachta indica*), Kadamba (*Anthocephalus kadamba*) and many others.

The Muslim revere the Date Palm (*Phoenix sylvestris*), Mulberry, Frangipani, Mehandi (*Lawsonia inermis*), Olive, Pomegranate (*Pumica granatum*), Fig, etc. The Buddhists consider Peepal (Indian 'Bodh'), *Ginkgo biloba* (Chinese 'Bodhi), Chandan (*Santalum album*), Bhojpatra (*Betula utilis*), Sevan, Rayan and Sal (*Shorea robusta*) among many others as sacred. The Christians have X-mas tree, Holly Sweet Bay, Persimmon, Poplar Willow, Oak, Olive, Aspen, Birch, etc. as their religious and sacred trees. The Sikhs treat Vat (*Ficus benghalensis*), Aritha, Sisam, Ber (*Zizyphus indica*), etc., as very sacred solely due to their close association with the Gurus. The Jain, in the same way, venerate Saptaparni (*Alostonia scholaris*), Shirish, Sal, Behada, Tendu (*Diospyros melanoxylon*), Jamun, Bakul (*Mimusops elengi*), etc.

Out of a very large number of tree species in the world why only a few have been considered 'sacred" is a matter for deep study and research. However, it is suggested that these religious trees properly labelled be used extensively for wasteland plantations and in social forestry.

The 'sacred' trees and groves should help preserve greenery in every place in any part of the country.

■■■■

Chapter - 20

Role of *Casuarina* in soil conservation along sea coast of Gujarat

Shalini Ojha

Department of Botany, Faculty of Science,
The Maharaja Sayajirao University of Baroda, Vadodara-390 002 India
E mail: shivalika02@rediffmail.com

ABSTRACT

Most coastlines are naturally dynamic, and cycles of erosion are often an important feature of their ecological character. Wind, waves and currents are natural forces that easily move the unconsolidated sand and soils in the coastal area, resulting in rapid changes in the position of the shoreline. Efforts are made to manage coastal erosion problems and to restore coastal capacity to accommodate short-and long-term changes induced by human activities. The erosion problem becomes worse whenever the counter measures applied are inappropriate and improperly designed.

Besides mangroves there are many trees which help in coastal areas to reduce soil erosion like *Alder, Sesbania, Thespesia, Anacardium* and many more but most research is still needed for other coastal forests and trees such as *Pandanus, Casuarina, Pinus*, which do not directly interact with waves during normal conditions, but do so during extreme events such as major storms and tsunamis. *Casuarina* roots also have symbiotic relationships with three other types of microorganism. The mycorrhizal fungi enter the root and facilitate the uptake of minerals, notably phosphorus and some trace elements. Actinorhizal plants showed increased nodulation caused by *Frankia*. Sand dune soils have high leaching rates, with a resulting low level of nitrogen. Frankia acts as a nitrogen fixers and helps in coastal area restoration. Nodule formation and physiology is described in present paper.

Keywords: *Casuarina*, Conservation, Coastal area, Gujarat

20.1 INTRODUCTION

The removal of top soil by water currents is called soil erosion, which is caused by soil disturbances and altered drainage patterns. Trees are good solution to many erosion problems. Trees reduce the rate of erosion by various means like protecting the soil from the impact of rain, transpiring large amount of water which counteracts very wet soil and binding soil to sloping land with their roots.

A well-developed wind break or shelter belt can have considerable influence in reducing wind velocity at the soil surface, where the barriers is at the right angles to the wind direction. Forest and trees provide some coastal protection and that the clearing of coastal forests and trees has increased the vulnerability of coast to erosion. Coastal erosion is a complex process that need to be investigated from the various angles like wind wave, tidal current, beach dynamics and human activities along the coastline. Scientific investigations on how coastal vegetation provides a measure of shoreline protection have been conducted. Many field and laboratory trials shows that mangrove forest and other coastal vegetation of certain density can reduce wave height considerably and protect the coast from erosion, as well as effectively prevent coastal sand dune movement during strong winds.

In the Nagapattinam coast of India, *Casuarina* shelterbelt planting is believed to have acted as a barrier and reduced the impact of the giant tsunami waves (Ganesan 2005). Under normal circumstances, most sand removed by storms from beaches in coastal areas is stored offshore, and moves back to the beach via wave transport. On the beach, *Casuarina* roots retain the sand. Stabilizing of shifting sand dunes is based on the principle of reducing the threshold velocity of wind by the tree-belt. Sand dune stabilization with plant species is more permanent than mechanical mulching and chemical fixation techniques (FAO 1989). *Casuarina equisetifolia* is a popular fast growing low-maintenance species, used extensively for windbreaks and coastal stabilization. *Casuarina* has root suckering habit, extensive foliage which arrests wind velocity and ability to reclaim soil under excess salinity conditions (Torrey 1976). Although *Casuarina* is not a legume, it does have the ability to form root nodules and fix atmospheric nitrogen (National Research Council 1984).

The definition of a coastal shelterbelt is highly complex because it needs to define the relationship of trees to other elements of the coastal and urban environment. It is a challenge to understand the dynamic nature of the coastal shelterbelt and its anticipated environmental, social and economic changes of urban environment. There is a strong relation between major coastal erosion problems and degradation of the protective function of coastal systems such as coastal forest and trees – particularly mangrove forest. Artificial and natural agents that induce mangrove loss and make coastal areas more susceptible to coastal erosion include anthropogenic factors such as excessive logging, direct land reclamation for agriculture, aquaculture, salt ponds, urban development and settlement, and to a lesser extent fires, storms, hurricanes, tidal waves and erosion cycles owing to changing sea levels (Kovacs, 2000). More scientific investigation and quantification of the physical processes and dynamic interaction of the system is needed to understand how and under what circumstances mangrove forests and other coastal vegetation effectively protect the shoreline from erosion.

Fig 1. Trees growing in sea coast of Tithal in Gujarat. A tree showing growth of *Lorenthus* Root Nodules caused by *Frankia*

Land degradation from water induced soil erosion is a serious problem in India, especially in Gujarat where several kilometers of area along sea coast is out of any vegetation, only fragmentary information on factors affecting soil erosion is available. Main factors which are responsible for the soil erosion in coastal belt are-Coastal geomorphology: Coastline and sensitivity to coastal processes Wind: The main force in wave generation, under the right environmental conditions, wind may transfer sediments from the beach environment landward on all open coastlines Waves: they are the most important forces for soil erosion. They introduce energy to the coast and also a series of current that move sediment along the shore. There are various human activities which have bad influences on these things.

A key component that guarantees stability of coastal sand dunes is vegetation. Network of endomycorrhizal hyphae can intermesh sand grains to form aggregates that subsequently help in stabilize loose sand. The endomycorrhizae definitely be beneficial from their aggregating ability to play a significant role in stabilizing coastal sand dunes, thus synergistically increasing the effectiveness of Casuarina planted as shelterbelts to conserve soil for other agricultural practices. Plantation of such type of trees can reduce wind speed by 30-50% and soil loss up to 80%. This planting would also contribute substantially to achieving other objectives of sustainable agriculture.

Among these species which helps coastal areas in several ways are exotic trees such as are Casuarina are widespread in tropical and subtropical zones where they play an important role due to their symbiotic relationship with mycorrhizal fungi and Frankia bacteria. These microorganism increase plant growth and development (Yang and Paszkowski, 2011). They also improve nutrient availability-particularly P and N (Xh and Critchley, 2008). Casuarinaceae species are actinorhizal plants which originated from Australia. The term Casuarina comes from the Malay word 'kasuari' due to resemblance of the twigs and the plumage of the cassowary bird (Boland *et al.*, 1994). These are widely established around the world with the exception of Antarctica.

Casuarina is a popular rapid-growing low-maintenance often planted in sandy soil as windbreaks. I has the potential to form root nodules and fix atmospheric nitrogen (National Research Council 1984). The coastal shelterbelt generally moderates local climate, slowing wind and storm water, cooling and cleaning the air, providing shade, increasing scenic value and enhancing coastal tourism and recreation.

20.2 *CASUARINA:* CHARACTERISTICS OF PLANTS

Casuarina is the most important plant of the tropics. *Casuarina* based farming is very popular in the coastal regions of India as it buffers the economy against crop failure and draught. The family Casuarinaceae comprises about 90 species of trees and shrubs and is the only family included under the order Casuarinales. The family comprises four genera (*Allocasuarina, Casuarina, Ceuthostoma and Gymnostoma*) and approximately 86 species and 13 subspecies (Steane *et al.*, 2003; Bisby *et al.*, 2007).

Casuarina is the most important actinorhizal genus of tropics and subtropics (Mansour and Baker, 1994). They are woody, angiosperms evergreen trees. Leaves are scale like, branchlets are needle like, bark is brown rough fibrous and exfoliating in longitudinal a trips. (Jose *et al.*, 2002). The *Casuarinas* are monoecious or dioecious and occasionally both conditions exist within one species.

Casuarina is a large fast growing tree. The tree attains height upto 40mt with diameter of 60 cm (180 cm girth). Variation in flowering and fruiting may occur within localities. Along the coastal regions, where *Casuarina* thrives, the temperature is extreme sometimes extending to 47°C. It grows well both southwest and North East monsoons. *Casuarina* grows best in loose, fine coastal sands, It tolerates lateritic and red soils and also saline, alkaline and acidic conditions. *Casuarina,,* which is a tree of major family that has been disseminated throughout the tropics owing to their ability to grow in adverse conditions (Echhab *et al.* 2007).

Casuarina is an actinorhizal species that is able to grow in nutrient poor soils, mainly because of it's ability to fix atmospheric N_2 in the root nodule through a mutualistic symbiosis with the actinomycete *Frankia*, and by it's capacity to develop mycorrhizae (Gordon 1998). Because of this *Casuarina* plantations may largely modify soil characteristics, especially in coastal dunes. It has been estimated that *Frankia* fixes atmospheric nitrogen up to 362 Kg N/ha/year, which is an essential nutrient for all plant metabolic activities and growth.

Casuarina roots also have symbiotic relationships with three other types of microorganism. Two form of mycorrhizal fungi enter the root and facilitate the uptake of minerals, notably phosphorus and some trace elements. Furthermore a range of unidentified microorganism interact with the plant root to produce dense mats of "proteoid roots the large surface area of which probably also help to absorb phosphorus and other vital minerals. Actinorhizal plants showed increased nodulation by *Frankia,* in presence of soil bacteria *Pseudomonas cepacia* (Knowlton *et al.*, 1980). It is due to the ability of bacterium to cause root hair deformation which is essential for Frankial infection. With the combination of all four symbiont, it seems hard to imagine plants better suited than *Casuarinas* to the reforestation of low fertility lands.

Sand dune soil have high leaching rates, with a resulting low level of nitrogen. When nitrogen fixers form large stands, local nitrogen cycling may be altered significantly (Vitousek 1986), resulting in higher phosphorus and nitrogen concentrations in leaves as well as enriched soil-nutrient levels (Versfeld and Van Wilgen 1986). This enriched soil has been observed in *Casuarina* stands too. Leaf and branch litter beneath *Casuarina* trees can be 5-10 cm thick (Fernald and Barnett 1991), with organic matter accumulating at rates that are similar to those recorded in boreal and subalpine forests (Mailly and Margolis 1992).

20.3 *CASUARINA:* PHYSIOLOGICAL AND BIOCHEMICAL ASPECTS

Casuarina jhunghuhniana was found with high salinity tolerance as compared to those of *Casuarina equisetifolia*. Antioxidants enzyme activity were significantly high in *C. junghuhniana*, which is positively correlated with high salinity tolerance.

Thick Mesophyll and high palisade/spongy parenchyma ratio are anatomical indicates of drought tolerance (Castro and Jauregui, 1996). Drought tolerant species has sunken stomata whereas drought intolerant species have flat stomata.

20.4 BENEFITS OF *CASUARINA*

All species of *Casuarina* showed high water logging tolerance due to large proportion of aerenchyma in the roots (Vander Moezel *et al.*, 1989). The nitrogen fixing ability of *Casuarina* depends on *Frankia* strains, host genotype, soil factor, nutrient status, temperature and moisture. (Nicodemus *et al.,* 2001). *C. equisetifolia* is important as a multipurpose tree for agroforestry, as wind breaks, shelterbelts in coastal tracts, in wide- row inter-planting systems, and in croplands (Viswanath *et al.,* 2001). *C. equisetifolia* is also known by various names such as Australian pine, beef wood, whistling pine and swamp oak.

20.5 *FRANKIA:* A ROOT SYMBIONT

These are Gram positive, nitrogen fixing soil actinomycetes capable of forming actinorhizal symbiosis (Bond, 1967). Studies also show that members of genus *Frankia* can survive and remain infective on soils that are devoid of host planta. (Hahn *et al.,*1999). The *Frankia* has possibility to infect approximately 194 species in 24 genera within 8 families (Benson and Silvester, 1993). Owing to their capacity for nitrogen fixation, nodulated species can grow and improve soil fertility in disturbed sites, and are used in recolonization and reclamation of eroded sand dunes, erosion control and in agroforestry (Wheeler and Miller, 1990). *Frankia* avoids desiccation in arid soil environments by surviving as vegetative hyphae deep within the soil (Burleigh and Dawson, 1994).

20.5.1. Root Nodule Symbiosis

Actinorhizal plants can be infected by root hair mediated mechanism (Callaham *et al.*, 1979) or by direct penetration of the root through the inter cellular spaces of the epidermis (Miller and Baker, 1985). In both cases Frankia hyphae penetrates the host cells As a result of root infection, a root meristem is induced from the pericycle near the site of infection. Symbiosis results in the formation of deformed root with dichotomous branching which is the actinorhizal nodule. Frankia grows as endosymbiont occupying the major part of living host cell. The morphology of endophyte is host dependent.

20.5.2 Morphology of *Frankia*

In culture, three morphological forms are characteristic for Frankia: hyphae (or filaments), spores, and vesicles. Hyphae are typically 0.5 µm thick, septate and branched while vesicles are spherical, about 1–5 µm in diameter and septate

(Newcomb and Wood 1987). Vesicles are typically formed under N limited conditions and are the site of the N_2-fixing enzyme dinitrogenase (Meesters 1987). A striking feature of the vesicles is the surrounding envelope which consists of multiple layers of bacterial steroid lipids, hopanoids (Berry *et al.* 1993), which presumably assist in regulation of oxygen tension near dinitrogenase. The envelope is continuous, but much thinner around hyphae. Multilocular sporangia with spores are formed terminally or in an intercalary position on the hyphae. Spores will germinate into hyphae.

In root nodules, hyphae are always present and vesicles are found in nodules of all studied genera except Casuarina and Allocasuarina. Vesicle shape and size, presence or absence of septa in vesicles, and the spatial distribution of vesicles within an infected plant cell differ among symbioses and are determined by the host (Huss-Danell 1997). Vesicles are the site of dinitrogenase (Huss-Danell and Bergman 1990). However, in Casuarina and Allocasuarina dinitrogenase must be localized in hyphae. Frankia may or may not form spores within nodules, and according to spore formation nodules are designated as Sp+ or Sp–, respectively (Van Dijk and Merkus 1976). Presence of spores has been studied mainly in Alnus, Comptonia and Myrica (Schwintzer 1990; Dijk and Merkus 1976) and depends on the Frankia strain and environmental factors (Simonet *et al.* 1994).

20.6 INFECTION MECHANISM AND NODULE DEVELOPMENT

Among actinorhizal symbioses, three of the eight host families (Myricaceae, Betulaceae and Casuarinaceae) are nodulated by Frankia via the intracellular infection pathway. In five of the families (Elaeagnaceae, Rhamnaceae, Rosaceae, Datiscaceae and Coriariaceae), early nodule initiation occurs, (only indirect evidence for the families Rosaceae, Datiscaceae and Coriariaceae), via intercellular colonization (Wall and Berry 2008).

Mechanisms are briefly described as follows in the intracellular infection, or root-hair infection pathway, Frankia induces deformation of the root hairs and in a deeply-folded region of these root hairs, Frankia filaments transit from the root-hair surface to the inner part of the root hair forming a structure within the root hair that is analogous to the infection thread found in legume-rhizobial symbioses. Frankia penetration of the root hair triggers cell divisions in the root cortex subadjacent to the infected root hair, forming a zone called the prenodule. Some of these newly-divided cells expand and subsequently become infected by Frankia. In Casuarina glauca it has been shown that the prenodule cells display the same differentiation as the corresponding nodule cells (Laplaze *et al.*, 2000).

In the intercellular infection pathway, Frankia filaments invade the cortex of young roots by growing in the middle lamella between adjacent epidermal cells and cortical cells. Root hair deformation and cortical cell division to form prenodules are not induced. The host cells secrete extracellular material, creating an expanded intercellular zone.

Once Frankia has started the infection of the root, either intracellularly or intercellularly, a nodule lobe primordium is initiated in the root pericycle. As the nodule lobe primordium expands, the nodule cortex becomes infected intracellularly by Frankia filaments transiting through the infected area of the root cortex (either the prenodule or the intercellular infected tissue, depending on infection strategy) into the base of the primordium. Finally, the mature nodule develops from the apical meristem, Frankia vesicles differentiate within the cortical cells, and N_2 fixation is expressed. Nodule aeration can be facilitated by lenticels or nodule roots. Nodule roots contain large air spaces in their cortex and grow geotropically from tips of nodule lobes. They show determinate growth. The length of nodule depends on the oxygen tension.

The organization of infected cells in the nodule cortex differs among actinorhizal plant genera. In most cases they are interspersed with uninfected cells. In *Casuarina,* vacuole filled with phenolics form continuous files from the base of the apex to the base of the nodule lobe, separating layers of infected and uninfected phenolic free cortical cells. These files are organized in concentric layers. In each layer *Frankia* grows acropetally without crossing the files of phenolic-filled cells except in the youngest part of the nodule where not all cells of the files are already filled with phenolics (Pawlowski and Sirrenberg, 2003).

20.7 FACTORS AFFECTING ACTINORHIZAL *FRANKIA* SYMBIOSIS

Actinorhizal plants can be infected by root hair mediated mechanism (Callaham *et al.,* 1978) or by direct penetration of hair root through the intercellular spaces of the epidermis (Miller and Baker, 1985).

The symbiotic system is more sensitive to temperature stress than does plant growth. Multiplication of *Frankia* in rhizosphere and infection and nodule development process are sensitive to low temperature which is ultimately reflected through delayed infection and reduced nodule mass (Reddell and Bowen, 1985). Most actinorhizal trees have very little nodule activity below 20ºC. The optimum temperature for actinorhizal symbiosis is 25ºC. Asoil temperature of 15ºC inhibits nodule formation but increasing soil emperature, increased nodule number and nodule weight and shoot weight with seedlings of *Casuarina cunninghamiana* (Reddell *et al.,* 1985) which is very helpful aspect for this tree to grow in coastal belt.

Most *Casuarinas* are known to grow well in saline environments. They have mechanism to overcome salt stress for their survival. *C. glauca* and *C. equisetifolia* have been reported to be highly salt tolerant, surviving up to 500mM sodium chloride concentration. Nodule activity is affected by solutes in several ways. There is wide variation in the salt tolerance of host plants as a well as their associated *Frankia* (Dawson and Gibson, 1987). Many *Casuarina* are known to grow well in saline environments. They have mechanism to overcome salt stress for their survival.

Soil moisture has an important role in symbiosis. Kant and Narayana (1978) reported the decreased number and size of the nodules on *C. equisetifolia* with increasing severity of moisture stress. Reduction of soil moisture content below 25% markedly reduced the acetylene reduction activities (Hopemans *et al.*, 1983). Actinorhizal trees have a deeper root system that helps to extract moisture from deeper root system that helps to extract moisture from deeper layers and nodulated well. Highly lignified nature of actinorhizal nodule facilitates N_2 fixation to function even when soil is very dry.

Thus owing to theirs symbiosis, actinorhizal plants are capable of growing in marginal soils and therefore used in erosion control, soil reclamation, dune stabilization, pulp and timber production and in inter cropping systems. *C. equisetifolia* is found to be very salt tolerant. Though they are very important tree but research on the has been hampered, as they are mostly trees and shrubs which are not very amenable to molecular biological analysis and due to their long generation time they are unsuited for genetic analysis (Pawlowski and Sirrenberg, 2003).

20.8 PROBLEMS IN THE SOIL CONSERVATION

Despite years of study and substantial investment in remediation and prevention, soil erosion continues to be a major environmental problem with regard to land use in India and elsewhere around the world.

Gujarat has a long coastline and two out of three gulfs of the Country, namely Gulf of Kutchh and Gulf of Khambhat. They are the regions of rich biological diversity. In fact, on the inter-tidal zone, mangrove vegetation are found which provide large-scale food and good habitat for marine life. Restocking of old mangrove areas and Afforestation of the new inter tidal mud flats would offer excellent resting, roosting and breeding sites for aquatic birds. The high biological activity associated with mangroves is also responsible for food provided for the hundreds of thousands of migrant birds that visit our region for wintering. Ramifying root systems of mangrove act as effective silt traps and the filtering action prevents the silt from getting deposited on coral formations. Thus mangrove ecosystem would improve the ecology as well as economic development of the area. They are of immense value for the growth of estuarine fisheries and allied occupations. The herding cattle find good fodder in mangrove particularly during scarcity periods. Gujarat has presently 940 Sq.km of mangrove forests. Recently potential area for raising mangrove has been identified and estimated to be approximately 1400 sq.km. Along with the plantation of new areas, restocking of existing mangrove forest is also necessary. Plantation programme is undertaken to address these issues.

The area as a whole faces problems common to most of the coastal areas of Gujarat. These include low soil fertility, increasing soil salinity, scanty vegetation cover, soil erosion, scarcity of fresh water, overgrazing and overexploitation of fuel wood, depletion of fish stocks due to mangrove destruction, and discharge of industrial waste into the sea, which has really need attention to grow. Mangroves buffer the destructive power of cyclones and other storms, thus reducing coastal erosion. They are the feeding and breeding habitats for a large number of commercially important fish and marine organisms. They are also the nesting sites for a large number of water birds. In the Saurashtra and Kutchh regions of Gujarat, mangroves are a vital source of fodder for the migratory camel herders (Jats). Mangroves have numerous uses such as fuel wood, fodder, green manure, boat building, fish traps and fishing floats, tannin for the preservation of fishing nets, raw material for paper production, food, drugs, honey, beverages and household items.

20.9 CONCLUSION

The tree control erosion and stabilize the littoral by holding sediments and building upland. A green belt of trees will effectively reduce the force of devastating storm surges and waves that accompany cyclones and tsunamis. It will provide an amenity and a source of food and material for coastal community. Growing such type of trees are beneficial to biodiversity and can create habitat corridors for wildlife. Soil conservation is very important because soil is crucial for many aspects of human life as it provides food, filters air and water and helps to decompose biological waste into nutrients for new plant life. Soils can be drained away or contaminated, destroying it for use.

REFERENCES

Benson, D.R. and Silvester, W.B. (1993). Biology of *Frankia* strains, actinomycete symbionts of actinorhizal plants. Microbiol Biol Rev. 57 (2): 293-319

Berry, A. M., Harriott O.T., Moreau R.A., Osman S.F., Benson D.R., and Jones. A.D. (1993). Hopanoid lipids compose the Frankia vesicle envelope, presumptive barrier of oxygen diffusion to nitrogenase, *Proceedings of the National Academy of Sciences* 90, (13): 6091-6094.

Bisby, F. A., Y. R. Roskov, M. A. Ruggiero, T. M. Orrell, L. E. Paglinawan, P. W. Brewer, N. Bailly, and J. van Hertum, J. (2007). Species 2000 and ITIS catalogue of life: 2007 annual checklist. Digital resource at www.catalogueoflife.org/annual-checklist/2007/Species 2000." *Reading, UK* (2007).

Boland, D J, M I H Brooker, G M Chippendale, N Hall, B P M Hyland, R D Johnston, D A Kleinig, and J D Turner (1994). Forest tree of Australia, CSIRO, Melbourne, Australia

Bond, G. (1967). Nitrogen fixation in some non-legume root nodules. Phyton 24: 57–66.

Burleigh, S.H., Jeffrey O. and Dawson J.O. (1994). Occurrence of Myrica-nodulating *Frankia* in Hawaiian volcanic soils." *Plant and Soil* 164(2): 283-289.

Callaham, D., Delitredici, P. and John G. Torrey J.G. (1978). Isolation and cultivation *in vitro* of the actinomycete causing root nodulation in *Comptonia*. *Science* 199, (4331): 899-902.

Dawson, J. O. and Gibson, A.H. (1987) Sensitivity of selected *Frankia* isolates from *Casuarina*, *Allocasuarina* and North American host plants to sodium chloride." *Physiologia Plantarum* 70(2): 272-278.

Dijk, C. V, and Merkus, E. (1976) Amicroscopical study of the development of a spore-like stage in the life cycle of the root nodule endophyte of *Alnus glutinosa* (L.) Gaertn. *New Phytologist* 77.1 (1976): 73-91.

Echhab H (2007), Successful nodulation of Casuarina by *Frankia* in axenic conditions. *J. Appl. Microbiol.* 103: 1728–1737

FAO (1989), Arid zone forestry: A Guide for Field Technicians; Publications Division, Food and Agriculture Organization of the United Nations, Rome, Italy.

Fernald, R.T. and B.S. Barnett (1991) Establishment of native hammock vegetation on spoil islands dominated by Australian pine (*Casuarina equisetifolia*) and Brazilian pepper (*Schinus terebinthifolius*). Pp 131–150

Ganesan, S. (2005), 'Shelter belt plantations take on killer waves'; The Hindu Online edition of India's National Newspaper, India, Sunday, 2 January 2005, p. 3. Loughborough, W.K. (undated), A study of the seasoning and allied characteristics of *Casuarina equisetifolia* .

Gordon, D.R. (1998), Effect of invasive, non-indigenous plant species on ecosystem processes: lessons from Florida. Ecological applications, Wiley Online Library

Hahn, D, Nicke, A.L. and Dawson J. (1999). Assessing *Frankia* populations in plants and soil using molecular methods." *FEMS microbiology ecology* 29,(3): 215-227.

Hopmans, P., Douglas, L.A. and Chalk P.M. (1983) Nitrogen fixation associated with *Acacia dealbata* Link. seedlings as estimated by the acetylene reduction assay." *Australian Journal of Botany* 31 (4): 331-339.

Huss-Danell K. and Bergman B. (1990), Nitrogenase in Frankia from root nodules of Alnus incana (L.) Moench: immunolocalization of the Fe- and MoFe-proteins during vesicle differentiation. *New Phytologist*, 116 (3): 443-455

Huss-Danell (1997), Actinorhizal symbioses and their N^2 fixation, New Phytologist, 136,(3)

Kant S and Narayana HS (1978). Effect of water stress on growth, nodulation and nitrogen fixation in *Casuarina quesetifolia* . *Ann.Arid Zone* 17: 216-221

Knowlton, S. Berry, A. and Torrey J.G. (1980) Evidence that associated soil bacteria may influence root hair infection of actinorhizal plants by *Frankia.*" *Canadian journal of microbiology* 26, (8): 971-977.

Kovacs, J.M. (2000), Perceptions of environmental change in a tropical coastal wetland. *Land Degradation & Development,* 11: 209–220.

Laplaze, L., Ribeiro A., Franche C., Duhoux E., Auguy F., Bogusz D. and Pawlowski K. (2000) Characterization of a *Casuarina glauca* nodule-specific subtilisin-like protease gene, a homolog of *Alnus glutinosa* ag12." *Molecular plant-microbe interactions* 13, (1): 113-117.

Lee, T.M. 2005. Monitoring, FAO (1989), Arid zone forestry: A Guide for Field Technicians; Publications Division, Food and Agriculture Organization of the United Nations, Rome, Italy.

Mailly D. and Margolis, H.A. (1992), Forest floor and mineral soil development in *Casuarina equsetifolia* plantations on the coastal sand dunes of Senegal, Forest ecology and management, Elsevier

Mansour, S. R., and Baker D.D. (1994) Selection trials for effective N2-fixing Casuarina-Frankia combinations in Egypt. *Soil Biology and Biochemistry* 26(5): 655-658.

Meesters T.M. (1987), Localization of nitrogenase in vesicles of Frankia sp. Cc1.17 by immuno goldlabelling on ultrathin cryosections, *Archives of Microbiology*, 146 (4): 327-331

Miller I. M. and Baker, D.D. (1985), The initiation, development and structure of root nodules in *Elaeagnus angustifolia* L. (Elaeagnaceae), *Protoplasma* 128 (2): 107-119.

National Research Council (1984), *Casuarinas*: Nitrogen-fixing Trees for Adverse Sites', National Academy Press, Washington, DC

Newcomb, W. and Wood S. M. (1987) Morphogenesis and fine structure of *Frankia* (Actinomycetales): the microsymbiont of nitrogen-fixing actinorhizal root nodules." *International review of cytology* 109: 1-88.

Pawlowski K. and Sirrenberg A. (2003). Symbiosis between *Frankia* and actinorhizal plants: Root nodules of non-legumes 1-339, NISCAIR Online Periodicals Repository

Reddell, P. and Bowen G. D. (1985) *Frankia* source affects growth, nodulation and nitrogen fixation in *Casuarina* species." *New Phytologist* 100 (1): 115-122.

Rouchelle R. S., Mascarenhas A. and Jagtap T.G. 2011 An evaluation of flora from coastal sand dunes of India: Rationale for conservation and management." *Ocean & coastal management* 54(2): 181-188.

Knowlton S. and J O Dawson, (1983), Effect of *Pseudomonas cepacia* and cultural factors on the nodulation of *Alnus rubra* roots by *Frankia*, *Can. J. Bot.*, 61:2877-2882

Schwintzer C.R. (1990), Spore -positive and spore-negative nodules, The biology of Frankia and actinorhizal plants.

Simonet, P., Marco B., Chrystelle C., Moiroud A., and Normand P. (1994) Molecular characterization of *Frankia* microsymbionts from spore-positive and spore-negative nodules in a natural *Alder* stand." *Applied and environmental microbiology* 60 (4): 1335-1341.

Steane, D. A., Karen L. W. and Robert S. Hill R.S. (2003) "Using matK sequence data to unravel the phylogeny of Casuarinaceae." *Molecular phylogenetics and evolution* 28(1): 47-59.

Torrey, J.G. (1976), Initiation and development of root nodules of Casuarina, American Journal of Botany, 63(2): 335–344.

Urban Development Authority (2006), Basic Guide to Design and Planting of Coastal Green Belt in Sri Lanka; Urban Development Authority, Colombo, Sri Lanka.

Van der M., Walton, P.G.C.S. Pearce-Pinto, G.V.N. and Bell, D.T. (1989) Screening for salinity and waterlogging tolerance in five *Casuarina* species." *Landscape and Urban Planning* 17(4): 331-337.

Versfeld DB and van Wilgen BW (1986), Impact of woody aliens on ecosystem properties. In: Macdonald IAW, Kruger FJ and Ferrar AA (eds) The Ecology and Management of Biological Invasions in Southern Africa, pp 239-246. Oxford University Press, Cape Town, South Africa

Viswanath, S., Manivahakam, P. and George, M. (2001), Casuarina equisetifolia in agroforestry practices, 187-192: In *Casuarina Improvement and Utilization,* Gurumurthi, K., Nicodemus, A. and Siddappa (eds.) Institute of Forest Genetics and Tree Breeding, Coimbatore, India.

Vitousek,P.M. (1986), Biological invasions and Ecosystem processes: Can species make a difference? Ecology of Biological Invasion of North America and Hawai, Ecological studies, 58: 163-176

Wall, L. G., and Berry. A.M. (2007) Early interactions, infection and nodulation in actinorhizal symbiosis." In *nitrogen-fixing actinorhizal symbioses*, pp. 147-166. Springer Netherlands.

Wheeler, C. T., and Miller.I.M. (1990) Current and potential uses of actinorhizal plants in Europe." *The biology of Frankia and actinorhizal plants* (1990).

Wolanski, E. (1992), Hydrodynamics of mangroves swamp and their coastal waters. *Hydrobiologia,* 247: 141–161.

Wu, Y., Falconer, R.A. and J. Struve. (2001). Mathematical modelling of tidal currents in mangrove forests. Environmental Modelling & Software 16 (2001) 19–29

XH HE and C Critchley, (2008), Frankia nodulation, mycorrhization and interactions between *Frankia* and mycorrhizal fungi in *Casuarina* plants, Springer

Yang S.Y. and Paszkowski, U. (2011), Phosphate import at the Arbuscule: Just a Nutrient? Molecular Plant-Microbe Interactions, 24(11): 1296-1299.

■■■■

Chapter - 21

Essential oils as next generation fungicides: Contrivances and challenges

Akash Kedia[1], Abhishek Kumar Dwivedy[2] and Nawal Kishore Dubey[2]

[1]Department of Botany, Government General Degree College at Mangalkote, P.O. Khudrun, Burdwan-713174, W.B., India

[2]Center of Advanced Study in Botany, Banaras Hindu University, Varanasi-221005, U.P., India

E mail: nkdubeybhu@gmail.com

ABSTRACT

Essential oils (EOs), the volatile aromatic oily liquids from higher plants are easily extracted by steam/hydro distillation and contain many low-molecular-weight terpenes and phenolics. Many EOs showed a broad spectrum of antifungal activity against various plant pathogenic fungi promoting their future application as plant based fungicide. Recent investigations pointed that some chemical constituents of these oils inhibit the synthesis of ergosterol in fungi, making them effective antifungal agents. As this target site is confined to fungi and is not shared with other groups, most EO chemicals are relatively non-toxic to mammals in toxicological tests, fulfilling the criteria for "reduced risk" fungicides. Further, many EOs and their components are already registered as safe and are widely used as flavoring agents in foods and beverages. This special regulatory status combined with the wide availability of EOs, can be convinced into fast-track commercialization of EO-based fungicides. Meanwhile, the issue of resistance development against many synthetic fungicides can further be minimized by using EOs owing to the complex mixture of constituents of EO. Ultimately, these "green fungicides have the greatest impact in development of "next generation fungicides" due to their safety to non-target organisms and the environment.

Keywords: Antifungal, essential oil, fungicide, safe.

21.1 INTRODUCTION

Fungi cause enormous loss to agricultural products both in the field and storage conditions throughout the globe. Success in reducing the damage of agricultural properties caused by fungi depends greatly on the timely application of fungicides. Chemical control through synthetic fungicides is the most prevalent method applied mainly throughout the world and large numbers of synthetic fungicides such as organochlorines, organophosphates, organomercurals, dithiocarbamic acids, systemic fungicides etc. are continuously used against plant pathogenic fungi. The history of synthetic fungicide dates back to 1885 when Bordeaux mixture was discovered by Millardet in France. It was the landmark in the history of chemical disease control and is categorized as 'first generation of fungicides', which include inorganic chemicals. The 'second generation of fungicides' including organic chemicals, dates back to 1934 with the discovery of dithiocarbamates along with quinones, captan and related compounds. 'Third-generation fungicides' are also organic but systemic in action i.e. penetrate the plant tissue and also useful to control established infections. These fungicides include 2-aminopyrimidines, benzimidazoles, carboxamides, phenylamides, azoles and related compounds (Ragsdale and Sisler, 1991).

This method of application of synthetic fungicides, however, has many negative concerns due to their post application side effects to non-target organisms, pest resistance and residual toxicity leading to threat in food security (Brent and Hollomon, 1998). Due to their side effects, many of these synthetic pesticides have been banned in the developed countries and the addition is continuously going on. As for example, Lindane, a well known organochlorine, become highly toxic if consumed along with food. Organophosphates such as Malathion, Dichlorvos cause residual toxicity in food. Pyrethroids such as Permethrin, Deltamethrin have carcinogenic properties. Many Dithiocarbamates have mutagenic, carcinogenic and teratogenic properties when consumed along with the food material (Heil, 2008; Fishbein 1976; OHS, 1991, USEPA, 2006). As plant disease control heavily depends on synthetic fungicides to combat the wide variety of fungal diseases and due to serious ill effects of synthetic fungicides, there is a quick need of some safer alternatives to protect the plants from the noxious fungal pathogens. Hence, in present period, the search for 'fourth and next generation of fungicides' is widely researched throughout globe having the compounds that is safe for human and environment, having multiple modes of action, no side-effects on non-target organisms than those produced by conventional fungicides.

21.2 ESSENTIAL OILS

Plants contain a number of defensive chemicals viz. terpenoids, phenolics, alkaloids etc. being used by them since antiquity for their protection against different pathogens without causing any ill effect to the nature. This valuable defensive chemistry can easily be exploited as plant based fungicide against fungal pathogen. These plant based formulations being biodegradable and eco-friendly would be a better alternative to synthetic fungicides. Essential oils (EOs), being a mixture of various terpenoids, phenolics and other defensive molecules and due to their simple process of extraction, emerging as better alternative of synthetic fungicides all over the world. EOs are usually stored in oil ducts, resin ducts, glands or trichomes (glandular hairs) of the plants (Baser and Demirci, 2007) and may be obtained from different plant parts viz. flowers, seeds, leaves, twigs, barks, fruits and roots through the processes like hydro-distillation, steam-distillation, cold pressing, expression, fermentation, effleurage or extraction etc (Rajendran and Sriranjini, 2008). For commercial production, steam or hydro-distillation is the most common method of extraction (Van de Braak, and Leijten, 1999). EOs may contain 20-100 different plant secondary metabolites belonging to a wide variety of chemical classes of which terpenoids and phenylpropanoids form the major constituents. Additionally, few aromatic and aliphatic constituents are also present (Carson *et al.*, 2006). The bioactivities of a particular EO are decided by either one or two of its main components or overall synergistic activities of all of its major and minor components (Isman *et al.*, 2008). Further, the chemical composition of a particular EO however, can vary according to many factors such as genetic variation, plant variety, plant nutrition, application of fertilizers, geographical and seasonal variation, stress factors, age of plant, post harvest drying and storage, method of extraction etc. and thereby decides the characteristic biological properties of EO (Hussain *et al.*, 2008). Hence, standardization of EO in terms of its chemical composition is mandatory before recommendation as plant based fungicide.

Around 3000 EOs have been produced by around 2000 plant species, out of which 300 are commercially important. Plants producing EOs belong to around 60 families of which the plants from families Alliaceae, Apiaceae, Asteraceae, Lamiaceae, Myrtaceae, Poaceae and Rutaceae show well ability to produce EOs of medicinal and industrial value (Vigan, 2010). Commonly, EOs are used as flavouring agents in food products, perfumaries, pharmaceuticals, cosmetics and use of EOs in traditional systems of medicine is being practiced since ancient times (Burt, 2004). Currently researchers from all over the world exploring the wide application of EOs as they are showing a wide range of biological properties including antimicrobial, antiviral, antimutagenic, anticancer, antioxidant, antiinflammatory, immunomodulatory and antiprotozoal activities (Raut and Karuppayil, 2014). The antifungal property of EOs against a wide range of fungi is a well researched area as a plethora of literature is available promoting their future applications as 'next generation fungicide'.

21.3 MODE OF ANTIFUNGAL ACTION OF ESSENTIAL OIL

Recent investigations pointed that some chemical constituents of EO inhibit the synthesis of ergosterol in fungi, making them effective antifungal agents. As this target site is confined to fungi and is not shared with other groups, most EO chemicals are relatively non-toxic to mammals in toxicological tests, fulfilling the criteria for non-toxic to non-target organisms. The mode of antifungal action of EOs on fungal cell is mainly attributed to the interruption of sterol biosynthetic pathways resulting in reduced ergosterol biosynthesis as similar with the antifungal action of azole antifungal drugs (Kelly *et al.*, 1995; Pinto *et al.*, 2009, Romagnoli *et al.*, 2005). Ergosterol is the fungal specific sterol in cell membrane responsible for maintaining the function and integrity of fungal cell membrane (Rodriguez *et al.*, 1985). In addition, being lipophilic, EOs target the plasma membrane and the membranous organelles of the fungal cell by either crossing or accumulating in the membrane bilayer. This results in membrane swelling, change in fluidity and increased passive flux of protons which ultimately leads to cell death as supported by the transmission electron microscopic study (Kedia *et al.*, 2015; Tian *et al.*, 2012). Further, the increased leakages of Ca^{+2}, K^{+} and Mg^{+2} ions from EO treated cells of *Aspergillus flavus* also supports the membrane dysfunction (Helal *et al.*, 2007; Kedia *et al.*, 2015). Some of the phenolic EO components viz. thymol, carvacrol, eugenol etc. have been well studied for their antifungal actions. Thymol and Carvacrol have been reported to cause depletion of intracellular ATP concentration and finally cell death. Eugenol disrupts cell membrane by dissipating H+ and K+ ion gradients causing leakage of vital cellular constituents resulting in water imbalance (Helander *et al.*, 1998).

21.4 SAFETY PROFILE OF ESSENTIAL OIL

Regarding safety of EOs on human consumption, most of the EOs are already in use in pharmaceuticals from ancient times and as a flavoring agent in foods without any side effects. Most of the EOs are already on the 'Generally Recognised as Safe' (GRAS) list approved by the Food and Drug Administration (FDA) and Environment Protection Agency (EPA) in USA, favoring their applications as safe fungicide without any ill effects on human health (Burt, 2004; Tripathi and Dubey, 2004). In last few years, a number of EOs have been tested for their safety limit profile in mammalian system and the results showed non-mammalian or less mammalian toxicity in most of the cases. Further, EO treated seeds can be used for sowing purpose as no ill effects on seed viability was observed on EO treated seeds (Zabka and Pavela, 2013).

21.5 POTENTIAL OF ESSENTIAL OILS AS NEXT GENERATION FUNGICIDE

Based on different studies, EOs showed a great potential for the formulation of plant based green fungicide against wide range of plant pathogenic fungi. Recently, progress in formulation of EOs in micro/nano capsules providing a boon for their use as fungicide as it will reduce the amounts, will increase the duration of effectiveness by controlled volatilization, will increase the activity of EOs at the same concentration and will slow down the rate of degradation of EOs (Dayan *et al.*, 2009). Some of the EOs worked in detail for their fungicidal effects having better efficacy to be develop as next generation fungicide is discussed below.

Ajowan oil – Kedia *et al.* (2015) reported the efficacy of *Trachyspermum ammi* fruit EO (ajowan oil) as plant based preservative in food system based on its detail study against *Aspergillus flavus*. The major component was found as cymene (76.27%) followed by thymol (13.30%), dl-limonene (3.23%), 1,8-

cineole (2.58%) and γ-terpinene (1.68%) as analyzed through Gas Chromatography-Mass Spectrometry (GC-MS) analysis. The EO showed minimum inhibitory concentration at 0.8 μL/mL against a wide range of food deteriorating fungi including *A. flavus*. The components showed synergistic effects having multiple modes of action as the highest antifungal action was showed by thymol. Electron microscopic observations of EO treated *A. flavus* cell showed the mode of action on fungal plasma membrane with reduced ergosterol quantity and increased leakages of ions. Further the EO showed less toxicity (LD_{50} 6620.43 μL/kg) when tested for oral-toxicity on mice and high antioxidant activity (IC_{50} 0.467 μL/mL) when tested through 2,2-diphenyl-1-picrylhydrazil (DPPH) free radical analysis. Based on its high antifungal activity, multiple modes of action and mammalian safety, the ajowan EO may thus be formulated as safe plant based fungicide for large scale application.

Basil oil – Prakash *et al.* (2011) reported the *Ocimum gratissimum* EO as plant based preservative and recommended its application as a nontoxic antimicrobial agent against fungal contamination of spices. The major components of the EO were identified as Methyl cinnamate (48.29%) and γ-terpinene (26.08%) through GC-MS analysis The EO was found to exhibit better efficacy as fungitoxicant than the synthetic fungicide Wettasul-80 against the fungal isolates from spices. The minimum inhibitory concentration of the EO against *A. flavus* was found at 0.6 μL/mL concentration. The EO showed better antioxidant activity through DPPH free radical scavenging and β-carotene-linoleic acid bleaching assay and showed non-mammalian toxicity (LD_{50} 11622.67 μL/kg) during oral toxicity on mice. Kumar *et al.* (2011) presented *O. basilicum* EO as a plant-based preservative against fungi. At 1 μL/mL concentration, the EO exhibited complete inhibition of *A. flavus*. The EO showed broad fungitoxic spectrum when used as fumigant in closed storage containers. Kumar *et al.* (2010), further showed the efficacy of *O. sanctum* EO and its major component eugenol against food contaminating fungi. The minimum inhibitory concentration of EO and eugenol was found at 0.3 and 0.2 μL/mL concentrations respectively against common food deteriorating fungi. The EO and eugenol also exhibited better efficacy than some prevalent synthetic antifungals. The EO was found non-mammalian toxic (LD_{50} 4571.43 μL/kg) on mice. The findings of these studies revealed the possible exploitation of basil oils as safe alternative to synthetic fungicides.

Citronella oil – Nakahara *et al.* (2003) elucidated the chemical composition and antifungal activity of *Cymbopogon nardus* EO (citronella oil). The major components as revealed through GC and GC-MS were geraniol (35.7%), trans-citral (22.7%), cis-citral (14.2%), geranyl acetate (9.7%), citronellal (5.8%) and citronellol (4.6%). The EO significantly suppressed the growth of several species of *Aspergillus*, *Penicillium* and *Eurotium* at 250 mg/L air dose following the vapor-agar contact method. Among the components, citronellal and linalool were found as the most active compounds as they completely inhibited the growth of all tested fungi at 112 mg/L air dose showing the synergistic effects of all components for the overall bioactivity of EO. Li *et al.* (2013) explored the antifungal effect of citronella oil with *Aspergillus niger* conidia through poisoned food technique, broth dilution method and disc volatility method. Through various experiments, they found that 0.5% (*v/v*) EO concentration was a threshold to kill the conidia thoroughly. With light and electron microscopy, they found that EO could destruct the cell wall, pass through the cell membrane, act on the main organelles leading to irreversible alteration of the hyphae and conidia and subsequently kill the conidia. Hence, citronella oil provides a potential of being a safe and eco-friendly fungicide in the future.

Clove oil – Pinto *et al.* (2009) studied the composition and antifungal activity of *Syzygium aromaticum* EO (clove oil). The GC and GC-MS analysis showed a high content of eugenol (85.3%). Further, from detailed study, it was confirmed that the bioactivity of EO was due to its major component eugenol as the EO and eugenol both showed strong inhibitory activity against *Candida*, *Aspergillus* and dermatophyte clinical and American Type Culture Collection strains through standard protocols. The mode of antifungal action was performed through flow cytometry and ergosterol synthesis inhibition test on yeasts and filamentous fungi. The EO and eugenol caused extensive lesions in cell membrane and reduction in the quantity of ergosterol. Below their minimum inhibitory concentrations, they also inhibited the germ tube formation by *Candida albicans*. This study argued the potential of clove oil and eugenol as green fungicide as they showed considerable antifungal activity against clinically relevant fungi, including fluconazole-resistant strains. A similar finding was also observed by Bullerman *et al.* (1977) where eugenol was found as the major active antifungal ingredient of clove oil favoring its exploitation as natural fungicide.

Cumin oil – Kedia *et al.* (2014a) reported the potential of *Cuminum cyminum* (cumin) seed EO as a plant based shelf life enhancer against fungal contamination of food items. The EO showed efficacy as a preservative in stored wheat and chickpeas against a wide range of food spoiling fungi including *Aspergillus flavus*. The major component of the EO was identified as cymene (47.08%) through GC-MS analysis. The EO showed toxicity against a broad spectrum of food borne fungi and the minimum inhibitory concentration was found at 0.6 μL/ml concentration. The mode of antifungal action of EO on EO treated *A. flavus* cells showed reduced ergosterol content. Further, the EO showed strong antioxidant potential (IC_{50} 0.092 μL/mL) through DPPH free radical analysis and provided sufficient protection of seed samples against fungal association without affecting seed germination. Hence, in view of the antifungal nature against a wide range of fungi, free radical scavenging potential and efficacy in food system, cumin oil may be able to develop as plant based fungicide for wider applications.

Dill oil – Tian *et al.* (2012) demonstrated the *Anethum graveolens* (Dill) seed EO as a potential source of an eco-friendly antifungal agent. The mode of antifungal action of EO was investigated on the plasma membrane and mitochondria of *Aspergillus flavus* through flow cytometry and then verified through the ergosterol synthesis inhibition. The EO at 4 μL/mL concentration caused morphological changes in the cells, reduction in the ergosterol quantity, elevation of mitochondrial membrane potential, suppression of the glucose-induced decrease in external pH, decreased ATPase and dehydrogenase activities and Reactive Oxygen Species accumulation in *A. flavus* cells. The same oil was further experimented for its *in vivo* antifungal activity as a potential food preservative (Tian *et al.*, 2011). The minimum inhibitory concentration of EO was obtained at 2.0 μL/mL concentration against *A. flavus*, *A. oryzae*, *A. niger* and *Alternaria alternata*. The Scanning electron microscopic examination of EO treated *A. niger* cells showed distorted conidial heads, decreased hyphal diameters, shriveled hyphal aggregates, and swelling of the hyphal wall. The EO significantly protected decay of cherry tomatoes in *in vivo* condition favoring its use to control food spoilage from fungi.

Geranium oil – Džamić *et al.* (2014) described the chemical composition, antifungal and antioxidant activity of *Pelargonium graveolens* EO (geranium oil). The main components were identified as citronellol (24.54%), geraniol (15.33%), citronellyl formate (10.66%) and linalool (9.80%) through GC and GC-MS analysis. The antifungal test was performed using microdilution and macrodilution methods against *Alternaria alternata*, *Aspergillus flavus*, *A. niger*, *A. ochraceus*, *A. terreus*, *A. versicolor*, *Aureobasidium pullulans*, *Candida albicans*, *Cladosporium cladosporioides*, *C. fulvum*, *Fusarium sporotrichoides*, *F. tricinctum*, *Mucor mucedo*, *Penicillium funiculosum*, *P. ochrochloron*, *Phoma macdonaldii*, *Phomopsis helianthi*, *Trichoderma viride* and *Trichophyton menthagrophytes*. The synthetic antifungal bifonazol was taken as a control. The EO at 0.25-2.5 mg/mL concentration showed fungicidal activity to all tested fungi except *Mucor mucedo* and *Aspergillus* species whereas it showed minimum inhibitory concentration below 1.00 mg/mL against all tested fungi except *A. flavus*. The EO showed strong antioxidant activity (EC_{50} 0.802 mg/mL) evaluated by means of the DPPH free radical assay. Hence, due to its wide range of fungitoxic activity, the geranium oil can be formulated for the generation of plant based fungicide.

Lemongrass oil – Mishra and Dubey (1994) screened 14 EOs for their antifungal activities against *Aspergillus flavus* and found *Cymbopogon citratus* (lemongrass) EO to exhibit better fungitoxicity. At 1 μL/mL concentration, the EO showed fungistatic nature, wide fungitoxic spectrum, non-phytotoxicity and superiority over synthetic fungicides viz. Agrosan G. N., Thiride, Ceresan, Dithane M-45, Agrozim, Bavistin, Emison, Thiovit, wettable sulfur and copper oxychloride. The fungitoxicity of EO remained unaltered for 7 months of storage as well as at high doses of inoculum. Further, the EO was thermostable within 5 to 100°C temperature. Hence, the EO indicated the possibility of exploitation as an safe alternative to synthetic fungicide.

Mentha oil – Kedia *et al.* (2014b) reported the chemically characterized *Mentha spicata* (Spearmint) EO tested as a plant-based pesticide against food-deteriorating molds. The GC-MS analysis of EO depicted the presence of 13 compounds of which carvone was found as major component (59.6%). The EO significantly inhibited growth of 19 food-deteriorating molds at 1.0 μL/mL concentration. The EO showed non-phytotoxicity as 100% germination of EO-treated chickpea seeds was observed and low mammalian toxicity (LD_{50} 8342.33 μL/kg) on mice. Kumar *et al.* (2009) evaluated the antifungal potential of *Mentha arvensis* EO to develop it as eco-friendly substitute of synthetic preservatives. The EO at 400 ppm showed minimum inhibitory concentration against *A. flavus* and exhibited broad fungitoxicity against 14 storage fungi. Further, the EO also found to be superior than some prevalent synthetic fungicides viz. Benomyl, diphenylamine, zinc dimethyl dithiocarbamate (Ziram), Carbendazim, Mancozeb, Bavistin, Wettasul-80, Agrosan GN, Dithane M-45 and Zineb. Hence, Mentha EO showed efficacious for formulation as plant based fungicide for future development of next generation fungicides.

Origano oil – Daouk *et al.* (1995) evaluated the antifungal activity of *Origanum syriacum* EO against *Aspergillus niger, Fusarium oxysporum,* and *Penicillium* species. The chemical composition of EO using gas liquid chromatography characterized thymol and carvacrol as the major components. The EO showed strong inhibitory action against the tested fungi tested at 0.1 μL/mL concentration. Similarly, Paster *et al.* (1994) evaluated the efficacy of oregano oil as a fumigant for protection of food items from fungi during storage. The EO was applied for 24 h as fumigants against *A. flavus, A. niger, A. ochraceus* and natural microflora of wheat grains. The data obtained from this study underline the toxicity of oregano EO as fumigants against fungi attacking stored grain and fortify the possibility of using it as an alternative to synthetic chemicals.

Rosemary oil – Prakash *et al.* (2015) explored the antifungal potency of *Rosmarinus officinalis* (Rosemary) EO and its major compounds along with its mode of antifungal action practical efficacy as preservative in food system. The GC-MS analysis of EO depicted 16 compounds of which α-pinene, 1,8-cineole and camphor were found as the major compounds. The minimum inhibitory concentration of EO, α-pinene, 1,8-cineole and camphor against *A. flavus* was found at 1.5, >5.0, 4.0 and 3.0 μL/mL concentrations respectively showing the synergistic action of EO components. The targeted site of EO on fungal cell was confirmed at plasma membrane as observed through ergosterol measurement test and Transmission Electron Microscopic analysis. Further, the EO significantly protected food sample against mould infestation upto 6 months of storage making it a suitable plant based fungicide.

Thyme oil – Kumar *et al.* (2008) recommended *Thymus vulgaris* (thyme) EO as effective antifungal after screening of 14 EOs from higher plants. The thyme oil completely inhibited the mycelial growth of A. flavus at 0.7 μL/mL concentration and at this concentration showed broad fungitoxic spectrum against different food deteriorating fungi viz. *Fusarium oxysporum*, *Cladosporium herbarum*, *Curvularia lunata*, *Aspergillus terreus*, *A. niger*, *A. fumigatus*, *Alternaria alternata* and *Botryodiploidia theobromae*. The EO was also found superior than most of the prevalent synthetic fungicides. The non-mammalian toxicity (LC_{50}

7142.85 μl/kg) of EO on mice further strengthening its safe exploitation for eco-friendly control of fungal pathogen.

Vetiver oil – *Vetiveria zizaniodes* (khas or vetiver) is a well known plant for curing many diseases in Ayurvedic system of medicine from a long time. The oil from vetiver is used traditionally to repel moths as many people place vetiver root among their clothes to keep insects away. Further it also showed strong antifungal effects against a wide range of fungi (Dikshit and Hussain, 1984; Singh *et al.*, 1978). Dubey *et al.* (2010) examined the antifungal efficacy of vetiver oil from two different geographical regions and observed significant fungitoxicity against phytopathogenic fungus *Rhizoctonia solani*. Hence, vetiver oil could be exploited as effective fungitoxicant for the formulation of plant based fungicide for future applications.

21.6 COMMERCIALIZATION OF ESSENTIAL OIL

Commercially, some of the EO based insecticides, miticides, weedicides and fungicides formulated by different agricultural industries are already available in developed countries for organic farmers. Among fungicides some examples are: E-Rase™ containing jajoba (*Simmondsia californica*) oil, Sporan™ containing rosemary (*Rosemarinus officianalis*) oil, Promax™ containing thyme (*Thymus vulgaris*) oil etc (Dayan *et al.*, 2009). However, these EO based fungicides are used only in developed countries in organic farming and there is a strong need of their promotion and large scale trial by both private and government agricultural organizations to develop them as future fungicide.

21.7 CHALLENGES

The major issue of using EO as fungicide is its chemical variation in different parameters viz. ecological and geographical variations, age of the plant, harvesting time and different methodology of their extraction etc. which influence its bioactivity (Burt, 2004). Hence, it is always advisable to standardize EO before its large scale trial. Another issue is the strong aroma of EO even at low concentrations, which may adversely affect the organoleptic property of food items on which it is applied. Hence, interactions of food matrix components with EO should be investigated before its *in vivo* application. Further, being volatile, there is a chance of accidental loss of EO during its applications and transportation. Some EO components are susceptible to degradation even in the closed containers (Isman, 2000). These limitations can be defeated by modern encapsulation technique in which EO loaded micro/nanocapsules are prepared for their controlled release and protection, however, there is a quick need for standardization of a common and cost effective protocol for production of EO loaded microcapsules in large scale.

21.8 CONCLUSION

Out of immense report of various EOs as antifungal agent, only handful of EO formulations is commercially available. The ability of EOs as fungicide through its biorational mode of action along with plant origin would be helpful in achieving next generation fungicide. The antifungal effects of EOs against a wide range of fungi, fungal specific mode of action, safety for non target organisms, complex composition and wide availability could play an important role in integrated pest management (IPM) programmes in future after developed into plant based 'next generation fungicide'. There is a huge opportunity to biodiversity rich countries for leading in global market in near future if they properly standardize the fungicidal effects of their wild plant material.

REFERENCES

Baser KHC and Demirci F (2007). Chemistry of essential oils. In: Berger RG, ed. Flavours and Fragrances: Chemistry, Bioprocessing and Sustainability. Springer, Berlin, Germany, 43-86.

Brent KJ and Hollomon DW (1998). Fungicide resistance: the assessment of risk. Monograph no. 2, FRAC Global Crop Protection Federation, Brussels.

Bullerman LB, Lieu FY and Seier SA (1977). Inhibition of growth and aflatoxin production by cinnamon and clove oils, cinnamic aldehyde and eugenol. J. Food Sci. 42(4):1107-1109.

Burt S (2004). Essential oils: their antibacterial properties and potential applications in foods – a review. Int. J. Food Microbiol. 94:223-253.

Carson CF, Hammer KA and Riley TV (2006). *Melaleuca alternifolia* (Tea Tree) oil: a review of antimicrobial and other medicinal properties. Clin. Microbiol. Rev. 19:50-62.

Daouk RK, Dagher SM and Sattout EJ (1995). Antifungal activity of the essential oil of *Origanum syriacum* L. J. Food Prot. 58(10):1147-1149.

Dayan FE, Cantrell CL and Duke SO (2009. Natural products in crop protection. Bioorg. Med. Chem. 17(12):4022-4034.

Dikshit A and Husain A (1984). Antifungal action of some essential oils against animal pathogens. Fitoterapia 55:171-176.

Dubey N, Raghav CS, Gupta RL and Chhonkar SS (2010). Chemical composition and antifungal activity of vetiver oil of North and South India against *Rhizoctonia solani*. Pestic. Res. J., 22(1):63-67.

Džamić AM, Soković MD, Ristić MS, Grujić SM, Mileski KS and Marin PD (2014). Chemical composition, antifungal and antioxidant activity of *Pelargonium graveolens* essential oil. J. Appl. Pharm. Sci. 4(3):1-5.

Fishbein L (1976). Environmental health aspects of fungicides. I. Dithiocarbamates. J. Toxicol. Environ. Health 1(5):713-735.

Heil A, Humphreys EH and Janssen S (2008). Outcomes of the California ban on pharmaceutical Lindane: Clinical and ecologic impacts. Environ. Health Persp.116:297-302.

Helal GA, Sarhan MM, Abu Shahla ANK and Abou El-Khair EK (2007). Effects of *Cymbopogon citratus* L. essential oil on the growth: morphogenesis andaflatoxin production of *Aspergillus flavus* ML2-strain. J. Basic Microbiol. 47:5-15.

Helander IM, Alakomi HL, Latva-Kala K, Mattila-Sandholm T, Pol I, Smid EJ, Gorris LGM and Von Wright A (1998). Characterization of the action of selected essential oil components on Gram-negative bacteria. J. Agric. Food Chem. 46:3590-3595.

Hussain AI, Anwar F, Hussain Sherazi ST and Przybylski R (2008). Chemical composition, antioxidant and antimicrobial activities of basil (*Ocimum basilicum*) essential oils depends on seasonal variations. Food Chem. 108:986-995.

Isman MB (2000). Plant essential oils for pest and disease management. *Crop Prot.* 19:603-608.

Isman MB, Wilson JA and Bradbury R (2008). Insecticidal activities of commercial rosemary oils (*Rosmarinus officinalis*) against larvae of *Pseudaletia unipuncta* and *Trichoplusia ni* in relation to their chemical compositions. Pharmaceut. Biol. 46:82-87.

Kedia A, Prakash B, Mishra PK, Chanotiya CS and Dubey NK (2014b). Antifungal, antiaflatoxigenic, and insecticidal efficacy of spearmint (*Mentha spicata* L.) essential oil. Int. Biodeter. Biodegr. 89:29-36.

Kedia A, Prakash B, Mishra PK and Dubey NK (2014a). Antifungal and antiaflatoxigenic properties of *Cuminum cyminum* (L.) seed essential oil and its efficacy as a preservative in stored commodities. Int. J. Food Microbiol. 168-169:1-7.

Kedia A, Prakash B, Mishra PK, Dwivedy AK and Dubey NK (2015). *Trachyspermum ammi* L. essential oil as plant based preservative in food system. Ind. Crops Prod. 69:104-109.

Kelly SL, Lamb DC, Corran AJ, Baldwin BC and Kelly DE (1995). Mode of action and resistance to azole antifungals associated with the formation of 14α-methylergosta-8, 24(28)-dien-3β, 6α-diol. Biochem. Biophys. Res. Commun. 207:910-915.

Kumar A, Shukla R, Singh P and Dubey NK (2010). Chemical composition, antifungal and antiaflatoxigenic activities of *Ocimum sanctum* L. essential oil and its safety assessment as plant based antimicrobial. Food Chem. Toxicol. 48:539-543.

Kumar A, Shukla R, Singh P, Prakash B and Dubey NK (2011). Chemical composition of *Ocimum basilicum* L. essential oil and its efficacy as a preservative against fungal and aflatoxin contamination of dry fruits. Int. J. Food Sci. Technol. 46:1840-1846.

Kumar A, Shukla R, Singh P, Prasad CS and Dubey NK (2008). Assessment of *Thymus vulgaris* L. essential oil as a safe botanical preservative against post harvest fungal infestation of food commodities. Innov. Food Sci. Emer. Technol. 9:575-580.

Kumar A, Shukla R, Singh P, Singh AK and Dubey NK (2009). Use of essential oil from *Mentha arvensis* L. to control storage moulds and insects in stored chickpea. J. Sci. Food Agric. 89:2643-2649.

Li WR, Shi QS, Ouyang YS, Chen YB and Duan SS (2013). Antifungal effects of citronella oil against *Aspergillus niger* ATCC 16404. Appl. Microbiol. Biotechnol. 97(16):7483-7492.

Mishra AK and Dubey NK (1994). Evaluation of some essential oils for their toxicity against fungi causing deterioration of stored food commodities. Appl. Environ. Microbiol. Apr.:1101-1105.

Nakahara K, Alzoreky NS, Yoshihashi T, Nguyen HTT and Trakoontivakorn G (2003). Chemical composition and antifungal activity of essential oil from *Cymbopogon nardus* (Citronella grass). Japan Agric. Res. Quarter. 37(4):249-252.

OHS (Occupational Health Services), MSDS for Dichlorvos; (1991). Available from www.bvsde.ops-oms.org.

Paster N, Menasherov M, Ravid U and Juven B (1995). Antifungal activity of oregano and thyme essential oils applied as fumigants against fungi attacking stored grain. J Food Prot. 58(1):81-85.

Pinto E, Vale-Silva L, Cavaleiro C and Salgueiro L (2009). Antifungal activity of the clove essential oil from *Syzygium aromaticum* on *Candida*, *Aspergillus* and dermatophyte species. J. Med. Microbiol. 58:1454-1462.

Prakash B, Kedia A, Mishra PK, Dwivedy AK and Dubey NK (2015). Assessment of chemically characterised *Rosmarinus officinalis* L. essential oil and its major compounds as plant-based preservative in food system based on their efficacy against food-borne moulds and aflatoxin secretion and as antioxidant. Int. J. Food Sci. Technol. 50:1792-1798.

Prakash B, Shukla R, Singh P, Mishra PK, Dubey NK and Kharwar RN (2011). Efficacy of chemically characterized *Ocimum gratissimum* L. essential oil as an antioxidant and a safe plant based antimicrobial against fungal and aflatoxin B1 contamination of spices. Food Res. Int. 44:385-390.

Ragsdale NN and Sisler HD (1991). The nature, modes of action, and toxicity of fungicides. In: Pimental D, ed. Handbook of Pest Management in Agriculture. Boca Raton, FL: CRC Press, 461-496.

Rajendran S and Sriranjini V (2008). Plant products as fumigants for stored product insect control. J. Stored Prod. Res. 44:126-135.

Raut JS and Karuppayil SM (2014). A status review on the medicinal properties of essential oils. Ind. Crops Prod. 62:250-264.

Rodriguez RJ, Low C, Bottema CD and Parks LW (1985). Multiple functions for sterols in *Saccharomyces cerevisiae*. Biochim. Biophys. Acta 837:336-343.

Romagnoli C, Bruni R, Andreotti E, Rai MK, Vicentini CB and Mares D (2005). Chemical characterization and antifungal activity of essential oil of capitula from wild Indian *Tagetes patula* L. Protoplasma 225(1-2):57-65.

Singh G, Singh BS and Kumar BRV (1978). Antimicrobial activity of essential oils against keratinophilic fungi. Indian Drugs 16(2):43-45.

Tian J, Ban X, Zeng H, He J, Chen Y and Wang Y (2012). The mechanism of antifungal action of essential oil from Dill (*Anethum graveolens* L.) on *Aspergillus flavus*. Plos One 7(1):1-10.

Tian J, Ban X, Zeng H, Huang B, He J and Wang Y (2011). *In vitro* and *in vivo* activity of essential oil from dill (*Anethum graveolens* L.) against fungal spoilage of cherry tomatoes. Food Cont. 22:1992-1999.

Tripathi P and Dubey NK (2004). Exploitation of natural products as an alternative strategy to control postharvest fungal rotting of fruit and vegetables. Postharvest Biol. Technol. 232:235-245.

USEPA (United States Environment Protection Agency), Permethrin facts, Reregistration Eligibility Decision (RED) Fact Sheet; 2006.

Van de Braak SAAJ and Leijten GCJJ (1999). Essential oils and oleoresins: a survey in the Netherlands and other major markets in the European Union. CBI, Centre for the Promotion of Imports from Developing Countries, Rotterdam.

Vigan M (2010). Essential oils: renewal of interest and toxicity. Eur. J. Dermatol. 20:685-692.

Zabka M and Pavela R (2013). Antifungal efficacy of some natural phenolic compounds against significant pathogenic and toxinogenic filamentous fungi. Chemosphere 93:1051-1056.

■■■■

Chapter - 22

Threatened and Endemic plant diversity of India

M. Daniel and P.S. Nagar

Department of Botany, Faculty of Science,
The Maharaja Sayajirao University of Baroda, Vadodara 390002

ABSTRACT

India is one of the most diverse hotspot in the world in terms of floristic diversity. It is largest producer of medicinal herbs and is known as the botanical garden of the world. About 54% of the country's land is under cultivation for food, ornamentals and medicinal plant crops and approximately 19% area has varying degree of forest vegetation, though as per Biodiversity vision we wish to attain 33% of forest vegetation, though as per Biodiversity vision we wish to attain 33% of forest vegetation by 2012. There are about 45,000 plant species in India which includes 17000 flowering plants, 23,000 fungi, 2500 algae, 1600 lichens, 1800 bryophytes and 30 microorganisms. Gujarat contributes approximately 3000 species of flowering plants. Still in term of floristic diversity of Gujarat state extensive surveys are required.

Keyword: Threatened, Endemic, plant diversity , Gujarat, India

22.1. PLANT DIVERSITY: INTRODUCTION

The plant diversity is needed for human survival in many ways (i) it serves and offers a great potential as a source of food, feed, medicine, fuel, fibre pesticides and other products of day to day importance, (ii) it constitute a source of genetic resource for improvement of future crop plants, (iii) plays an important role in regulating hydrological cycle and climate, (iv) protects soil erosion, and (v) perform essential ecological functions associated with structural stability. The magnitude of erosion of the genetic resource base is exemplified by the global awareness of what is known as species extinction crisis or biological diversity crisis. According to the estimates of Reid and Miller (1989), (source Heywood and Stuart, 1990), roughly 5 to 10 per cent of the world's species will be lost per decade. According to a recent estimate, about 19,077 plant species are known to be endangered, rare or otherwise threatened (Pandit *et al.* 1994).

Biodiversity change as projected for the 21st Century by models or extrapolations are based on experiments and observed trends. The term "biodiversity" is used in a broad sense as it is defined in the Convention on Biological Diversity to mean the abundance and distribution of an interaction between genotypes, species, communities, ecosystems and biomes. This synthesis pays particular attention to the interactions between biodiversity and ecosystem services and to critical "tipping points" that could lead to large, rapid and potentially irreversible changes. Comparisons between models are used to estimate the range of projections and to identify sources of uncertainty. Experiments and observed trends were used to check the plausibility of these projections. In addition the studies have identified possible actions at the local, national and international levels that can be taken to conserve biodiversity. Wide range of scientists were called upon to participate in this synthesis, with the objective to provide decision makers with messages that reflect the consensus of the scientific community and that will aid in the development of policy and management strategies that are ambitious, forward looking and proactive (CBD, 2010). Being a Biodiversity year we should meet at national level to rethink on the strategies adopted by us to conserve the biodiversity. Following are the key factors for studying the biodiversity of the various regions.

(1) Projections of global change impacts on biodiversity show continuing and, in many cases, accelerating species extinctions, loss of natural habitat, and changes in the distribution and abundance of species and biomes over the 21st century.

(2) Thresholds, amplifying feedbacks and time lag effects leading to "tipping points" are widespread and make the impacts of global change on biodiversity hard to predict, difficult to control once they begin, and slow and expensive to reverse once they have occurred.

(3) For many important cases the degradation of ecosystem services goes hand-in-hand with species extinctions, declining species abundance, or widespread shifts in species and biome distributions. However, conservation of biodiversity and of some ecosystem services, especially provisioning services, is often at odds.

(4) Strong action at international, national and local levels to mitigate drivers of biodiversity change and to develop adaptive management strategies could significantly reduce or reverse undesirable and dangerous biodiversity transformations if urgently, comprehensively and appropriately applied.

22.2 THREATENED AND ENDEMIC SPECIES AT NATIONAL LEVEL

It is now realized all over the world that several species of plants and animals are threatened, many are critically rare and a few already extinct. Studies in some parts of the world have shown that on an average about 0% of vascular plants fall into one or the other category of threatened species. Many countries in the world have taken stock of their rare plants and have developed provisional or fairly accurate lists of their threatened plants. Botanical Survey of India initiated work on rare and endangered species in India way back in 1980 and published several lists of rare and endangered species. "An Assessment of Threatened Plants of India" (Jain and Rao, 1983) brought to light several hundred rare species from different parts of the country. This was followed by publication of 4 volumes of Red Data Book of India which included nearly 620 threatened plants falling under various threat categories of ICUN as below:

Table 1. Threatened and endangered plants and families in India

Category	Families	Species
EX	22	62
EN	43	128
V	50	110
R	67	247
I	37	71
K	4	5
Total	223	623

Source: Nayar & Sastry (1988-90) EX: Possibly/Presumed extinct: En: Endangered: V: Vulnerable: R: Rare: I: Indeterminate: K: Insufficiently known.

22.3 CATEGORIES OF THREATENED PLANTS

The International Union for Conservation of Nature and Natural Resources (IUCN, 1995) has recognized following updated categories of threatened plants on the basis of geographical range, populations and fragmentation of population.

Extinct (EX): A taxon is Extinct when there is no reasonable doubt that the last individual has died.

Extinct in the Wild (EW): A taxon is Extinct in the wild when it is known only to survive in cultivation.

Critically Endangered (CR): A taxon is Critically Endangered when it is facing an extremely high risk of extinction in the wild in the immediate future (80% decline in the last 10 years).

Endangered (EN): A taxon is Endangered when it is not Critically Endangered but is facing a very high risk to extinction in the wild in the near future (50% decline in the last 10 years).

Vulnerable (VU): A taxon is Vulnerable when it is not Critically Endangered or Endangered but is facing a high risk of extinction in the wild in the medium term future (50% decline in last 20 years)

Vulnerable (VU): A taxon is Vulnerable when it is not Critically Endangered or Endangered but is facing a high risk of extinction in the wild in the medium term future (50% decline in last 20 Years)

Conservation Dependent (CD): Taxa which are the focus of a continuing taxon-specific or habitat specific conservation programme targeted towards the taxon in question, the cessation of which would result in the taxon qualifying for one of the threatened categories above within a period of five years.

Data Deficient (DD): A taxon is Data Deficient when there is inadequate information to make a direct or indirect assessment of its risk of extinction based on its distribution and/or population status.

Low Risk (LR): A taxon is said to be at low risk when it has been evaluated, does not satisfy the criteria for any of the categories. Critically Endangered. Endangered, Vulnerable. Conservation Dependent or Data Deficient.

Not Evaluated (NE): A taxon is not evaluated when it has not yet been assessed against the criteria.

22.4 CURRENT STATUS OF ENDANGERED PLANTS IN INDIA

In India following the general criteria laid down by IUCN (1978) the botanical Survey of India has brought out several assessments of rare and threatened species in the country (Jain & Rao, 1984) these lists are fundamentally based on herbarium and partly field studies. It is estimated that about 3000 species of flowering plants out of 17000 species fall in one or the other category of threatened plants which also include several medicinal plants. These lists have also formed the basis for the publication of the four volume of Red Data Books of India (Jain & Sastry, 1984: Nayar & Sastry, 1987-1990) wherein 620 species are included, of which 550 species are endemic including some valuable medicinal species (A list of species included in the Red Data Book and also list of species supposed extinct is appended at the end).

These lists are however not without any drawbacks or limitations, Detailed field studies throughout the entire distribution range of these species to estimate the size and number of populations of particular species, their specific ecological niches, reproductive behaviour and demographic studies are lacking. Future studies should also highlight the correct status of threatened and endemic plants at State/Regional level based on extensive field surveys.

22.5 ENDEMISM IN THE INDIAN FLORA

An excellent account of endemism and endemic plants of the Indian subcontinent has been provided by Nayar (1996) which is based on several earlier masterly accounts on the topic. The information provided in the following few paragraphs has been largely drawn by this work.

What are endemic plants?

Generally, when a taxonomic unit is found to occur in a restricted area which is isolated by geographical or temporal barriers it is called endemic. Conventionally an endemic taxon is one which occurs in a small restricted area i.e an island, a peninsula, a phytogeographical area or a mountain range.

Endemics can be classified as below:

i. **Palaeoendemics** which are taxonomically isolated, show no variations and occur in isolated refugia.

ii. **Schizoendemics** which are produced by gradual specification having common origin, but isolated in different ecological niches: they usually have identical chromosome.

iii. **Patroendemics** or parent endemics are diploids which give rise to polyploids.

iv. **Apoendemics** which are polyploids usually of hybrid origin, arising from widely distributed diploids.

Endemic Plants in Indian Flora

By virtue of its varied physiognomy and climate, India harbours a rich flora, with a fairly high degree of endemism. Being bounded on the north by the Himalaya and with it peninsular region surrounded by oceans, the isolation of the Indian flora to a large extent help in the development of endemism.

The flowering plants of India have about 17000 species of which 5725 are broadly considered as endemics. The endemics represent 33.5% of the flora. The present study shows there are about 3471 endemics in the Himalaya, 2015 in the Penninsular India and 239 in the Andaman & Nicobar islands. The three mega centres of Endemism are the Eastern Himalaya, Western Ghats, and Western Himalaya. The Central Himalaya not only includes the fringe endemics of Western and Eastern Himalaya, but it is a crucible of speciation.

According to Nayar (1980) there are 14 endemic genera in India. Acanthaceae and Poaceae have the highest number of endemic genera (17 each). Pteracanthus and Niligirianthus have the number of endemic species (20 each). Since Nayar's (1980) publication there are five additional endemic genera recorded from India. About 114 endemic species which represent monotypic. There are about 5725 endemic species which represent 33.5% of Indian flora (17000 species). As many as 15 genera are found to be endemic although by and large spread throughout India as a wide. The genera, restricted to India as a whole are:

Amischophacelus Rao & Kamm. (Commelinaceac); Bentinkia Berry ex Roxb, (Palmae); Caesulia Roxb. (Compositae); Colebrookea Sm. (Labiatae); Glossocardia Cass, (Cmpositae); Hoppea Wild. (Gentianaceae); Indochloa

Indopiptadenia Brenan (Mimosaceae); Lophopogon Hack. (Poaceae); Octotropis Bedd. (Rubiaceae); Oueinia Benth (Papilionaceae); Plesmonium Schott (Araceae); Porteresia Tateoka (Poaceae); Trochiscus O.E. Schulz (Cruciferae).

Table 2. Families having more than 150 species and the percentage of Endemism in India

Family	Total No. of taxa	Total Endemic taxa	% of Endemic
Poaceae	1205	376	31.2%
Fabaceae	1152	298	25.9%
Orchidaceae	1100	259	23.5%
Asteraceae	1052	390	37.0%
Rubiaceae	500	205	41.0%
Cyperaceae	446	108	24.2%
Labiatae	420	109	25.0%
Acanthaceae	380	224	58.9%
Scrophulariaceae	354	87	24.5%
Rosaceae	250	182	72 8%
Umbelliferae	209	165	78.8%
Liliaceae	203	114	55.6%
Ranunculaceae	180	135	75.0%
Balsaminaceae	180	123	68.3%
Ericaceae	168	80	47.6%
Primulaceae	165	80	48.4%
Lauraceae	163	133	80.9%

Endemic plants in India are concentrated mainly in the Himalayas Peninsular India and in Andaman & Nicobar Island

22.6 ENDEMISM IN THE HIMALAYA

The distribution of high altitude Himalayan flora was described by Rau (1974) as one of migration of floras, survival of relicts, evolution of new flora by an intermixing of different of floras and acclimatization of species from lower altitudes. A fairly good degree of endemism is found in all the altitudinal zones of the Himalayan vegetation. About 71 genera are found restricted to the Himalayas. Many genera with restricted distribution extend to Tibet, S.W. China, N.E India-Yunnan Burma area. Kedarnath, Keraymonia (Umbelliferae) Brachycaulos (Rosaceae) are three new additions to the Himalayan endemic genera. About 42 genera are confined to the Eastern Himalayas while 12 genera are confined to the W. Himalayas.

Jaeschkea and *Parajaeschkea* are examples of these endemic pairs with species compliments adapted to different climatic zones of the Himalayas. The monotypic genera *Drimycarpus* and *Parrotiopsis* represent arborescent tree genera. *Listrobanthes, Megacodon, Pseudaechmanthera, Pseudostachyum, Pteracanthus, Sympagis* and *Triaenacanthus* represent the other arborescent genera with shrubby elements. Genera with undershrub elements are: *Catamixis, Lacaitaea, Indopolysolena, Pottingeria, Roylea* and *Tarphochlamys*. The rest of the endemic genera comprise herbs, twining climbers and epiphytes. The genus *Catamixis* is on the verge of depletion (Ahmedullah and Nayar, 1986).

At the species level there are about 3471 endemics restricted to the Himalayas. A large number of these are threatened. It is seen that neoendemics generally have herbaceous and shrubby forms where there are climatic and environmental stresses.

Table 3. List of families of Himalaya which have high degree of endemics

Family	Total taxa in India	Total endemic taxa in the Himalaya	Percentage
Berberidaceae	79	77	98.7%
Saxifragaceae	117	108	92.3%
Ranunculaceae	180	131	72.7%
Rosaceae	250	176	70.4%
Umbelliferae	209	144	68.8%
Caryophyllaceae	121	79	65.3%
Gentianaceae	147	84	57.1%
Primulaceae	165	80	48.4%
Lauraceae	163	78	47.9%

22.7 ENDEMISM IN PENINSULAR INDIA

Detailed studies on the endemism in the Peninsular Indian flora were carried out recently (Ahmedullah and Nayar, 1987). According to this study there are 58 genera which are endemic to Penninsular India. Of these 49 are monotype. The families Poaceae, with 13 genera and Acanthaceae with 9 genera have the highest representation of endemic genera in Peninsular India.

Table 4. List of Peninsular Indian families which have high degree of endemics.

Family	Total taxa in India	Total endemic taxa in Peninsular India	Percentage
Melastomataceae	81	46	56.7%
Balsaminaceae	180	80	44.4%
Acanthaceae	380	146	38.6%
Asclepiadaceae	209	67	32.4%
Rubiaceae	500	150	30.0%
Lauraceae	163	43	26.3%

The four large families having more than 1000 species are Poaceae (1205) Orchidaceae (1100), Fabaceae (1152) and Asteraceae (1052). However the family Orchidaceae has 23.5% endemics, Poaceae has 31.2% endemics, Asteraceae has 37% endemics and Fabaceae has 25.9% endemics. The other families having high degree of endemices are Rosaceae (182 endemics out of 250), 72.8% endemics; Umebelliferae (165 endemics out of 209) 78.8% endemics; Ranunculaceae (135 endemics out of 180) 75% endemics; Balasminaceae (80 endemics out of 123) 68.3% endemics; Saxifragaceae (108 endemics out of 117), 92.3%; Berberidaceae (68 endemics out of 79), 86%. It is interesting to note that the palaetropic family Lauraceae (133 endemics out of 163) has 80.9% endemics.

The four large families have the following representation of endemics in the Himalayas; Poaceae (154 endemics out of 1205). Fabaceae (183 endemics out of 1152), Orchidaceae (130 endemics out of 1100), Asteraceae (302 endemics out of 1052). There is high representation of endemics in the family Berberidaceae (98.7%), Saxifragaceae (92.3%) Ranunculaceae (72.7), Rosaceae (70.4%), Umbelliferae (68.8%), Caryophyllaceae (65.3%), Gentiacaceae (57.1%).

The four large families which have more than 1000 species each and their percentage of endemics in peninsular India are as follows: Poaceae (219 are endemics out 1205), 18% endemics; Orchidaceae (129 are endemics out of 1100), 11.8% Fabaceae (112 are endemics out of 1152), 9.7% Asteraceae (87 are endemics out of 1052); 8.2%.

The families having high degree of endemism in peninsular India, are Melastomataceae (46%), Acanthaceae (38.6%), Rubiaceae (30%), The genus Impatiens (Balsaminaceae) is herebaceous and it has 80% endemics in the Peninsular India. Andaman & Nicobar islands number of endemics in the family Annonance in Peninsular India and Andaman and Nicobar islands.

At the species level Peninsular India harbours as many as 1932 taxa, including 1788 taxa at species level and 144 taxa at intraspecific level (Ahmedullah and Nayar, 1987). The major concentration of these endemics are found in the W. Ghats region, while only 76 taxa are endemics to the Eastern Ghat region. Here endemic *Cycas beddomei* comes under endangered category

(Ahmedullah and Nayar, 1986). There are 2015 endemic taxa 1425 belonging to the monocotyledons. Families like Poaceae, Rubiaceae, Acanthaceae, Orchidaceae and Leguminosae have the largest representation of endemic taxa in the region. There are 60 genera which are endemic to Peninsular India. Of these 49 are monotypic. The families Poaceae, with 13 genera and Acanthaceae with 9 genera have the highest representation of endemic genera in Peninsular India.

The endemic genera of Peninsular India comprise of herbs, succulent scapigerous plants, shrubs, under shrubs and trees. Trees species are mainly represented by the genera *Poeciloneuron, Blepharistemma, Pseudoglochidon, Meteroronmyrtus*, *Otonephelium* and *Erinocarpus*. These woody life forms are generally considered to be relictual in nature. The genus *Poeciloneuron*, with two trees species *P. indicum* and *P. parviflorum* occurs in the evergreen forests of southern W. Ghats. The mangrove member *Blephaistemma* with single species (*B. serratum*) is especially interesting endemic genus; this inland mangrove species had adapted itself to the low land forest conditions of the S.W. Ghats. *Carvia, Nilgirian, Phlebophyllum, Taeniandra, Xenacanthus* are represented by shrubs that form undergrowth components of the tropical rain forests of W. Ghats.

Genera like *Moullava, Lamprachaenium, Nanothamnus* and *Carviahave* successfully established themselves. These are examples of endemics that show successful adaption for survival and spread (Ahmedullah and Nayar, 1987). As many as 29 genera are commonly endemic to peninsular India and Sri Lanka reflecting their past geological connections. They are as follows: *Brysophyllum* Hook.f., *Chloroxylon* DC; *Coelachropsis* Bor; *Cottonia* Wt; *Cullenia* Wt.; *Didyplosandra* Wt.; *Eleiotes* DC; *Farmeria* Willis; *Fergusonia* Hook f,; *Kendrickia* Hook, f.; *Lagenandra* Dalz; *Leptacanthus* Nees; *Lopholepis* Decne; *Mackenziea* Nees; *Meyenia* Nees; *Mischodon* Thw.; *Neurocalyx* Hook; *Pommereulla* L.f.; *Sirhookera* Kuntze; *Stenosiphonium* Nees; *Terniola* Tul; *Thelepaepale* Bremek; *Theriophonum* B1,; *Thraulococcus* Radik.; *Veterial* Lim, *Zenkeria* Trin.

22.8 ENDEMISM IN ANDAMAN & NICOBAR ISLANDS

Andaman and Nicobar islands in the Bay of Bengal are considered as continental islands, the mountain tops of the Arakan yoma Range forming an arc to Sumara and they consist of 349 islands stretching for 590 km, with a total land area of 8249 sq. km. Of the 2200 estimated higher group of plants about 239 taxa are endemics. Some of the endemics occur in small islands like Narcondam, Rutland, Kamorta, Katchal, Nancoury, which are on the road to extinction.

The genus Bentineka (palmae) has interesting disjunct distribution. *Bentinckia nicobaricca* occurs in Nicobar islands, while its counter part *Bentinckia condapanna* occurs in the southern W. Ghats especially Agasthaimalai hills. Important endemc flagship species which are of timber value are *Dipterocarpus andamanicus*, *Dysoxylum andamanicum*. Some of the endemic plants of genetic resources are *Pandanus leram*, *Paramignya andamanica*,

Calamus nicobaricus, *Pinanga manii*, *Oryza indandamanica*, *Vanilla andamanica*, *Dioscorea vexans*, *Myristica connariodes*, *Magnifera andamanica* *Phyllanthus andamanica*. Some of the locally used medicinal endemic plants are *Canarium* mannii, *Toxocarpus kurzii*, *Orophea katschallica*, *Strychnos andamaensis*.

22.9 CENTRES OF ENDEMISM

Based on General Principles of Location of Endemism the following centres are identified:

Table 5. List of endemic species in different regions

Regions	No. of Endemic Species
Andaman group Islands	132
Nicobar group islands	45
Agsthymalai hills (Ariankavu Pass to Aramboli Pass)	189
Anamalai and High Ranges	94
Palni Hills	43
Shimoga-Kanara	58
Mahabaleshwar Khandala	63
Konkan Raigad	50
Marathawada Satpura	27
Tirupati-Cudapa-Nallamalai hills	27
Vizagapatnam-Ganjam-Leeward side	23
Southern Deccan Leeward side	43
Chotonagpur Plateau	15
Kathiawar- Kutch	08
Rajasthan – Aravalli hills	20
Khasia-Jaintia hills	37
Assam	04
Arunachal Pradesh Himalaya	26
Nilgiri Silent valley-Wyanad-Kodagu	150
Sikkim Himalayas	183
Garhwal- Kumaon Himalayas	44
Lahul –Himachal Pradesh	20
Kashmir Himalaya	80
Nepal Himalaya	40

REFERENCES

Heywood V.H. and Stuart S.N. (1990) Species extinction in tropical forests. Paper presented at Triennial Congress of IUCN.

Jain, S.K. and Rao R.R. (1983) An Assessment of Threatened plants of India. Bot. Surv. India, Calcutta.

Jain, S.K. and Sastri, A.R.K. (1980) Threatened Plants of India. A state of the Art Report, Department of Science and Technology, New Delhi. P.48.

Kothari, M. J. and Hajra, P. K. (1983) Materials for a catalogue of Threatened Plants of India. Botanical Survey of India, Howrah. P. 45.

Pandit M.K., Sulaiman I.M. and Babu C.R. (1994) Conservation of plant diversity: Herbarium and systematics. Bull. Bot. Surv. India 36(1-4):61-67

Sabnis, S. D. and Rao K.S.S. (1983) Observation on some rare or endangered endemics of South-eastern Kutch. An Assessment of threatened plants of India. Bot. Surv. India, Calcutta. P.71-77.

Chapter - 23

Endangered Animals of Gujarat and Their Conservation

Ankita Bhatt and Neha Singh

Department of Environmental Studies, Faculty of Science,
The Maharaja Sayajirao University of Baroda, Vadodara, Gujarat-390002, India

ABSTRACT

The Gujarat state is unique in its geographical features. It has four out of ten biogeographical zones in India, which is the highest in any state of the country. The topography is flat with a very large cost line. Biodiversity provides the rich resources on which the humanity is dependent. Industrialization and human activities has resulted into loss of biodiversity. Some of the species like Asiatic lion (*Panthera leo persica*) and wild ass does not occur anywhere, else in the country. Conservation of rare and endangered biodiversity and implement action plan to prevent extinction is of prime importance. A study was undertaken jointly by the departments of Botany and zoology to find out status of rare plants and animals. Few results are discussed in this chapter.

During an intensive and extensive ambles in the forests of JWLS, 18 species of mammals and 132 species of birds were recorded. Amphibians were observed during environmental education camps at Kada reservoir and Jhand spring by Dr. Rajendra Vyas, Assistant Curator, Baroda Zoo and Shri Mihir Dave, Landscape Designer, Baroda. Reptiles were listed over a period of years during camps undertaken by experts and several naturalists.

Keywords: Endangered, Biodiversity, Asiatic lion, Wild ass, Great Indian Bustard. Black Mahseer

23.1 INTRODUCTION

According to IUCN's studies, the number of endangered fauna species in the country had increased from 648 in 2013 to 665 in 2015. Similarly, the number of endangered mammals had risen from 95 to 98, birds from 80 to 88, reptiles from 52 to 53, amphibians from 74 to 75, fishes from 213 to 216 and mollusks from six to seven in the last two years. The wildlife habitats were being destroyed due to mining and other human interventions.

India is home to different types of animals, birds and fishes which include some important farm animals like goats, poultry, cows, buffaloes, pigs etc. The country is also a habitat for wild animals like Bengal tigers, deer, wolves, pythons, Indian lions, bears, snakes, monkeys, many types of bison, Asian elephants and antelope species. India is one of the mega diverse countries out of a total of seventeen mega diverse countries in the world. These seventeen mega diverse countries, including India, are the habitats of around 60–70% of the world's biodiversity. The Western Ghats, The Eastern Himalayas and Indo–Burma are the three biodiversity hotspots out of total 34 in the whole world. India has 6.5% of the world's total wildlife species according to a report which was published by the United Nations Office on Drugs and Crime (UNODC) that includes 7.6% of all mammals and 12.6% of all bird species. According a report issued by the International Union for Conservation of Nature (IUCN) Red List in 2014, 15 species of birds, 12 species of mammals, and 18 species of reptiles and amphibians have joined the critically endangered list.

As many as 36 animal and plant species in Gujarat have been categorized as endangered by the International Union of Conservation of Nature (IUCN). The vast area is surrounded by three mountain range Aravalli, Vindhya and Western ghats. The animal species endangered include Black Mahseer, Golden Mahaseer, Leatherback Sea Turtle, Green Sea Turtle, Indian White-backed Vulture, Long-billed Vulture, Red-headed Vulture, Steppe Eagle, Greater Adjutant-Stork, Great Indian Bustard, Lesser Florican, Sociable Lapwing, Spotted Greenshank, Forest Spotted Owlet, Dhole, Caracal, Blue Whale, Fin Whale and Indian Wild Ass.

According to IUCN Red List the critically endangered species are at a highest risk of extinction. There are basically five ways to determine whether the particular species are endangered or not. When the species have a limited geographical range, Very limited or small population of less than 50 adult individuals was recorded. Whether the population has decreased or will decrease by more than 80% for the last three generations or 10 years. If the population is less than 250 individuals and is continuously declining at 25% for the last one generation or three years. There is a high possibility of extinction in the wild. Indian elephant, Bengal tiger, Indian lion, Indian Rhino, Gaur, lion tailed macaque, Tibetan Antelope, Ganga river dolphin, the Nilgiri Tahr, snow leopard, dhole, black buck, great Indian bustard, forest owlet, white – winged duck and many more are the most endangered animals in India.

Reptiles

Among the terrestrial vertebrates, Reptiles form a link between the amphibians on one hand and birds and mammals on the other. The number of living reptile species is about 6300, half of which consists of lizard species. Reptiles are characterized by their dry scalp skin, habit of laying eggs on land, nature of eggs and their thermo-regulation ability. One species of fresh water turtle, 11 species of lizards and 7 species of snakes have been enumerated. The scheme of Classification, common English and zoological names with author' names have been adopted from The Book of Indian Reptiles by (Daniel, 1989). The local names with identification features, habitat and observations have been given for each species. 132 species of birds have been enumerated based on the field studies and personal communications with the natives. Birds were abundant in certain localities – Jhand, Lambhiya, Kathiari, Moto Raska, Targol (Sankheds taluka); Katkuva, Borkach, Vav, Malhar, Narukot, Dhanimataji hill, Keva, Laphni (Jambughoda taluka): Koba, Rinchhbar (Halol taluka).

23.2 REASONS FOR BEING ENDANGERMENT

1. Loss of habitat is one of the primary reasons for the endangerment of species. Today, human intervention plays a major role in the destruction of the natural landscape. Human activities like removal of trees that provide both food and shelter for innumerable number of species, mining and agriculture, the land can be used in a variety of ways, and it threatens wild life.
2. Over hunting and poaching has a very destructive and catastrophic effect on the number of animals and fishes all over the world, various efforts have failed in this direction.
3. Pollution like air pollution, water pollution and waste pollution, especially in the form of plastic plays a very dominant role in the endangerment of animal species. Pollution not only causes health hazards for humans, but it affects the animals also.
4. In a robust and hearty environment there is always an accurate balance between the number of predators and their prey animals. The predators who are natural enemies of their prey animals choose the old and sick preys as they cannot keep up with their group. In this scenario the relationship between them is totally healthy as the predators only eat these prey animals that are already nearing the end of their life. But the problems become more apparent when the predators wander in such an area where they will get only a few numbers of their prey animals.
5. To save the animals from poaching and hunting they are often kept in a sanctuary and shelter. Though for some animals it has proved to be very much beneficial, there are other animals too who suffer and are in trouble for being endangered. The main two reasons are overcrowding and overgrazing. Usually there are too many animals that are enclosed in small areas. These animals often eat the same grass and trees in a limited area whereas in natural surroundings the grazing animals keep on changing the landscape while eating and keep on moving for most of the time. But in an enclosed and small area they eat from the same plants repeatedly which finally destroys the plants due to overstress.

23.3 RARE AND ENDANGERED ANIMALS OF GUJARAT

Panthera sondaica (Temminck, 1844)

The Javan Tiger likely became extinct in the mid-1970s (Seidensticker 1987). On the basis of morphology Mazak and Groves (2006) classified the Javan Tiger as a distinct species, *Panthera sondaica*. Classically it is considered to be a subspecies of Tiger *Panthera tigris* (Nowell and Jackson 1996).

Pangolins

When the door opened to Lucky's cage, he didn't appear to be alive.

He was just this nautilus of rust-colored scales, all spiraled up in a ball inside that underground metal box, which was connected by a concrete pipe to his cage, or "pangolarium," as it's called here at Cuc Phuong National Park.

I'd arrived at the park late that morning after driving about 3½ hours from Hanoi, past enormous gumdrop mountains and flooded rice paddies that reflected perfect rectangles of sky when the sun peeked through the rain. Phuong Quang Tran, a researcher who introduced me to Lucky, met me at the gate of the Carnivore and Pangolin Conservation Program, just outside the park. The setting reminded me of the TV show "Lost," with person-sized ferns and aging concrete buildings covered in moss. The gate at Pangolin Rehab, as I'll call it, is topped with barbed wire. Hidden cameras perch on the property like birds of prey.

With pangolins, it turns out, you can't be too careful.

These pangolins have avoided a major cause of pangolin death -- poaching -- because they were found in shipments that were intercepted by the Vietnamese government, bound for markets in Hanoi or, by road, to China. Here, they face new threats. Maybe it's the food that kills them, Phuong said. There's not been enough research on what pangolins eat in the wild to know for certain what their caregivers should feed them in captivity. Pangolin Rehab has been relatively successful with a combination of frozen ants (25%), soy (35%) and silkworm larvae (50%).

Yet they still have occasional digestive problems. And the stress of captivity alone can be fatal. It's one reason you rarely -- almost never, actually -- see them in zoos. (The San Diego Zoo is the only one in North America that has a pangolin. His name is Baba. The zoo used to have two pangolins, but one died of digestive problems, according to Jenny Mehlow, a spokeswoman for the zoo.) It's hard to overstate the amount of stress that trafficking routes put on pangolins. They're not happy travelers. Often they haven't had food or water for days and are perilously dehydrated. Forty percent die within a day or two of arriving at the center, Phuong told me. The rest are injected with hydrating fluids and kept in quarantine until they can be moved to a larger cage.

All of this made me a little nervous to meet Lucky. After standing there for a minute, I saw him breathing. It was hard to notice at first, but the hexagonal scales on his back rose and fell in a slow, oceanic motion. He must have smelled us, because his slender, toothless snout started to peek out of the ball like a cobra rising from a snake charmer's basket. Soon, he was looking at us with curious blueberry eyes, bobbing his head and sniffing the air.

"He's performing," Phuong said with a smile. After Lucky had a few dozen photos taken, a worker from Pangolin Rehab picked up the pangolin -- wait, what? -- and passed him over to me. I tried to refuse at first, but I could tell this was happening whether I wanted it to or not.

He felt like a lump of bricks in my arms, and a squirmy one. He's one of the biggest pangolins in the world in captivity, Phuong told me. He weighs about 8 kilos, or 17 pounds. Lucky wrapped his girthy pangolin body around my forearm, scales and claws pinching my skin, and held on tight. His stomach was warm and hairy, a somewhat awkward reminder that this otherwise reptilian-looking thing is actually a mammal. His body clamped around my arm like a giant snap bracelet.

"Careful, he's very strong," Phuong said. I could tell. This animal must have the abs of Pink.

And, this is embarrassing, but I kind of wanted him off of me. Just think how much I could stress out a pangolin. After meeting Lucky, I sat in the office at Pangolin Rehab with Phuong and rattled off what I thought would be a ridiculously softball list of questions about pangolins:

1. How long do they live?
2. How big do they get?
3. What's their gestation period?
4. What do they eat in the wild, and how much?
5. Are they solitary or do they live in pairs?

Phuong, who has talked about pangolins at international conferences, couldn't answer any of these with specificity. No one can. It's known that pangolins can live up to 20 years in captivity, for example, but it's not clear exactly how long they live in the wild. Pangolins eat ants, termites and various larvae, but it's also possible they consume bee larvae, flies, worms and crickets, according to a 2007 technical review of pangolin diets and husbandry. Estimates on their weight range from 4 to 72 pounds. Pangolins are a near-complete scientific mystery.

And it's not hard to see why if you keep poking around.

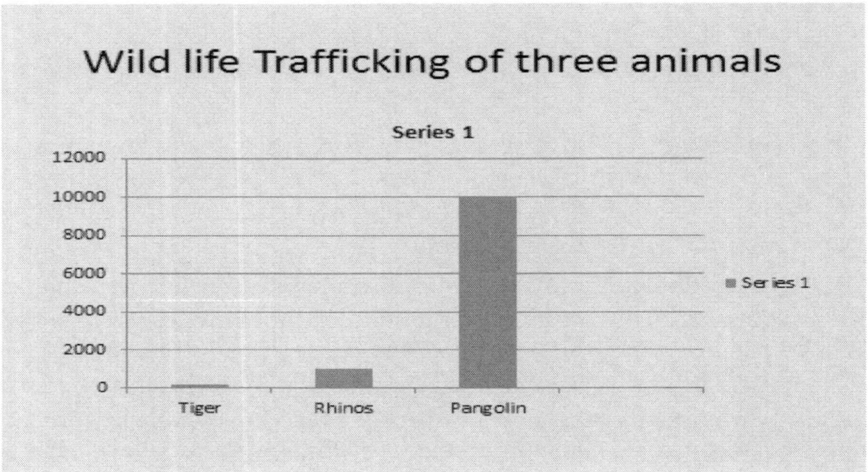

Sources: TRAFFIC, South Africa and IUCN. Note: Rhino numbers are for South Africa only.

***Chelonia mydas* (Frazier)**

An olive green and brown turtle, with brown and black spots. Carapace length is 85 cm. Carapace shields do not overlap. A pair of prefrontal shields on head is a distinguishing character. Lays eggs well above high water mark and nesting activity occurs at night. Known to migrate long distances. Found in temperate, tropical and sub tropical seas. It nests at the coast of Gulf of Kachchh, Mangrol and Okha and also, to a small extent, along the entire Saurashtra coast. Endangered due to sand mining at nesting sites and hunting for flesh eggs.

***Geochelone elegans* (Schoeff)**

A largely Crepuscuar tortoise which spends most of the time under cover. Upper shell (carapace) length of full grown tortoise may be about around 30 cm. Carapace has a yellow central area with yellow central area with yellow lines radiating from it and giving star like appearance. Lower shell has black rays on a yellow background. Head and limbs are brownish yellow with darker markings. Feeds on fallen fruits, grass and also on snails. Eggs are laid in pits dug by females. Inhabits dry scrub forest in Kachchh and dry evergreen forests in Panchmahals, Saurashtra, North and South Gujarat. Vulnerable due to loss of habitat and exploitation for trade.

***Kaloula pulchra* (Gray)**

A medium sized frog with short and rounded head. Colour pattern distinctive and comprises of blackish brown area bordered by two red bands on lateral sides and spotted with red brown patches. Fingers are without webs; first finger is shorter than the second and fingertips are dilated into discs. Floats on water with the help of vocal sacs. Feeds on ants and termites. Breeds from June to July. Mainly lives in burrows and tree hollows in hilly, moist, deciduous forests; aquatic margins, temporary water pools and on trees in Shoolpaneshwar Wild Life Sanctuary. Threatened due to loss of habitat and deforestation.

Ardeotis nigricops (Vigors)

A vulture sized, heavy ground bird, endemic to Indian subcontinent and has black crested crown. Males are about 1.2 m tall while females are smaller. Body colour is deep buff above and white below with road black gorget on lower side. Legs and feet are yellowish. Feeds on insects, snakes, grass seeds, wild berries etc. Breeds from March to September. Normally seen in pairs or parties. Habitat comprises of grasslands and open scrublands in Kachchh and Saurashtra.

Endangered due to habitat degradation from overgrazing. *Prosopsis* invasion and poaching.

Equus hemionous Pallas Khur Lesson

An endemic ass which thrives in a unique ecosystem of Rann of Kachchh. About 1.2 m high at shoulder. Colour varies from reddish grey to brown to pale chestnut above and white below.

Grazes on grass between dusk and sunrise and moves solitary or in troops. Mating occurs between August and October, gestation period is of eleven months. Inhabits saline desert and grassland and shrublands in Little Rann of Kachchh, but is also found in surrounding districts like Surendranagar, Banaskantha, Mehsana and Kachchh districts fringing Little Rann.

Endangered due to loss of habitat from expanding agriculture and salt pans encroachments.

23.4 WAYS TO SAVE ENDANGERED ANIMALS

- If pollution can be controlled all across the globe then it can have a major positive impact on animals, fish and birds all over the world.
- To save the endangered animals from extinction, a number of breeding programmes have been introduced. The government, NGOs and the other corporate bodies should come forward for this noble cause as this programme involves dedicated and special people and of course a lot of money.
- Reintroducing the endangered animals to the wild once their numbers increase has become successful in some cases though all species have not done well.
- If hunting and poaching can be controlled then there can be a significant change in the numbers of the endangered animals.

23.5 STEPS TAKEN BY THE GOVERNMENT FOR WILDLIFE PROTECTION

- All the states, excluding Jammu and Kashmir (it has its own Act), adopted Wildlife (Protection) Act in 1972 which prohibits any type of trade for endangered and rare species.
- The Central Government provides every type of financial assistance to the State Governments for the conservation and protection of the endangered species.
- National ban was imposed on tiger hunting in 1970 and Wildlife Protection Act became effective in 1972. According to the latest tiger census (2015), there is an overall increase of 30% in tiger population. In 2010, there were 1700 tigers left in India according to tiger census which has gone up to 2226 tigers in 2015.
- Innumerable numbers of National Parks, Wildlife Sanctuaries, Parks etc have been established by the government.
- In 1992, Central Zoo Authority (CZA) was initiated for the supervision of the management of zoological parks in the country.
- In 1996, the Wildlife Advisory Committee and Wildlife Institute were established for seeking advice on different features of Wildlife conservation and matters related to it.

There are several other initiatives which have been taken up by the government to save the endangered species of India.

India is part of five main International Conventions which are associated to wildlife conservation. They are (i) Convention on International Trade in Endangered Species (CITES), (ii) Coalition Against Wildlife Trafficking (CAWT), (iii) International Whaling Commission (IWC), (iv) United Nations Educational, Scientific and Cultural Organization – World Heritage Committee (UNESCO – WHC) and (v) the Convention on Migratory Species (CHS).

Numerous positive steps are being taken for saving the endangered species of India. But that is not enough. There is a dire need for more NGOs and the private corporate sectors to come forward for this noble cause.

■■■■

Chapter - 24

"Yagna" an Ancient Indian tradition for keeping Environment Clean

Arun Arya and Chitra Arya

Department of Botany, Faculty of Science, The Maharaja Sayajirao University of Baroda, Vadodara-390002 Gujarat, India

E Mail: sarojarun10arya@rediffmail.com, caarya@yahoo.co.in

ABSTRACT

Daily agnihotra is the basic form of yagna tuned to the biorhythm of sunrise and sunset. When specific substances or *Samagri* are burnt along with a wood from definite tree in a copper / iron pyramid vessel with a specific mantra a specific effect is guaranteed to occur. Tremendous amounts of energy is gathered around the yagna kund just at the agnihotra time. This energy field neutralizes the negative type of energy and reinforces the positive pattern of energy. When the earth is rotating on its axis, it is also revolving around the sun. For completing one full circle on its axis, the earth takes a specific time. This specific time is called the 'Rhythm of Nature'. Wooden pieces often termed samidha can be used of the following trees, *Butea monosperma*, (Palas) *Prosopis spicigera* L. (Shami), *Ficus religiosa* L.(Peepal), *F. benghalensis* (Vad), *F. glomerata* (Gular) *Mengifera indica* L.*(mango)*, (Bilva) *Aegle marmelos* (L.) Corrêa, (Chandan) *Santalum album*. *Cedrus deodara* (Roxb.) G.Don Guggul Agar Tagar Somlata Tejapatra Jao Kala Til Makhana

Keywords: Yagna, Time, Fire and Yagnakund.

24.1 INTRODUCTION

At the time of exact sunrise, the place where the sun is said to have risen, the atmosphere of the area is tremendously activated by electrical energies. Performance of agnihotra exactly synchronizing this momentarily activated period results in attracting a stream of vital life sustaining energies of the solar system towards the earth.

At sunrise, many rays, electricity, and more subtle energies emanating from the sun extend all the way to the earth and produce a flood effect at those coordinates, where the sun is said to rise. This torrent of life sustaining energies causes all life so rejoices. At sunrise that is much heard. The morning agnihotra mantra is the essence of that music. It is quintessential sound of that flood. At sunset, the flood recedes (from light towards divine path) (Raju 2007).

There are five basic things necessary in the performance of a Yagna: (1) Definite timings (sunrise /sunset), (2) Wood pieces of different holy trees. (3) Pyramid shaped copper or iron pot, (4) Offering of Hawan samagri (Table 1) with cow's pure Ghee and (5) Vedic mantras. The yagna involves chanting of vedic mantras. The fire helps to clean the atmosphere. Various plant products when burnt produce different gases which kill the harmful bacteria and germs causing diseases. It has been found that even tuberculosis and diseases like swine flue can be prevented by continuous performance of yagna. The scientists have tried to prove that after pouring ghee in fire the produced gases gets converted into formaldehyde like substances when sprinkling of water is done all around the yagna kund. This may lead to killing of germs present around the yagna place. An increase in the yield of crops has been reported by scientists by application of burnt ash. Efforts should me made to find out explanation for various activities recommended during daily yagna. Paper reports review work done by different workers in this field.

24.2 RHYTHM OF NATURE: TIME OF AGNIHOTRA

Agnihotra is the basic form of yagna tuned to biorhythm of sunrise sunset. Sunrise and sunset are the timings of Agnihotra. When specific substances are burned in a copper pyramid with a specific Mantra at exact timings of sunrise/ sunset, a specific effect is guaranteed to occur. Tremendous amounts of energy are gathered around the Agnihotra copper pyramid just at the Agnihotra time. This energy field net realizes the negative type of energy and reinforces the positive pattern of energy. When the earth is rotating on its axis, it is also revolving around the sun. For completing one full circle on its axis, the earth takes a specific time. This specific time is called the 'Rhythm of Nature'. Revolving earth's half portion comes in front of the sun and the remaining half is going away from the sun. the place where the rays the last days vanish, that state is called 'Sunrise' and the place where the last rays vanish, that state is called 'Sunset'. This activity of sunrise and sunset takes place once in every 24 hours. The moment when changes are occurring in the nature exactly at sunrise and again at sunset is known as the 'Rhythm of Nature' the smallest rhythmic cycle of sunrise and sunset is known as 'Circadian Rhythm'. At the time of exact sunrise, the place where the sun is said to have risen, the atmosphere of the area is tremendously activated by electrical energies. Performance of Agnihotra exactly synchronizing this momentarily activated period results in attracting a stream of vital life sustaining energies of the solar system towards the earth. Agnihotra attracting a stream of vital life sustaining energies of the solar systems towards the earth. Agnihotra attracts these energies and they get landed at the time and 'aura energy field' gets created of its own, around the plant life in the area. The vital life sustaining energies coming from the solar system get entangled in this 'aura energy field' through the mediula of copper pyramid. At sunset time, the growth of pathogenic bacteria is alarming and hence Agnihotra performed at this time has the bacteriostatic effect on the atmosphere. All such bacteria vanishes.

At sunrise, the many rays, electromagnetic radiations, others and more subtle energies emanating from the sun extend all the way to the earth and produce a flood effect at those coordinates where the sun is said to rise. The flood enlivens and purifies everything is its path destroying what is impure in its wake. This torrent of life sustaining energies causes all life so rejoices. At sunrise that much and heard. The morning Agnihotra mantra is the essence of that music. It is quintessential sound of that flood. At sunset, the flood recedes (from Light towards Divine path). Agnihotra is the basic form of yagna tuned to biorhythm of sunrise/sunset.

Definite timings : Agnihotra is performed at the sunrise and sunset timings every day. The exact time of sunrise and sunset is very important. The sunrise/sunset timing can be had from the Radio, television, local news paper or can be obtained from Agnihotra centre. Agnihotra performed at the correct time activated the copper pyramid so that it constantly attract a stream of solar energies throughout the day. Similarly, after the sunset Agnihotra, the pot attract all through the night lunar and other planetary energies. The inducement of these energies create a beneficial, tranquil atmosphere in the house.

24.3 FIRE AND YAGNAKUNDA

Fire: can be produced in two ways Wooden pieces that is called Samidha can be used of the following trees: (Palas) *Butia monospermum*, (Shami) *Prosopis spicigera,* (Peepal) *Ficus religiosa,* (Bad) *F. benghalensis*, (Gular) *F. glomerata* (Amb) *Mangifera indica*, (Bilva) *Aegle marmelos*, (Chandan) *Santalum album.* In another way fire can be generated by dried cow dung cakes. Cow dung is the essential ingredient. Fire is prepared using dried cow dung cakes in the pyramid shaped pot. The disinfectant values and the medicinal of cow dung due to the presence of chemical like menthol, ammonia, phenol, formalin are well known.

Fire should be prepared 8-10 minutes before Agnihotra times with camphor or cotton wicks duly soaked in cow ghee, pieces of dry cow dung are arranged in the pot in such away so that air can circulate and proper combustion is possible. On a piece of cow dung pieces or wood pieces place in a pyramid shaped pot, a spoonful gugul is spread and a piece of Kapur (Camphor) is placed on that and Kapur is lighted to create fire. Fire should be ready at Agnihotra time. To help ignite the fire in the time, use of dry twigs of prescribed trees is advised (Sanskarvidhi, 1981).

Pyramid shaped copper pot: Copper pyramid measurement 14.5 cm × 14.5 cm at the top, 5.25 × 5.25 cm at the bottom and 6.5 inch height is recommended. The pyramid shape of the pot is very important, for the fact that the shape of the pyramid accumulates inside it electromagnetic waves or cosmic waves and defines all known laws of science and electronics. Preferably this pot should be of copper. Clay pot can also be used. The most important aspect is the shape and size of this pot. The energy concentration is optimum at the geometric centre of the pyramid. The fumigation gasses in the pot are sequentially transformed from pot. The energy rich fumes enter into body through nose and by inhalation directly get mixed in the blood stream and supply energy to the cells wherever required, with metabolic action is complete. Electromagnetic forces are present inside the pyramid and these forces prevent any object kept inside from decomposing. Pyramid energy will prove to be more revolutionary than the solar energy. Spoon and other article used should be of Silver or wooden.

24.4 OFFERINGS

Offerings : Hawan samagri and cow ghee without preservatives of chemicals is necessary. The Hawan samagri contains mixture of Kesar, Agar, Tagar, Chandan, Elaechi, Jayphal, Javitri, fruits, grains, full Rice grain, wheat, barley, Udad, Til, Kheer, Sugar, Honey, Somlata, Giloy.

The oblations should be offered to the fire very attentively. After the offering, concentrate the fire, sit as long as you feel comfortable, in these moments of complete concentration, you will feel pleasantly relaxed, peaceful and vigorously happy. This state of mind remains so till the next Agnihotra time and, therefore, this morning evening cycle should not be disturbed.

After the Agnihotra offerings are made, the fire should not be disturbed the in the pot. Let the fire cool down on its own. The cold agnihotra ash remaining in the pot is very useful. It is used to preparing effective homa therapy medicine. It is also used as a fertilizer and pesticide for the plants.

24.5 CHANTING OF MANTRAS

After the offerings are made, one can sit and try to concentrate or mediate for as many minutes as one feels like and can spare. Agnihotra is the basis of wholistic healing. The mantras are original words in Sanskrit. The mantras should be uttered perfectly. The mantras cause specific vibrations useful for the mind and atmosphere. Jean Le Mee in his introduction to the book 'Hymns from the Rig-Veda writes, each verse is a formula, precisely measure which carries the direction, the forces, and the illumination required for the work. Hence the importance of the meters, which governs the rhythms of the verse is symbolic of the cosmic rhythm itself. There are three such basic meters which by combination give the seven meters representing the sever rhythms, the seven pranes governing the whole process of life. The sound that these rhythms articulate are the vibrations traveling through the four substances, the four levels of being of speech, from the absolute state, unaffected by anything, to the causal state, where meaning are universal and undifferentiated, to the mental state, where they are formed into separate thoughts, and finally to the physical state of utterance where they are heard by ear. It is in second stage, the casual, that the vision takes place; this is where the mantra is perceived by seet. With the third stage metal comes the rhythm and with the forth, the articulated sound itself. The arrangements of these articulated sounds are the ceasing of the Rishis, while the ideas and their truth are simply seen and heard by them'. The Agnihotra mantras are from the basic Vedic text handed down right from the time of creation. Agnihotra sets up a life pattern based on bio physiological techniques given through Vedas. The vibration, the words create has varied effect on the body, mind and prana. The qualities of vibration therapy and all the beneficial qualities of chromopathy to the present in the Agnihotra Mantra.

When Cow's ghee is mixed with rice grains and Hawan samagri, the Mantras uttered. Agnihotra obtained are offered to the fire, four types pf gasses are produced with the burning of oblations. They are Ethylene oxide, Propylene oxide, Formaldehyde and Butapropiolactons. After the oblations are offered to the fire, cow's ghee produces acetylene, the energy of excessive heat which sucks in the polluted air and purifies it. The gases that are produced from burning of the cow's ghee posses the wonderful capacity of removing mental tensions and curing many diseases (Ansari 2005).

24.6 EFFECT OF AGNIHOTRA ON HUMAN BODY

1. Carbonation environment induces rhythmic action in heart due to vibration effect.
2. The tone of nervous system is improved.
3. Coronary blood flow improves.
4. Pyramid effect ensures ample supply of energy to body cells.
5. Pyruvates supply instant energy to cells.
6. Surrounding atmosphere would be clear due to disinfectant gases.
7. Virus gases like formalin protect the Agnihotra from any infection at sunset, when return home after hard work and sit down for Agniohotra.

Following things happen

1 At sunset, body is in a state of transition from exogenous to endogenous condition.

2 Carbonation effervescent improves the blood supply due to vagotropic effect there is more intake of oxygen in blood.

3 Pyruvates and oxygen give necessary energy to strained cells and they relax.

4 The 'Idam Na Mama' relives the superficial tension in mind and it be comes quiet and relaxed.

5 Relaxed mind relaxes the strained nerves faster.

6 In the process of Agnihotra all the five senses come into action and concentrate on some activity. This enhances the response of nerves and relaxes them faster

i. Eyes watches the offerings.

ii. Nose smells the fragrance of offerings.

iii. Mouth chants the mantras.

iv. Ears are hearing the sound of mantra.

v. Hand are engaged in offerings.

The cumulative result of simultaneous action is "Total Relaxation", which is achieved much faster.

24.7 INVESTIGATIONS ON AGNIHOTRA

(a) *Field of Medicine*: Col. Golecha observed that Hawan cures drug addictions. Dr. Selvamurthy observed that yagna atmosphere has beneficial effects in the form of mental tranquility. Dr. A.G. Mondkar, practicing microbiologist from Bombay studies Hawan effect on bio-energetic system of individual microorganism and concluded that yagna ash has antibacterial effect against well known pathogenic bacteria and helps in wound healing. It is observed by several earlier investigators that Yagna ash has curative effect in shin gung us, scabies, dhobi itch, raches, pain in kidney, gastritis, migraine, tonsillitis, wound healing, warts etc., it is observed that hawan has a wonderfully caming effect on hyper active children and gives relief from bronchial asthma, bronchial troubles, psoriasis etc., Folk medicines are prepared from hawan ash and its application in various ailments is proving to be very beneficial.

(b) *Field of Agriculture*: Homa therapy farming is considered to be useful:

1) To heal and improve the lands that cultivate rather than burdening it with pollution of chemicals.

2) To grow superior crops without the use of chemical fertilizers, pesticides and herbicides.

3) To protect and heal the environment.

(c) *Effects on Microbes*: Tiwari and Mishra 1996 found that after hawan 40%decrease in total number of colonies isolated and a decrease of over 50% in the colonies of *Aspergillus flavus, A. fumigatus, A. niger,* the three most common species of *Aspergillus* found in working place. The fungicidal properties of Mango and Sandal are known, Schmutterer (1995).most of the ingredients of Hawan samagree are known for their medicinal properties and a number of of them are used as medicines by practitioners of the Ayurveda and Unani method of treatment. Studies shows that hawan is more effective in reducing the fungal load (Tiwari and Mishra 1997).

Hawan ash is found to be useful and handy for preservative food grains. Hawan ash can be useful for pest control. Yagna was found to have the effect on germination of seeds, increase in crop yield, food grain preservation Dr. B.G. Bhujbal observed that Yagna atmosphere and hawan ash induced Dr. Ramasharaya Mishra of Kanpur observed that hawan ash treated seeds with dusting of ash gave a higher and better yield as compared to traditional methods.

24.8 CONCLUSIONS

Everyday in the evening and morning and exactly sunset and sunrise, oblations of a little quantity of two pinchful, hawan samagri with cow's pure ghee duly applied to it, to be offered, accompanied by chanting of mantras, in the fire specially prepared from cow dung cakes and mango wooden chips for the purpose, in a pyramid shaped copper or iron pot, two oblations with mantras at sunrise and two oblations with mantras at sunset are to be offered. This is in brief yagna. To perform it, we need very little ingredients, a very short time and cost wise, it is very economical. There are no taboos concerning personal habits etc. Yagna strengthens everything, anybody, any human being can practice it. Yagna is the best care taker. Yagna is a wonder weapon to counter pollution. Yagna heals the atmosphere. Yagna is a practicable, result oriented scientific practice.

Table 1. Ingredients (plant origin) of Hawan samagri used in Yagna

S/No	Sanskrit/Hindi Name	Botanical name
1	Guggul	*Commiphora mukul*
2	Agar	*Aquilaria agallocha*
3	Tagar	*Valeriana wallichii*
4	Kapoor	*Cinnamomum camphora*
5	Somlata	*Ephedra gerardiana*
6	Giloy	*Tinospora cardifolia*
7	Elaychi	*Elateria cardamom*
8	Tejapatra	*Cinnamomum macrocarpum*
9	Jaiphal	*Myristica fragrans*
10	Makhana	*Euryale ferox*
11	Jao	*Hordium vulgaris*
12	Dhan	*Oryza sativa*
13	Udad	*Phaseolous mungo*
14	Dakh	*Vitis vinifera* (Dry)
15	Kala Til	*Sesamum indicum*

S/No	Sanskrit/Hindi Name	Botanical name
	Samidha (Fire wood)	
1	Palas	*Beutia monospermum*
2	Shami	*Prosopis cineraria*
3	Pipal	*Ficus religiosa*
4	Vad	*Ficus benghalensis*
5	Gular	*Ficus glomerata*
6	Amb	*Mengifera indica*
7	Bilva	*Aegle marmelos*
8	Chandan	*Santalum album*

Non plant origin ingredients of Samagri : 1 Cow's Ghee, 2 Cow's Milk, 3 Honey, 4 Sugar, 5 Mohan bhog and 6 Kheer

Butea monosperma: In Kerala, this is called 'plasu' and 'chamata'. Chamata is the vernacular version of Sanskrit word 'Samidha', small piece of wood that is used for 'agnihotra' or fire ritual. In most of the old namboodiri (Kerala Brahmin) houses, one can find this tree because this is widely used for their fire ritual.

It is used for timber, resin, fodder, medicine, and dye. The wood is dirty white and soft and, being durable under water, is used for well-curbs and water scoops. Good charcoal can be obtained from it. The leaves are usually very leathery and not eaten by cattle. The leaves are used by street food sellers to serve food placed on the leaves.

Aegeline (N-[2-hydroxy-2(4-methoxyphenyl) ethyl]-3-phenyl-2-propenamide) is a known constituent of the bael leaf and consumed as a dietary supplement for a variety of purposes. In 2013, the U.S. Food and Drug Administration (FDA), other federal regulators, and local health officials, investigated an outbreak of 97 persons with acute nonviral hepatitis that first emerged in Hawaii. Seventy-two of these persons had reported using the dietary supplement OxyElite Pro, produced by USPlabs.

The wood of *P. cineraria* is a good fuel source, and provides excellent charcoal plus firewood, fodder, green manure and goat-proof thorny fences.

Ficus benghalensis: Native to India and Pakistan, banyan is a type of strangling fig. The plant begins life growing on other trees and eventually envelops them completely. Aerial roots hang down from the branches and these eventually become trunks. This circle of trunks deriving from one original tree can reach an enormous size – 200 cm in diameter and 30 m in height. Their welcome shade has made them important gathering places. Known in Hindu mythology as 'the wish-fulfilling tree', banyans represent eternal life.

Ashwamedha yagna by Yudhisthir

Yagna performed by saints and family members

The *Ficus religiosa* tree is considered sacred by the followers of Hinduism, Jainism and Buddhism. In the Bhagavad Gita, Krishna says, "I am the Peepal tree among the trees, Narada among the sages, Chitraaratha among the Gandharvas, And sage Kapila among the Siddhas." Ethnomedical uses of *F. religiosa* are recorded throughout South Asia, where it has been used for about 50 types of disorders. Phytochemical research had led to the isolation of phytosterols, amino acids, furanocoumarins, phenolic components, hydrocarbons, aliphatic alcohols, volatile components and few other classes of secondary metabolites from *F. religiosa*. Fresh plant materials, crude extracts and isolated components of *F. religiosa* showed a wide spectrum of *in vitro* and *in vivo* pharmacological activities like, antidiabetic, cognitive enhancer, wound healing, anticonvulsant, anti-inflammatory, analgesic, antimicrobial, antiviral, hypolipidemic, antioxidant, immunomodulatory, antiasthmatic, parasympathetic modulatory, esterogenic, antitumor, antiulcer, antianxiety, antihelmintic, endotheilin receptor antagonistic, apoptosis inducer and hypotensive.

The botanical name, which is also the English common name, derives from the Sanskrit term devadāru, which means "wood of the gods", a compound of deva "god" and dāru "wood, tree".The bark of *Cedrus deodara* contains large amounts of taxifolin. The wood contains cedeodarin (6-methyltaxifolin), dihydromyricetin (ampelopsin), cedrin (6-methyldihydromyricetin), cedrinoside and deodarin (3′,4′,5,6-tetrahydroxy-8-methyl dihydroflavonol)(Sabnis and Daniel 1990).

Among Hindus, as the etymology of deodar suggests, it is worshiped as a divine tree. Deva, the first half of the Sanskrit term, means divine, deity, or deus. Dāru, the second part, is cognate with (related to) the words durum, druid, tree, and true. Several Hindu legends refer to this tree. For example, Valmiki Ramayan – Kishkinda khanda

Forests full of Deodar or Devadāru trees were the favorite living place of ancient Indian sages and their families who were devoted to the Hindu god Shiva. To please Lord Shiva, the sages used to perform very difficult tapasya (meditation) practices in deodar forests. Also the ancient Hindu epics and Shaivite texts regularly mention Darukavana, meaning a forest of deodars, as a sacred place.

REFERENCES

Ansari S.H. (2005). Alternative System of Medicine in Essentials of Pharmacognosy. Scientific Publisher Jodhpur Rajasthan pp. 489-494.

Dayanand Swami (1991) *Satyarth Prakash*. Arsh Sahitya Prachar Trust pp 271

Raju K.M.(2007). Veda and Yagna in Indian tradition. Indian J. Traditional Knowledge.6 (4):687-691.

Sabnis S.D. and Daniel M.(1990) a phytochemical approach to economic Botany. Kalyani Pub. Ludhiana 487 pp.

Schmutterer H,(1995) The neem Tree, *Azadirachta indica* A, Juss. and other Meliaceous plants. Source of unique natural products for integrated pest management, medicine, industry and other purposes. New York; VCH publishers.

Tiwari R.and Mishra J.K. (1996). Allergic potential of some Aspergilli on saw mill workers in Lucknow, India. Aerobiologia 12:229-32.

Tiwari R. and Mishra J.K. (1997). Hawan – An effective method to reduce fungal load at small work places. *Aerobiologia* 13: 135-138.

Chapter - 25

Surface Modified Magnetic Nanoparticles: A New Generation of Nanoadsorbents for Facile Remediation Protocols

Monika Yadav, Manita Das, Sonal Thakore*, Rajendrasinh N. Jadeja*

Department of Chemistry, The Maharaja Sayajirao University of Baroda, Vadodara

ABSTRACT

Adsorption has been considered to be a promising technology for water remediation due to its operation easiness and cost-effectiveness. Regardless of the fact that multiple adsorbents have been developed, it is still challenging to develop materials with high performance and easy-separation. To address these challenges, various new materials have been developed. Among these, magnetic nano adsorbents synthesized with various surface modifications have been extensively studied because of their small sizes, high surface area, and convenient separation. This chapter gives an insight about some of the magnetic nano-adsorbents which have been developed so far using various surface functionalizations for the removal of various pollutants present in water. The effect of a particular functional group on the mechanism of adsorption of a specific pollutant has also been discussed. These systems demonstrate that such novel fabricated nano-adsorbents could subsequently become potential materials for environmental remediation.

25.1 INTRODUCTION

Water is an important supply for maintaining life on this planet. Even though it is present in abundance but their chemical composition varies across the strata affecting its propriety for domestic and industrial purposes. Only 0.6% of the total available water resources contributes for groundwater (Mehta *et al.*, 2015).

Environmental pollution, especially water contamination, is one of the leading global concern creating greatest ecological issues of the day. The primary source of water contamination leading to soil pollution is industrial effluent. Among various pollution sources, wastewaters from the textile industry plays crucial role in environmental pollution. Serious health issues and ecological problems to human beings have been aroused due to release of textile effluents into surface water. Growth of plants and animals might be inhibited by colorants and they have propensity to chelate metal ions that outturn micro toxicity to fish and other organisms in natural streams (Mousavi, Shokoofehpoor, & Mohammadi, 2019).

Large amount of hazardous species have been discharged in environment due to industrial activities. Among different species, dye effluents discharged in water supplies was of more concern because of immense toxicity and excessive color which interferes with light penetration, reduces photosynthetic action in aquatic life and proved aesthetically unpleasant. Allergies, skin irritations, cancers and mutagenicity are different adverse impact of dyes on aquatic organisms and humans. (Pourjavadi, Abedin-Moghanaki, & Tavakoli, 2016). Synthetic dyes are used in the manufacturing of textiles, paper, leather, plastic, food, etc. 1.6 million tons of dyes are produced annually to meet the industrial demand and 10–15% of this volume is discarded as wastewater creating dyes as major water pollutant. Dyes are stable, resistant to degradation, toxic, carcinogenic and mutagenic because of their complicated chemical structure, they can also cause severe health problems such as allergy and dermatological diseases (Demarchi et al., 2016).

Global epidemic arise due to release heavy metals into the environment by anthropogenic activities. Wastewater, industrial activities and domestic effluents ends up releasing these species in natural water systems. Rapid industrialization deteriorates the quality of drinking water significantly. Natural sources of heavy metal include weathering, erosion from rock and soil, rainwater which is dominated by interfaces between solids, liquids and gases. Natural processes occur through mineral dissolution, precipitation, and sorption/desorption of chemical species, evolving it as pollutants in soils and groundwater (Ray & Shipley, 2015).

Various techniques like membrane separation process, coagulation, flocculation, filtration, adsorption and chemical treatment have been developed to treat water. Surface boundary phenomenon of accumulation of species onto a liquid or solid phase from a bulk phase is termed as adsorption. The development of isotherm models by Langmuir (Rooper et al., 1916) and Freundlich (Mukasa-Tebandeke et al., 2015) and the kinetic models developed by Lagregren and Ho were some pioneer significant contributions in this field. Higher efficiency and insensitivity to toxic substances in comparison to the other conventional methods of water purification has made adsorption much more prominent (Mehta et al., 2015). The advantages of adsorption are versatile; it does not require huge amount of energy, enormous amount of water or additional chemicals. Access to large amounts of power and financial resources might be an issue in developing countries thus simple and cheap process of adsorption would be a feasible option. The mechanical process of adsorption grants versatility in the design and use of the adsorbent. In batch method of adsorption the adsorbate was added to the sorbent in an aqueous suspension with continuous mixing or agitation. In column or flow method continuous flow of liquid through a fixed bed of the adsorbent was used by fixing the flow rate of liquid. The flow method is generally utilized in industrial settings (Burglsser, Cernk, Borkovec, Stlcher, & Ecology, 1993).

25.2 NANOADSORBENT

An excellent adsorbent is the one having high surface area and attains equilibrium time quickly. It is necessary that an adsorbent causes rapid removal of toxic impurities by faster adsorption and also generation of sludge is lowest. In this context many research groups have investigated the use of nanoparticles as adsorbent. Nanoparticles are particles in the range of 1-100 nm with high surface area and having novel size-and shape-dependent properties. Nanomaterials have been extensively used over a decade exhibiting higher efficiency and faster removal rates as compared to the conventional materials. Upon scaling down the particle size to nano-scale considerable changes occur in the chemical, physical, mechanical and optical properties of the adsorbent. The most important change is the increase in surface area, presence of specific functionalities and change in pore structure. These changes lead to higher adsorption capacity and faster transportation of the pollutants to the internal active sites of the adsorbent making the entire process of adsorption very rapid (Othman et al., 2018). Nanoadsorbents could be employed most effectively not only in a very low concentration range (approx. 1 ppm) of pollutant, but also in a very high concentration range (approx. 1000 ppm), where other techniques are ineffective, time-consuming, or costly.

In addition, it is the need of the hour to develop adsorbents for in-situ treatment approach over ex situ approach. In an in situ approach, treatment of contaminants can be done at any affected place. On the contrary ex situ approach requires transferring the contaminated material to a more convenient location, adding more cost and environmental impact to the process (e.g. pumping, transportation, instrumentations and treatment of contaminants). Nanoadsorbents are highly mobile in porous media as they are much smaller than the relevant pore spaces, so they can be transported effectively by the flow in porous media. Thus it becomes easy to employ the nanoadsorbents in situ, within the contaminated zone where treatment is needed (Nassar, 2010).

25.3 SYNTHESIS OF NANOPARTICLES

The metal salts are used as precursors to produce metal ions of interest. These ions, after reduction, serve as the small blocks that join to form nanoparticles. The nanoparticles are stabilized using a capping agent. The capping agent helps to prevent further agglomeration of nanoparticles either by electrostatic repulsion between the nanoparticles or by polymeric deterrents. As per recent trends nanoparticles are synthesized by using plant extracts. The plant extracts are composed of compounds, such as polyphenols that help in the nanoparticle synthesis by reducing the metallic precursor ion as well as by acting as capping agents to prevent agglomeration of particles to restrict particles to nanoscale (Tahir & Saad, 2019).

25.4 CHARACTERIZATION OF NANOPARTICLES

The commonly employed analytical techniques for characterization of magnetic nanoadsorbents are as per shown in **Table 1.**

Table 1: Characterization of Nanoparticles.

Techniques	**Property Characterization**
Scanning Electron Microscope (SEM)	Shape, surface morphology and assembly structure.
Transmission Electron Microscope (TEM)	Particle size, shape and crystallinity.
Thermo Gravimetric Analysis (TGA)	Thermal properties, thermal degradation.
X-Ray Diffraction (XRD)	Crystal structure.
Dynamic Light Scattering (DLS)	Hydrodynamic radius.
Infrared Spectroscopy (IR)	Surface functionalization, composition, and conjugation.
Nuclear Magnetic Resonance Spectroscopy (NMR)	Confirmation of the polymer structure.
Ultraviolet Visible Spectroscopy (UV-vis)	Light absorption, scattering and optical properties.
Brunauer-Emmett-Teller (BET) Analyser	Measurement of surface area, pore volume.

Iron Oxide Nanoparticles

The growing interest for use of iron oxide nanoparticles for waste water remediation is due to their simplicity and resourcefulness. They can be synthesized using a wide range of techniques like sol–gel, thermal decomposition, hydrothermal, solvothermal, and chemical vapor deposition methods. They can also be synthesized into different structural morphologies like nanochains, flowers, cauliflower-like structures, nanotubes, nanorods and nanoparticles.

Magnetism and Adsorption

The use of magnetism for water purification is an age old concept. Available texts predate the use of magnetism for water treatment to as early as 1873. Since then magnetism has been employed in various water treatment methods such as anti-scaling technique in boilers, pipelines in factories, coagulation, biological processes. The use of magnetism in an adsorption process is a relatively newer concept; one which is gathering increasing attention from the researchers day by day. Magnetic adsorbents are a new class of adsorbents where a base adsorbent is embedded with magnetic particles which are oxides of metals such as Fe, Co, Ni and Cu (Hu et al., 2010). On application of an external magnetic field, the magnetic adsorbent can be rapidly and easily separated from water due to the presence of the metal component in the adsorbent. Magnetic ion exchange resins (MIEX) were first used for the removal of natural organic matter in 1995 (Ambashta & Sillanpää, 2010). Since then magnetic particles modified with polymer (Albornoz, Sileo, & Jacobo, 2004; Philippova, Barabanova, Molchanov, & Khokhlov, 2011), carbon nanotubes (Hu et al., 2010) etc., have been used commercially.

Magnetic Nanoadsorbent

Numerous studies have been conducted to develop cheaper effective adsorbents containing natural biopolymers. Even though these adsorbents have proved to be very effective in the removal of pollutants from wastewater, they suffer one inherent disadvantage. It is difficult to separate them from wastewater in a continuous flow system. Hence, studies have been conducted to explore the use of magnetism as an effective means of separating these suspended adsorbents from water (Mehta et al., 2015).

Magnetic nanoparticles are expected to be an effective heavy metal adsorbent due to its high surface area and its magnetic recoverability, which overcomes the instinctive hardness of nanoadsorbents to separate from wastewater after using. Coating and modified magnetic nanoparticle with functional shells were the effective methods to decrease their agglomeration in wastewater and increase their adsorption affinity for heavy metals. Amino groups, carboxyl groups and thiol groups are the commonly used functional groups, but most of papers only report the adsorption properties of nanoadsorbents either with sole amino groups or with sole thiol groups (Ji, Chen, & Zhao, 2019). Magnetic nanoparticles are typically synthesized via cross-linking. The magnetites (Fe_3O_4) are used commonly as magnetic cores and many methods have been developed for Fe_3O_4 nanoparticles (X. Jin et al., 2016).

This chapter focuses on the use of various magnetic adsorbents for the removal of pollutants from wastewater. A summary of the data published pertinent to the topic is presented and the results of these findings have been discussed.

Functionalized MNPs

Coating and modified magnetic nanoparticles with functional shells were the effective methods to decrease their agglomeration in wastewater and increase their adsorption affinity for heavy metals. Amino groups, carboxyl groups and thiol groups are the commonly used functional groups (Ge, Li, Ye, & Zhao, 2012; X. Jin et al., 2016). The facile refluxing condition could employed for the synthesis of magnetic nanoparticles with simultaneous surface functionalization using modifiers composed of functional groups (amine, hydroxyl and carboxylic group) (Nonkumwong, Ananta, & Srisombat, 2016).

The most widely explored functional groups for functionalization of MNPs are amino and thiol groups. Thiol group functionalization helps to enhance the dispersibility and stability of the nanoparticles by controlling the degree of thionylation. Further, the choice of thiol functionalization provides an additional active site-"sulpho group" which aids to the removal of heavy metal ions via complexation. Thiol group can be introduced onto iron oxide using various precursors such as by grafting of AMPS branches (D. Lin et al., 2017), MPTMS (3-mercaptopropyl trimethoxysilane) (S. Lin et al., 2017) etc. Similarly amine functionalization can be introduced using ethanolamine (Nonkumwong et al., 2016), ethylenediamine (Liao et al., 2012) etc. The heavy metals that could successfully targeted for remediation using amine and thiol functionalization were Pb^{2+}, Ca^{2+}, Cd^{2+}, Zn^{2+}, Cu^{2+} and Ni^{2+}.

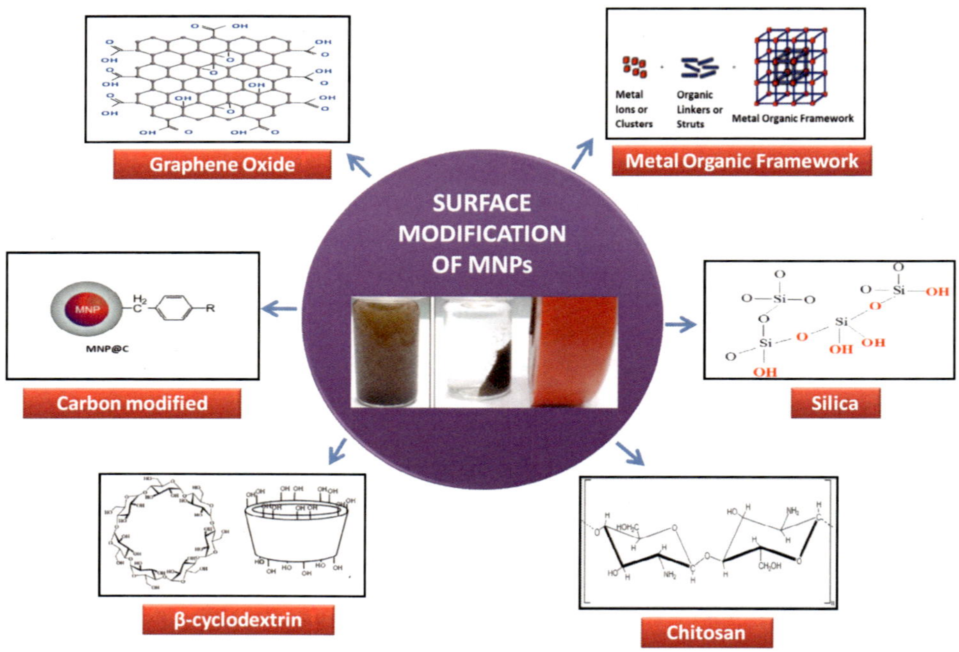

Figure 1: Surface modification of magnetic nanoparticles.

Efforts were made by Chen et al. to study the effects of coexistence of amino and thiol groups on adsorption of heavy metal. The sequential modification of iron oxide nanoparticle was done using (3-Chloropropyl) trimethoxysilan, polyetherimide, epichlorohydrin and thiourea. It was established that amino and thiol group play an important role in the adsorption of heavy metals. The maximum adsorption capacity for Pb^{2+}, Cd^{2+} and Cu^{2+} reached as high as 110.13 mg·g-1, 40.23 mg·g-1 and 29.37 mg·g-1, respectively. Apart from organic functional groups, presence of inorganic metal oxides like MnO_2 in a core-shell type of arrangement enhances the removal of lead (Chen et al., 2014).

In 2013, Roy et al. removed Cu(II), Zn(II) and Pb(II) from water using maghemite nanotubes (MHNT). They synthesized the maghemite nanotubes using a template free microwave irradiation and determined adsorption kinetics (Ray & Shipley, 2015).

25.5 SILICA MODIFIED MAGNETIC NANOADSORBENT

Silicon dioxide (SiO_2) is largely used in the industry, since is a semiconductor with high chemical resistance. Silica nanoparticles in turn, is a material generated from SiO_2, containing interesting characteristics, which enables several applications. In environmental area, the effect of nanosilica to decrease the salinity stress on plants was investigated. Furthermore, nanosilica is utilized to improve the resistance, durability and flame resistance of cements. The production of silica nanoparticles already presents improvements with respect to silica, but, it is still possible modify the nanosilica characteristics through alternative processes, such as microwave, hydrothermal, ultrasound or oxidative processes, in order to improve the potential of this material (Hosseini–Bandegharaei, Peres, Dotto, Cunha, & Slaviero, 2017). These nanostructures are capable of removing dyes as well as cationic pollutants from water. For instance a Zn^{2+} silica modified $CoFe_2O_4$ (CZFS) nanostructured composite, prepared by wet chemical method can adsorb cationic pollutants from water. The composite comprises cubic spinel crystallites of with amorphous silica clusters decorated on the crystallites-surface. CZFS efficiently removes MB as well as heavy metal ions (especially Pb^{2+}) from contaminated water (Sengupta, Rao, & Bahadur, 2017). On the other hand mercaptoamine-functionalised silica-coated magnetic nanoparticles (MAF-SCMNPs) and was effective only in the extraction and recovery of mercury and lead ions from wastewater (X. Jin et al., 2016). The results demonstrate the effect surface functionalization has on the categories of pollutants that can be removed by a particular adsorbent.

Metal Organic Framework

Metal-organic frameworks (MOFs), which consist of metal ions or clusters and organic ligands, have been recognized as alternative precursors for the synthesis of magnetic porous carbon materials. These materials show adsorption properties of organic dyes from aqueous solution. As per reports γ-Fe_2O_3/C obtained through the carbonization of MIL-53(Fe) using a microwave-enhanced high-temperature ionothermal method showed high dye adsorption capacity (499 mg g-1, 30°C) of malachite green (MG) (Shi et al., 2018). Magnetic porous carbon composites obtained from CNTs/Fe_2O_3 adsorbent with high nano-Fe_2O_3 loading (>50%) using Fenton's reagent method was capable of removing methyl orange (MO) (F. Yu et al., 2012).

It is important to take into consideration that use of magnetically modified MOFs are not only restricted to removal of dyes and metal ions but it can also be used for removal of drug molecules present as micropollutants. For instance magnetic copper based metal-organic frameworks (MOF) could be used successfully for removal of two widely used antibiotics ciprofloxacin (CIP) and norfloxacin (NOR) fluoroquinolone (FQAs) from aqueous solutions within a short time span of 30 minutes (G. Wu et al., 2018).

However, the preparation of most of MOF precursors in existing established works need time consuming building-up processes, high temperatures and the help of toxic and teratogenic organic solvents such as N,N-dimethylformamide and N,N-diethylformamide (DEF). Therefore, in this regard, the development of high performance magnetic carbon without the use of any toxic solvents and with relatively short reaction time is of great importance (L. Jin, Zhao, Qian, & Dong, 2018).

Polymer

Polymers are well established adsorbents. Well established polymers for modification of Iron oxide nanoparticles are m-aminobenzoicacid, Polyvinyl alcohol, Polyvinyl sulfate etc. In such type of modifications generally core-shell type materials are formed with the iron oxide nanoparticles forming the magnetic core and polymer comprises the shell. Conductive terpolymer of aniline/m-aminobenzoicacid/m-phenilendiamine shell self-doped by protonic acid of thiacalixarenetetrasulfonate pendent groups have been synthesized and applied for dye removal (Lakouraj, Norouzian, & Balo, 2015). Surface modification with a copolymer using 2, 4-diaminophenol and formaldehyde (DAPF) were capable in removing anionic dyes, namely, amaranth (AM), orange II (OR-II) and acid red 18 (AR-18), from aqueous solutions (Huo et al., 2018). Graft copolymerization is also a commonly used strategy to coat the nanoparticles with polymers. For instance magnetic nanoparticle coated Starch-g-poly (vinyl sulfate) nanocomposite was prepared graft copolymerization of vinyl acetate onto starch in the presence of magnetic nanoparticles, where the acetate groups were converted to hydroxyl groups followed by the sulfation of the hydroxyl groups using chlorosulfonic acid. Nanocomposite was used as an adsorbent for the removal of typical cationic dyes, methylene blue (MB) and malachite green (MG), from aqueous solutions (Pourjavadi et al., 2016). Polyvinylpyrrolidone (PVP)-coating had a potential in adsorption of a variety of organic micropollutants such as Tonalide, Bisphenol-A, Triclosan, Metolachlor, Estriol and Ketoprofen from the aqueous phase(Alizadeh Fard, Vosoogh, Barkdoll, & Aminzadeh, 2017). Recent advances for development of stimuli-sensitive polymers have provided a new insight for removal of radioactive wastes. A strippable coating material consisting of a magnetic field-sensitive adsorbent showed excellent ^{137}Cs adsorption. For the decontamination of ^{137}Cs-contaminated surfaces temperature sensitive elastic Poly vinyl alcohol (PVA)-borate hydrogel was developed. This type of methodology suggests a new eco-friendly volume-reduction method for radioactive waste after surface decontamination (Yang, Hwang, Park, & Lee, 2017).

Nitrogen doped magnetic Nanoadsorbent

Novel nitrogen doped magnetic carbons (NMC), in-situ synthesized through facile pyrolysis carbonization processes using $Fe(NO_3)_3$ and melamine as precursors, were demonstrated as excellent nanoadsorbents to remove Cr(VI) effectively. An unprecedented high adsorption performance of 2001.4 mg g_1 was observed which can be attributed to the incorporation of the nitrogen dopant, which increased the negative charge density on the surface of adsorbent and thereby enhanced the interaction between the adsorbents and Cr (VI) ions (Khasanov et al., 2016).

GO modified magnetic Nanoadsorbent

Graphene oxide (GO) and graphene are indeed the most notable carbon based materials after its magnificent discovery in 2004. It is a mono-atomic carpet fabricated by sp2 carbon. Due to inherent size and shape dependent attributes of GO, many applications such as sensing, photothermal, drug delivery and adsorption are explored (Khan et al., 2017). Graphene and graphene oxide (GO) based nanocomposites have attracted tremendous attentions due to their many excellent properties such as mechanical flexibility, chemical stability, and large surface area. For this reason, it has been reported that GO based nanocomposites can serve as a dye remover in aqueous solutions. Compared with conventional adsorbents, the merits of GO as adsorbents are their structure types and diverse compositions, tunable pore size, large surface area, and coordinatively unsaturated or saturated metal sites to regulate the adsorption ability (Fan, Luo, Sun, Qiu, & Li, 2013).

Magnetic nanoparticles can be anchored onto Graphene oxide by hydrothermal or sonochemical technique. Adsorbents prepared by interaction of only Graphene oxide- magnetic iron oxide nanoparticles were used as effective adsorbents for synthetic methylene blue (MB) dye removal. Adsorbents were prepared by incorporating metal oxides in the magnetic nanocomposite matrix, interestingly such systems were used as efficient adsorbents for removal of fluoride ions from aqueous system (Sahoo & Hota, 2018). Some researchers have used polymers like polyethylenimine to get GO/Fe3O4/PEI nanocomposites. The excellent complex ability of Polyethyleneimine, was explored to increase the removal efficiency of heavy metal ions like Cu^{2+} (W.W. Yu et al., 2014).

Chitosan modified magnetic Nanoadsorbent:

Chitosan (CS) is a natural polymer obtained from extensive deacetylation of chitin, the second most abundant polysaccharide on earth after cellulose. It is mainly composed of two kinds of structural units: 2-amino-2-deoxy-d-glucose and N-acetyl-2-amino-2-deoxy-d-glucose linked by a β (1 → 4) bond. It is biodegradable and biocompatible polymer which can be modified to prepare adsorbents for water remediation (Zamora-Mora et al., 2017). It is well known that chitosan is capable of efficiently adsorbing heavy metal ions due to its NH_2 or OH groups that can coordinate with heavy ions to form chelate compounds (Liu et al., 2014). The presence of abundant amino as well as hydroxyl groups on chitosan makes them important modifiers as versatile modifications can be achieved. Iron oxide nanoparticles fabricated through facile route using chitosan and having β-cyclodextrin and graphene oxide linked via the functional groups of chitosan act as excellent adsorbents for dye removal (Fan, Luo, Sun, Qiu, et al., 2013). Use of O carboxymethyl chitosan (O-CM) a derivative of chitosan did not hamper the adsorption but could successfully remove dye reactive red 120 (RR120) from aqueous solution(Demarchi et al., 2016). Chitosan attached to magnetic core can be crosslined using crosslinkers like glutaraldehyde. The resulting adsorbent was capable of methyl orange removal. Chitosan-modified ferrite nanoparticles having metals such as Mn or biologic reagent, α-ketoglutaric acid, were employed for Cu removal (He, Zhou, Branford-White, Zhu, & Nie, 2008; Liu et al., 2014). Further pyrolyzed chitosan, act as protective framework for magnetic nanoparticles (magnetite/ maghemite and/or metallic Fe), and was used using magnetic guidance for polycyclic aromatic hydrocarbons (PAHs) such as anthracene and napththalene removal. Detailed experimentation in this regard performed with a using mixture of eight PAHs demonstrated interesting results showing that PAH molecules with more hydrophobicity and highly extended aromaticity had stronger sorption capacities (Nisticò et al., 2017). Modification of chitosan-iron oxide naoparticles using epichlorohydrin and cysteine assisted in removal of rare earth metals like La(III), Nd(III) and Yb(III). As per a report magnetic composite of chitosan mixed with polymers like polyvinyl alcohol has the ability to uptake radioactive wastes like ^{60}Co and strontium metal ions from nuclear wastewater via using of adsorption and microfiltration system (Saad, Allam, Soliman, El-Khatib, & Mahmoud, 2017).

β-Cyclodextrin modified magnetic Nanoadsorbent:

Cyclodextrin (CD), a cyclic oligosaccharide consist of 6-8 glucose units, is commonly used as functional water-soluble units to construct polymer materials by various approaches. CD and its derivates contained a relatively hydrophobic cavity, therefore, it can be used as the molecular recognizer through host-guest interactions. A group of important contaminants such as polycyclic aromatic hydrocarbon (PAHs), phthalic acid esters (PAEs) and phenolic compounds could be formed inclusion complexes with cyclodextrin and its derivates. Based on these properties, the cyclodextrin polymer functionalized magnetic nanomaterials have been dramatically explored to efficiently adsorb and eliminate contaminants from waste water, which improve the inclusion properties of β-CD and allow for the exploitation of magnetic separation technology of MNPs(T. Gong et al., 2016). Cyclodextrin possesses the host–guest interaction, hydrophobic interaction or hydrogen-bond interaction and is widely used indrug delivery, molecular recognition, supramolecular self-assembly or catalysis. It has interactions with some environmental pollutants, such as aromatic molecules or heavy metals. As a result, β-cyclodextrin can be utilized in the cleanup of environmental pollutants(N. Wang et al., 2014). Some of the β-CD based magnetic nanoadsorbents developed are discussed herein. One of the most effective modifier for such systems is graphene oxide. A graphene oxide modified magnetic β-CD based adsorbent was capable of Malachite Green removal(D. Wang, Liu, Jiang, Yu, & Chen, 2015). A resin loaded with magnetic b-cyclodextrin bead and graphene oxide sheet (MCD-GO-R) prepared by one step methodology was effective for Hg(II) removal. Here the good ion exchange property and mechanical strength, of resins was used to increase the mechanical strength of the adsorbent and its adsorption performance (Cui et al., 2015). Sulfur groups introduced on such composites to stabilize the nanoparticles exhibited a high Pb (II) adsorption capacity (L. Yan, Kong, Yan, Qu, & Li, 2015).

Polymers such as poly (glycidylmethacrylate) can also be grafted as coatings on β-cyclodextrin magnetic Fe3O4 nanoparticles with using ordinary radical polymerization to removal of organic pollutants like bisphenol A and heavy metal ions like copper (N. Wang et al., 2014).

The conversion of β-cyclodextrin to its well-known derivative carboxymethyl-β-cyclodextrin yields a polymer which can be grafted on iron oxide nanoparticles. This adsorbent possess the fascinating features of superparamagnetism and adsorption properties for favorable removal of organic pollutants like bisphenol A and resorcin (T. Gong et al., 2016).

Carbon modified magnetic Nanoadsorbent:

Researchers have worked for the development of Carbon modified Fe_2O_3 nanoparticle for ultrasonic-assisted removal of methylene blue (MB) and malachite green (MG) (Asfaram, Ghaedi, Hajati, & Goudarzi, 2016). But carbon modified Fe3O4 nanoparticles prepared by hydrothermal method can be used for removal of large quantities of methylene blue only (R. Wu et al., 2014).

25.6 EFFICACY OF MAGNETIC NANOADSORBENTS

Efficacy of the adsorbent can be assessed by carrying out adsorption experiments under various conditions of pH, temperature, ionic strength, concentration etc. How these factors will affect the adsorption depends upon the interactions between adsorbate and adsorbent. For instance the removal of Pb (II) ions from aqueous solution can be done effectively in a pH range of 5-7 if the iron oxide nanoparticles are thiol or amine functionalized. On the other hand a formaldehyde functionalized magnetic nanoadsorbent was selective only in removal of anionic dye like Methyl Orange. A broader picture of adsorption can be viewed by carrying out adsorption isotherm modeling studies and kinetics studies. The magnetic nanoadsorbents reported so far exhibit high adsorption efficiencies. The values are as high as 898 mg/g for removal of dye (Malachite

Green) and 2002 mg/g for removal of metal ions (Cr^{6+}). These studies provide results and present an existing scenario of adsorption using magnetic nano-adsorbent. New adsorbents can be fabricated keeping in mind the reported surface functionalization and adsorption conditions in order to achieve higher removal efficiencies.

Table 2: Comparison table of magnetic nanoadsorbents.

S.No.	Adsorbent Name	Contact Time	Optimal pH	Concentration Range	Adsorbent Capacity	Adsorbate	Reference
1.	MNP modified with Fe-Mn binary oxide	90 min	7	200 µg/L	47.76 mg/g	As	(Shan & Tong, 2013)
2.	Fe_3O_4 and MnO2 modified Graphene oxide nanocomposite (3:8)	24 h	7	0.01-10 mg/L	48.57 mg/g	As	(X. Luo et al., 2012)
3.	γ-Fe_2O_3–TiO_2	10-720 min	7	0.1-50 mg/L	46.28 mg/g	As	(L. Yu et al., 2013)
4.	Carbon Nanotubes modified with Magnetite and Manganese Oxide	150 min	2	50–300 mg/L	150.8 mg/g	Cr	(C. Luo, Tian, Yang, Zhang, & Yan, 2013)
5.	Fe_3O_4 coated glycine doped polypyrrole magnetic nanocomposite	24 h	2	50-650 mg/L	238 mg/g	Cr	(Ballav, Choi, Mishra, & Maity, 2014)
6.	S. cerevisiae immobilised on chitosan-coated magnetic nanoparticles	2 h	4.5	40-500 mg/L	144.9 mg/g	Cu	(J. Zhang et al., 2009)
7.	Thiol-functionalised silicacoated magnetic nano particles	1 h	6	40-1000 mg/L	207.7 mg/g	Hg	(Hakami, Zhang, & Banks, 2012)
8.	Magnetic Hydroxyapatite nanoparticles	24 h	5 ± 0.1	10^{-4}–10^{-2} mol/L	1.964 mmol/g for Cd (II) 2.151 mmol/g for Zn (II)	Cd, Zn	(M. Yan et al., 2010)

S.No.	Adsorbent Name	Contact Time	Optimal pH	Concentration Range	Adsorbent Capacity	Adsorbate	Reference
9.	Humic acid-coated Fe_3O_4 nanoparticles	2 h	8.4	0.2-5 mmol/L	0.291 mmol/g	Methylene Blue	(X. Zhang et al., 2013)
10.	CNTs/k-Fe_2O_3	3 h	3-10	20-80 mg/L	28.8 mg/g	Methyl Orange	(F. Yu et al., 2012)
11.	Carbon nano-tubes and iron oxide composed nanocomposite	24 h	7	1.4-37.4 mg/L	MB- 15.74 mg/g NR- 20.33 mg/g BCB- 23.55 mg/g	Methylene Blue, Neutral Red, Brilliant Cresyl Blue	(J. L. Gong et al., 2009)
12.	magnetic-modified multi-walled carbon nanotubes	15 min	7	20 mg/L	CV- 227.7 mg/g JG- 250 mg/g Th- 36.4 mg/g MB – 48.1 mg/g	Crystal Violet, Thionine, Janus Green B, Methylene Blue	(Madrakian, Afkhami, Ahmadi, & Bagheri, 2011)
13.	Fe_3O_4 impregnated on Halloysite nanotube	8 h	7	150 mg/L	90.09 mg/g	Methyl Violet	(Duan, Liu, Chen, Zhang, & Liu, 2012)
14.	Fe_2O_3cross-linked chitosan composite (Mγ-Fe_2O_3/CSCs)	120 min	4	30 mg/L	29.5 mg/g	Methyl Orange	(Zhu, Jiang, Xiao, & Li, 2010)
15.	Carboxymethyl-β-cyclodextrin polymer grafted on Fe_3O_4	45 min	5.5	50–400 mg/L	Pb(II) – 64.5 mg/g Cd(II) - 27.7 mg/g Ni(II) – 13.2 mg/g	Pb, Cd, Ni	(Abu Zayed M Badrud-doza, Shawon, Tay, Hidajat, & Uddin, 2013)
16.	Chitosan/ Graphene Oxide composite (MCGO)	60 min	5		76.94 mg/g	Pb	(Fan, Luo, Sun, Li, & Qiu, 2013)
17.	Carboxymethyl-β- cyclodextrin modified Fe_3O_4	4 h	6	50-400 mg/L	47.2 mg/g	Cu	(A. Z M Badruddoza, Tay, Tan, Hidajat, & Uddin, 2011)

S.No.	Adsorbent Name	Contact Time	Optimal pH	Concentration Range	Adsorbent Capacity	Adsorbate	Reference
18.	Pectin coated Iron Oxide	24 h	5.5 ± 0.1	10-1000 mg/L	48.99 mg/g	Cu	(Deng et al., 2012)
19.	$CuFe_2O_4$ from PCB sludge	180 min	6	100 mg/L	17.54 mg/g @ 45 ◦C	Cd	(Tu, You, & Chang, 2012)
20.	Aminated β-cyclodextrin functionalized Fe_3O_4	4 h	7	5-20 mg/L	CBZ- 1.304 mg/g, NAP- 1.074 mg/g, BPA- 0.899 mg/g	Carbamazepine, Naproxen, Bisphenol A	(Ghosh, Badruddoza, Hidajat, & Uddin, 2013)

25.7 CONCLUSION

The book chapter attempts to identify and inspect a wide range of magnetic adsorbents for the removal of diverse type of pollutants such as heavy metals, nonmetals, dyes and organic pollutants from water. The use of magnetic adsorbents offers a significant advantage over other adsorbents which is the ability to separate them from an aqueous solution on application of a magnetic field. From the published literature it was observed that the mechanism and kinetics of the adsorption processes depended on the chemical nature of the adsorbent surface as well as the experimental conditions such as pH, adsorbent dosage, contact time, temperature and initial pollutant concentration. Further it was noticed that most of the adsorbents followed either of the Langmuir and Freundlich isotherm and these models could be used to determine the adsorption capacities of different adsorbents. Amongst the kinetic models, pseudo second order model was followed by the most number of adsorbents. Even though a significant amount of literature regarding magnetic adsorbents is available and is increasing at an immense pace every day, there still exists several gaps that need to be addressed. These include:

1. Economic studies of magnetic adsorbents to evaluate their feasibility on a commercial scale.
2. Detailed assessment of magnetic adsorbents in case of multipollutant solutions such as
    1. Heavy metals and Dyes,
    2. Heavy metals and Organic pollutants
    3. Organic pollutants and Dyes.
3. Investigations of the adsorption capability of different magnetic adsorbents with real industrial effluents.
4. Investigation of the removal capability of the adsorbents in a fixed bed column.

REFERENCES

Albornoz, C., Sileo, E. E., & Jacobo, S. E. (2004). Magnetic polymers of maghemite (γ-Fe2O3) and polyvinyl alcohol. *Physica B: Condensed Matter*, *354*(1–4 SPEC. ISS.), 149–153. https://doi.org/10.1016/j.physb.2004.09.038

Alizadeh Fard, M., Vosoogh, A., Barkdoll, B., & Aminzadeh, B. (2017). Using polymer coated nanoparticles for adsorption of micropollutants from water. *Colloids and Surfaces A: Physicochemical and Engineering Aspects*, *531*, 189–197. https://doi.org/10.1016/j.colsurfa.2017.08.008

Ambashta, R. D., & Sillanpää, M. (2010). Water purification using magnetic assistance: A review. *Journal of Hazardous Materials*, *180*(1–3), 38–49. https://doi.org/10.1016/j.jhazmat.2010.04.105

Asfaram, A., Ghaedi, M., Hajati, S., & Goudarzi, A. (2016). Synthesis of magnetic γ-Fe2O3-based nanomaterial for ultrasonic assisted dyes adsorption: Modeling and optimization. *Ultrasonics Sonochemistry*, *32*, 418–431. https://doi.org/10.1016/j.ultsonch.2016.04.011

Badruddoza, A. Z. M., Shawon, Z. B. Z., Tay, W. J. D., Hidajat, K., & Uddin, M. S. (2013). Fe 3O 4/cyclodextrin polymer nanocomposites for selective heavy metals removal from industrial wastewater. *Carbohydrate Polymers*, *91*(1), 322–332. https://doi.org/10.1016/j.carbpol.2012.08.030

Badruddoza, A. Z. M., Tay, A. S. H., Tan, P. Y., Hidajat, K., & Uddin, M. S. (2011). Carboxymethyl-β-cyclodextrin conjugated magnetic nanoparticles as nano-adsorbents for removal of copper ions: Synthesis and adsorption studies. *Journal of Hazardous Materials*, *185*(2–3), 1177–1186. https://doi.org/10.1016/j.jhazmat.2010.10.029

Ballav, N., Choi, H. J., Mishra, S. B., & Maity, A. (2014). Synthesis, characterization of Fe3O4@glycine doped polypyrrole magnetic nanocomposites and their potential performance to remove toxic Cr(VI). *Journal of Industrial and Engineering Chemistry*, *20*(6), 4085–4093. https://doi.org/10.1016/j.jiec.2014.01.007

Burglsser, C. S., Cernk, M., Borkovec, M., Stlcher, H., & Ecology, T. (1993). 1993-Buergisser_etal-Determination of nonlinear adsorption isotherms.pdf, *27*(5), 943–948.

Chen, J., He, F., Zhang, H., Zhang, X., Zhang, G., & Yuan, G. (2014). Novel core-shell structured Mn-Fe/MnO2magnetic nanoparticles for enhanced Pb(II) removal from aqueous solution. *Industrial and Engineering Chemistry Research*, *53*(48), 18481–18488. https://doi.org/10.1021/ie502967a

Cui, L., Hu, L., Du, B., Gao, L., Wang, Y., & Wei, Q. (2015). Removal of Hg(II) from aqueous solution by resin loaded magnetic β-cyclodextrin bead and graphene oxide sheet: Synthesis, adsorption mechanism and separation properties. *Journal of Colloid and Interface Science*, *456*, 42–49. https://doi.org/10.1016/j.jcis.2015.06.007

Demarchi, C. A., Chahm, T., Martins, B. A., Debrassi, A., Nedelko, N., Ślawska-Waniewska, A., & Rodrigues, C. A. (2016). Adsorption of reactive red dye (RR-120) on nanoadsorbent O-carboxymethylchitosan/γ-Fe2O3: Kinetic, equilibrium and factorial design studies. *RSC Advances*, *6*(41), 35058–35070. https://doi.org/10.1039/c6ra04249j

Deng, J.-H., Zhang, X.-R., Chen, L., Wang, X.-Y., Gong, J.-L., Niu, Q.-Y., & Zeng, G.-M. (2012). Copper (II) removal by pectin–iron oxide magnetic nanocomposite adsorbent. *Chemical Engineering Journal*, *185–186*, 100–107. https://doi.org/10.1016/j.cej.2012.01.050

Duan, J., Liu, R., Chen, T., Zhang, B., & Liu, J. (2012). Halloysite nanotube-Fe 3O 4 composite for removal of methyl violet from aqueous solutions. *Desalination*, *293*, 46–52. https://doi.org/10.1016/j.desal.2012.02.022

Fan, L., Luo, C., Sun, M., Li, X., & Qiu, H. (2013). Highly selective adsorption of lead ions by water-dispersible magnetic chitosan/graphene oxide composites. *Colloids and Surfaces B: Biointerfaces*, *103*, 523–529. https://doi.org/10.1016/j.colsurfb.2012.11.006

Fan, L., Luo, C., Sun, M., Qiu, H., & Li, X. (2013). Synthesis of magnetic β-cyclodextrin-chitosan/graphene oxide as nanoadsorbent and its application in dye adsorption and removal. *Colloids and Surfaces B: Biointerfaces*, *103*, 601–607. https://doi.org/10.1016/j.colsurfb.2012.11.023

Ge, F., Li, M. M., Ye, H., & Zhao, B. X. (2012). Effective removal of heavy metal ions Cd 2+, Zn 2+, Pb 2+, Cu 2+ from aqueous solution by polymer-modified magnetic nanoparticles. *Journal of Hazardous Materials*, *211–212*, 366–372. https://doi.org/10.1016/j.jhazmat.2011.12.013

Ghosh, S., Badruddoza, A. Z. M., Hidajat, K., & Uddin, M. S. (2013). Adsorptive removal of emerging contaminants from water using superparamagnetic Fe3O4 nanoparticles bearing aminated β-cyclodextrin. *Journal of Environmental Chemical Engineering*, *1*(3), 122–130. https://doi.org/10.1016/j.jece.2013.04.004

Gong, J. L., Wang, B., Zeng, G. M., Yang, C. P., Niu, C. G., Niu, Q. Y., … Liang, Y. (2009). Removal of cationic dyes from aqueous solution using magnetic multi-wall carbon nanotube nanocomposite as adsorbent. *Journal of Hazardous Materials*, *164*(2–3), 1517–1522. https://doi.org/10.1016/j.jhazmat.2008.09.072

Gong, T., Zhou, Y., Sun, L., Liang, W., Yang, J., Shuang, S., & Dong, C. (2016). Effective adsorption of phenolic pollutants from water using β-cyclodextrin polymer functionalized Fe3O4 magnetic nanoparticles. *RSC Advances*, *6*(84), 80955–80963. https://doi.org/10.1039/c6ra16383a

Hakami, O., Zhang, Y., & Banks, C. J. (2012). Thiol-functionalised mesoporous silica-coated magnetite nanoparticles for high efficiency removal and recovery of Hg from water. *Water Research*, *46*(12), 3913–3922. https://doi.org/10.1016/j.watres.2012.04.032

He, Z.-Y., Zhou, Y.-T., Branford-White, C., Zhu, L.-M., & Nie, H.-L. (2008). Removal of Cu2+ from aqueous solution by chitosan-coated magnetic nanoparticles modified with α-ketoglutaric acid. *Journal of Colloid and Interface Science*, *330*(1), 29–37. https://doi.org/10.1016/j.jcis.2008.10.026

Hosseini–Bandegharaei, A., Peres, E. C., Dotto, G. L., Cunha, A. M., & Slaviero, J. C. (2017). Microwave synthesis of silica nanoparticles and its application for methylene blue adsorption. *Journal of Environmental Chemical Engineering*, *6*(1), 649–659. https://doi.org/10.1016/j.jece.2017.12.062

Hu, J., Shao, D., Chen, C., Sheng, G., Li, J., Wang, X., & Nagatsu, M. (2010). Plasma-induced grafting of cyclodextrin onto multiwall carbon nanotube/iron oxides for adsorbent application. *Journal of Physical Chemistry B*, *114*(20), 6779–6785. https://doi.org/10.1021/jp911424k

Huo, Y., Wu, H., Wang, Z., Wang, F., Liu, Y., Feng, Y., & Zhao, Y. (2018). Preparation of core/shell nanocomposite adsorbents based on amine polymer-modified magnetic materials for the efficient adsorption of anionic dyes. *Colloids and Surfaces A: Physicochemical and Engineering Aspects*, *549*, 174–183. https://doi.org/10.1016/j.colsurfa.2018.04.021

Ji, J., Chen, G., & Zhao, J. (2019). Preparation and characterization of amino/thiol bifunctionalized magnetic nanoadsorbent and its application in rapid removal of Pb (II) from aqueous system. Journal of Hazardous Materials (Vol. 368). Elsevier B.V. https://doi.org/10.1016/j.jhazmat.2019.01.035

Jin, L., Zhao, X., Qian, X., & Dong, M. (2018). Nickel nanoparticles encapsulated in porous carbon and carbon nanotube hybrids from bimetallic metal-organic-frameworks for highly efficient adsorption of dyes. *Journal of Colloid and Interface Science*, *509*, 245–253. https://doi.org/10.1016/j.jcis.2017.09.002

Jin, X., Ning, P., Tang, L., Peng, J., Li, K., & Bao, S. (2016). Highly effective removal of mercury and lead ions from wastewater by mercaptoamine-functionalised silica-coated magnetic nano-adsorbents: Behaviours and mechanisms. *Applied Surface Science*, *393*, 457–466. https://doi.org/10.1016/j.apsusc.2016.09.098

Khan, M. S., Pandey, S., Bhaisare, M. L., Gedda, G., Talib, A., & Wu, H. F. (2017). Graphene oxide@gold nanorods for chemo-photothermal treatment and controlled release of doxorubicin in mice Tumor. *Colloids and Surfaces B: Biointerfaces*, *160*, 543–552. https://doi.org/10.1016/j.colsurfb.2017.09.001

Khasanov, A., Liu, J., Peng, X., Qiu, S., Cao, D., Young, D. P., … Peng, F. (2016). One-pot melamine derived nitrogen doped magnetic carbon nanoadsorbents with enhanced chromium removal. *Carbon*, *109*, 640–649. https://doi.org/10.1016/j.carbon.2016.08.035

Lakouraj, M. M., Norouzian, R. S., & Balo, S. (2015). Preparation and Cationic Dye Adsorption of Novel Fe<inf>3</inf>O<inf>4</inf> Supermagnetic/Thiacalix[4] arene Tetrasulfonate Self-Doped/Polyaniline Nanocomposite: Kinetics, Isotherms, and Thermodynamic Study. *Journal of Chemical and Engineering Data*, *60*(8), 2262–2272. https://doi.org/10.1021/acs.jced.5b00080

Liao, T., Hu, X., Li, T., Liu, Y., Lu, M., Peng, Q., & Wang, H. (2012). Removal of lead(II) from aqueous solution with ethylenediamine-modified yeast biomass coated with magnetic chitosan microparticles: Kinetic and equilibrium modeling. *Chemical Engineering Journal*, *214*, 189–197. https://doi.org/10.1016/j.cej.2012.10.055

Lin, D., Kong, L., Chen, K., He, J., Liu, T., Li, Y., … Liu, J. (2017). Removal of cadmium and lead ions from water by sulfonated magnetic nanoparticle adsorbents. Journal of Colloid and Interface Science (Vol. 494). https://doi.org/10.1016/j.jcis.2017.01.082

Lin, S., Yang, Y., Chen, G., Chen, X., Zhang, W., Xu, M., … Lin, K. (2017). Study on the influence of thiolation on the adsorption and magnetic recovery of superparamagnetic nanoadsorbents for Cd2+ removal. *Applied Surface Science*, *425*, 141–147. https://doi.org/10.1016/j.apsusc.2017.06.282

Liu, Z., Zeng, D., Meng, Y., Chen, D., Sun, Y., & Jiao, D. (2014). Adsorption of Cu2+ ions using chitosan-modified magnetic Mn ferrite nanoparticles synthesized by microwave-assisted hydrothermal method. *Applied Surface Science*, *324*, 745–750. https://doi.org/10.1016/j.apsusc.2014.11.028

Luo, C., Tian, Z., Yang, B., Zhang, L., & Yan, S. (2013). Manganese dioxide/iron oxide/acid oxidized multi-walled carbon nanotube magnetic nanocomposite for enhanced hexavalent chromium removal. *Chemical Engineering Journal*, *234*, 266–275. https://doi.org/10.1016/j.cej.2013.08.084

Luo, X., Wang, C., Luo, S., Dong, R., Tu, X., & Zeng, G. (2012). Adsorption of As (III) and As (V) from water using magnetite Fe 3O 4-reduced graphite oxide-MnO 2 nanocomposites. *Chemical Engineering Journal*, *187*, 45–52. https://doi.org/10.1016/j.cej.2012.01.073

Madrakian, T., Afkhami, A., Ahmadi, M., & Bagheri, H. (2011). Removal of some cationic dyes from aqueous solutions using magnetic-modified multi-walled carbon nanotubes. *Journal of Hazardous Materials*, *196*, 109–114. https://doi.org/10.1016/j.jhazmat.2011.08.078

Mehta, D., Mazumdar, S., & Singh, S. K. (2015). Magnetic adsorbents for the treatment of water/wastewater-A review. *Journal of Water Process Engineering*, *7*, 244–265. https://doi.org/10.1016/j.jwpe.2015.07.001

Mousavi, S. H., Shokoofehpoor, F., & Mohammadi, A. (2019). Synthesis and Characterization of γ-CD-Modified TiO2 Nanoparticles and Its Adsorption Performance for Different Types of Organic Dyes. *Journal of Chemical and Engineering Data*, *64*(1), 135–149. https://doi.org/10.1021/acs.jced.8b00656

Mukasa-Tebandeke, I. Z., Ssebuwufu, P. J. M., Ntale, M., Lugolobi, F., Andreas, S., Nyakairu, G. W., & Nyanzi, S. A. (2015). Adsorption Behavior of Acid-Leached Clays in Bleaching of Oil. *American Journal of Analytical Chemistry*, *06*(06), 495–512. https://doi.org/10.4236/ajac.2015.66049

Nassar, N. N. (2010). Rapid removal and recovery of Pb(II) from wastewater by magnetic nanoadsorbents. *Journal of Hazardous Materials*, *184*(1–3), 538–546. https://doi.org/10.1016/j.jhazmat.2010.08.069

Nisticò, R., Franzoso, F., Cesano, F., Scarano, D., Magnacca, G., Parolo, M. E., & Carlos, L. (2017). Chitosan-Derived Iron Oxide Systems for Magnetically Guided and Efficient Water Purification Processes from Polycyclic Aromatic Hydrocarbons. *ACS Sustainable Chemistry and Engineering*, *5*(1), 793–801. https://doi.org/10.1021/acssuschemeng.6b02126

Nonkumwong, J., Ananta, S., & Srisombat, L. (2016). Effective removal of lead(II) from wastewater by amine-functionalized magnesium ferrite nanoparticles. *RSC Advances*, *6*(53), 47382–47393. https://doi.org/10.1039/c6ra07680g

Othman, N. H., Alias, N. H., Shahruddin, M. Z., Abu Bakar, N. F., Nik Him, N. R., & Lau, W. J. (2018). Adsorption kinetics of methylene blue dyes onto magnetic graphene oxide. *Journal of Environmental Chemical Engineering*, *6*(2), 2803–2811. https://doi.org/10.1016/j.jece.2018.04.024

Philippova, O., Barabanova, A., Molchanov, V., & Khokhlov, A. (2011). Magnetic polymer beads: Recent trends and developments in synthetic design and applications. *European Polymer Journal*, *47*(4), 542–559. https://doi.org/10.1016/j.eurpolymj.2010.11.006

Pourjavadi, A., Abedin-Moghanaki, A., & Tavakoli, A. (2016). Efficient removal of cationic dyes using a new magnetic nanocomposite based on starch-g-poly(vinylalcohol) and functionalized with sulfate groups. *RSC Advances*, *6*(44), 38042–38051. https://doi.org/10.1039/c6ra02517j

Ray, P. Z., & Shipley, H. J. (2015). Inorganic nano-adsorbents for the removal of heavy metals and arsenic: A review. *RSC Advances*, *5*(38), 29885–29907. https://doi.org/10.1039/c5ra02714d

Rooper, C., Martin, M., Butler, J., Jones, D., Weber, T., Wilson, C., … Zimmermann, M. (1916). Comparison of three methods for estimating species and si e Comparison of three methods for estimating species and size iti f kfi h i ti f t l bl composition of rockfishes in acoustic surveys of untrawlable areas V Species Composition I Background and Nee, *252*. Retrieved from https://access.afsc.noaa.gov/pubs/posters/pdfs/pRooper07_comparison-three-methods.pdf

Saad, E. A., Allam, E. A., Soliman, M. A., El-Khatib, A. M., & Mahmoud, M. E. (2017). Adsorptive Removal of Zn(II),Co(II) and Their Radioactive Isotopes 65Zn, 60Co on The Surface of Sodium Nano Bentonite Coated with Oleyl-Amine. *Journal of Radiation and Nuclear Applications*, *2*(3), 87–93. https://doi.org/10.18576/jrna/020302

Sahoo, S. K., & Hota, G. (2018). Surface functionalization of GO with MgO/MgFe2O4binary oxides: A novel magnetic nanoadsorbent for removal of fluoride ions. *Journal of Environmental Chemical Engineering*, *6*(2), 2918–2931. https://doi.org/10.1016/j.jece.2018.04.054

Sengupta, A., Rao, R., & Bahadur, D. (2017). Zn2+-silica modified cobalt ferrite magnetic nanostructured composite for efficient adsorption of cationic pollutants from water. *ACS Sustainable Chemistry and Engineering*, *5*(2), 1280–1286. https://doi.org/10.1021/acssuschemeng.6b01186

Shan, C., & Tong, M. (2013). Efficient removal of trace arsenite through oxidation and adsorption by magnetic nanoparticles modified with Fe-Mn binary oxide. *Water Research*, *47*(10), 3411–3421. https://doi.org/10.1016/j.watres.2013.03.035

Shi, Z., Xu, C., Guan, H., Li, L., Fan, L., Wang, Y., … Zhang, R. (2018). Magnetic metal organic frameworks (MOFs) composite for removal of lead and malachite green in wastewater. *Colloids and Surfaces A: Physicochemical and Engineering Aspects*, *539*(2010), 382–390. https://doi.org/10.1016/j.colsurfa.2017.12.043

Tahir, H., & Saad, M. (2019). *Next-Generation Nanoadsorbents: Opportunities and Challenges. Composite Nanoadsorbents*. Elsevier Inc. https://doi.org/10.1016/b978-0-12-814132-8.00011-3

Tu, Y. J., You, C. F., & Chang, C. K. (2012). Kinetics and thermodynamics of adsorption for Cd on green manufactured nano-particles. *Journal of Hazardous Materials*, *235–236*, 116–122. https://doi.org/10.1016/j.jhazmat.2012.07.030

Wang, D., Liu, L., Jiang, X., Yu, J., & Chen, X. (2015). Adsorption and removal of malachite green from aqueous solution using magnetic β-cyclodextrin-graphene oxide nanocomposites as adsorbents. *Colloids and Surfaces A: Physicochemical and Engineering Aspects*, *466*, 166–173. https://doi.org/10.1016/j.colsurfa.2014.11.021

Wang, N., Zhou, L., Guo, J., Ye, Q., Lin, J. M., & Yuan, J. (2014). Adsorption of environmental pollutants using magnetic hybrid nanoparticles modified with β-cyclodextrin. *Applied Surface Science*, *305*, 267–273. https://doi.org/10.1016/j.apsusc.2014.03.054

Wu, G., Ma, J., Li, S., Guan, J., Jiang, B., Wang, L., … Chen, L. (2018). Magnetic copper-based metal organic framework as an effective and recyclable adsorbent for removal of two fluoroquinolone antibiotics from aqueous solutions. *Journal of Colloid and Interface Science*, *528*, 360–371. https://doi.org/10.1016/j.jcis.2018.05.105

Wu, R., Liu, J. H., Zhao, L., Zhang, X., Xie, J., Yu, B., … Liu, Y. (2014). Hydrothermal preparation of magnetic Fe3O4@C nanoparticles for dye adsorption. *Journal of Environmental Chemical Engineering*, *2*(2), 907–913. https://doi.org/10.1016/j.jece.2014.02.005

Yan, L., Kong, L., Yan, N., Qu, Z., & Li, L. (2015). β-Cyclodextrin stabilized magnetic Fe 3 S 4 nanoparticles for efficient removal of Pb(ii) . *Journal of Materials Chemistry A*, *3*(30), 15755–15763. https://doi.org/10.1039/c5ta03442f

Yan, M., Niu, Q.-Y., Gong, J.-L., Deng, J.-H., Niu, C.-G., Zeng, G.-M., … Feng, Y. (2010). Adsorption of Cd (II) and Zn (II) from aqueous solutions using magnetic hydroxyapatite nanoparticles as adsorbents. *Chemical Engineering Journal*, *162*(2), 487–494. https://doi.org/10.1016/j.cej.2010.05.049

Yang, H. M., Hwang, K. S., Park, C. W., & Lee, K. W. (2017). Polyvinyl alcohol-borate hydrogel containing magnetic adsorbent for surface decontamination. *Annals of Nuclear Energy*, *109*, 359–364. https://doi.org/10.1016/j.anucene.2017.05.044

Yu, F., Chen, J., Chen, L., Huai, J., Gong, W., Yuan, Z., … Ma, J. (2012). Magnetic carbon nanotubes synthesis by Fenton's reagent method and their potential application for removal of azo dye from aqueous solution. *Journal of Colloid and Interface Science*, *378*(1), 175–183. https://doi.org/10.1016/j.jcis.2012.04.024

Yu, L., Peng, X., Ni, F., Li, J., Wang, D., & Luan, Z. (2013). Arsenite removal from aqueous solutions by γ-Fe2O3-TiO2 magnetic nanoparticles through simultaneous photocatalytic oxidation and adsorption. *Journal of Hazardous Materials*, *246–247*, 10–17. https://doi.org/10.1016/j.jhazmat.2012.12.007

Yu, W. W., Liu, M., Wang, L., Wan, J., Xiao, H., Sui, N., … Sui, J. (2014). Polyethylenimine modified magnetic graphene oxide nanocomposites for Cu 2+ removal . *RSC Advances*, *5*(1), 746–752. https://doi.org/10.1039/c4ra11669k

Zamora-Mora, V., Fernández-Gutiérrez, M., González-Gómez, Á., Sanz, B., Román, J. S., Goya, G. F., … Mijangos, C. (2017). Chitosan nanoparticles for combined drug delivery and magnetic hyperthermia: From preparation to in vitro studies. *Carbohydrate Polymers*, *157*, 361–370. https://doi.org/10.1016/j.carbpol.2016.09.084

Zhang, J., Peng, Q., Yang, C., Xu, W., Liu, Y., & Zeng, G. (2009). Biosorption of copper(II) by immobilizing Saccharomyces cerevisiae on the surface of chitosan-coated magnetic nanoparticles from aqueous solution. *Journal of Hazardous Materials*, *177*(1–3), 676–682. https://doi.org/10.1016/j.jhazmat.2009.12.084

Zhang, X., Zhang, P., Wu, Z., Zhang, L., Zeng, G., & Zhou, C. (2013). Adsorption of methylene blue onto humic acid-coated Fe3O4 nanoparticles. *Colloids and Surfaces A: Physicochemical and Engineering Aspects*, *435*, 85–90. https://doi.org/10.1016/j.colsurfa.2012.12.056

Zhu, H. Y., Jiang, R., Xiao, L., & Li, W. (2010). A novel magnetically separable γ-Fe2O3/crosslinked chitosan adsorbent: Preparation, characterization and adsorption application for removal of hazardous azo dye. *Journal of Hazardous Materials*, *179*(1–3), 251–257. https://doi.org/10.1016/j.jhazmat.2010.02.087

■■■■

Subject Index

Related Books

ISBN: 9789389184662

ISBN: 9789389412833

ISBN: 9788172332990

ISBN: 9788172337551

ISBN: 9789387893887

ISBN: 9789388449267

ISBN: 9789387991941

ISBN: 9789389184112

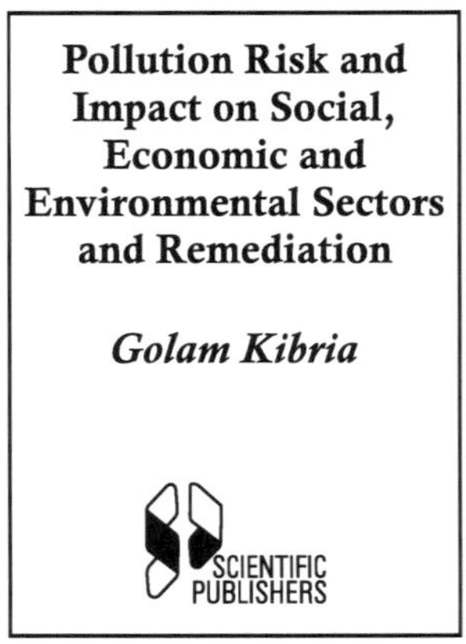

ISBN: 9789389184969